AF411434

Land, Ageing, Gender and Environment: Problems and Challenges from Different Disciplines

Land, Ageing, Gender and Environment: Problems and Challenges from Different Disciplines

Editors

Vanessa Zorrilla-Muñoz
María Silveria Agulló-Tomás
Eduardo Fernandez
Blanca Criado Quesada
Sonia de Lucas Santos
Jesus Cuadrado Rojo

MDPI • Basel • Beijing • Wuhan • Barcelona • Belgrade • Manchester • Tokyo • Cluj • Tianjin

Editors

Vanessa Zorrilla-Muñoz
Miguel Hernandez University
Elche, Spain

María Silveria Agulló-Tomás
University Carlos III of
Madrid
Getafe, Spain

Eduardo Fernandez
Miguel Hernandez University
and CIBER BBN
Elche, Spain

Blanca Criado Quesada
University Carlos III of
Madrid and University
Complutense of Madrid
Getafe, Spain

Sonia de Lucas Santos
Autonomous University of
Madrid
Madrid, Spain

Jesus Cuadrado Rojo
Basque Country University
Bilbao, Spain

Editorial Office
MDPI
St. Alban-Anlage 66
4052 Basel, Switzerland

This is a reprint of articles from the Special Issue published online in the open access journal *Land* (ISSN 2073-445X) (available at: https://www.mdpi.com/journal/land/special_issues/land_ageing_gender_enviroment_problems_challenges_differents_disciplines).

For citation purposes, cite each article independently as indicated on the article page online and as indicated below:

LastName, A.A.; LastName, B.B.; LastName, C.C. Article Title. *Journal Name* **Year**, *Volume Number*, Page Range.

ISBN 978-3-0365-8202-3 (Hbk)
ISBN 978-3-0365-8203-0 (PDF)

Cover image courtesy of María Silveria Agulló-Tomás

Contents

About the Editors

Vanessa Zorrilla-Muñoz

Vanessa Zorrilla-Muñoz holds a PhD in Mechanical Engineering (2012), PhD in Analysis and Evaluation of Social and Political Processes (2017) and MBA in Safety, Hygiene and Ergonomics. Currently, she is a Bioingenieering Institute-UMH Innovation agent. She is also the expert in European projects (EIC; EIC Acceletor), devices, robots and medical device testing and accreditacion. She has expertise in industrial safety and energy-saving projects and has developed her research in aging and old age; gender; ergonomics; land; rural development; emerging technologies; biomedical social engineering; climate change and emissions; health; social psychology and sociology (see her ORCID or other sites).

María Silveria Agulló-Tomás

María Silveria Agulló-Tomás is an Associate Professor at the Department of Social Analysis, Universidad Carlos III de Madrid, having held this position since 1995. She received a Ph.D. in Sociology and Political Science, Social Psychology (2000), at Universidad Complutense de Madrid. She holds a MA in Evaluation and Social Research, from the University of North London. She obtained a "IMSERSO Research Award 2000", and won other grants or merits (e.g., Pilares 2020). She has been a visiting postdoctoral academic (LSE, London, 2000, and University Berkeley, 2001). She is Deputy Director of the Social Analysis Department. She has been leading or collaborating in projects at the Institute of Gender Studies or CSIC (H2019/HUM-5698, ENGAGEn-CM, 2020-2023; CM:LEDYEVA, 2010-2013). Her research interests, publications and projects include work in the fields of aging and old age; land/environment; gender and feminism; habitat/rural spaces; evaluation of programmes; caring/work; climate/social change; health; technologies; social psychology and sociology (see her ORCID or other sites).

Eduardo Fernández

Eduardo Fernandez received a M.D. degree from the University of Alicante (1986) and a Ph.D. in Neuroscience with honors in 1990. He has been a visiting professor at the University of Utah (USA), Oldenburg (Germany), Beth Israel, Harvard Medical School (USA) and University of Vienna (Austria), and he is currently Full Professor and Chairman of the Department of Histology and Anatomy, and Director of the Neural Engineering Division of the Bioengineering Institute of the University Miguel Hernández (Spain). He is a qualified MD who combines biomedicine (molecular and cellular biology, biochemistry, anatomy, physiology and regenerative medicine) with the physical sciences and engineering (mathematics, physics and applied chemistry) to help visually disabled people and to better understand and safely interact with the nervous system. Currently he is the President of Instead Technologies. In recent years, he has been coordinating several projects to demonstrate the feasibility of a neuroprosthesis, interfaced with the occipital cortex, as a means through which a limited but useful sense of vision can be restored in profoundly blind people. His main interest is to develop innovative solutions to the problems raised by interfacing the human nervous system, and on this basis, to develop two-way direct communication with neurons and ensembles of neurons, which could be applied to more effectively enhance the life of people that are affected by visual impairments. Some of his present research pertains to the development of new aids for the blind, with novel therapeutic approaches for retinal degenerative diseases. Furthermore, he is also working on brain plasticity and reorganization in severe vision loss and developing non-invasive methodologies for the selection of appropriate candidates for the implantation of visual neuroprosthesis. He also has expertise in health in land; emerging technologies and interfaces related to rural development and land; aging and old age in land; biomedical engineering; bioengineering and neuroscience.

Blanca Criado Quesada

Blanca Criado Quesada is a program technician at the Network for Addiction Care (UNAD), having held this position since 2021. She holds a bachelor's degree in Sociology from Carlos III University of Madrid and a master's degree in Advanced Studies in Work and Employment from the Complutense University of Madrid. Currently, she is a Ph.D. candidate in Sociology at the Complutense University of Madrid, researching the relationship between older adults and new technologies from a gender perspective. Her interests are centered around aging and the elderly, gender, program evaluation, labor relations, and addictions. She has been involved in various social projects related to the elderly, such as "Age-Friendly Cities," where she conducted diagnostic reports, "Gender Equality Classroom for Seniors," where she delivered gender equality training in senior centers, "ENCAGE", a research project focused on aging, quality of life, and gender, and "The School for Seniors and Addictions," where she coordinated professional training. Additionally, she has various publications on various topics related to aging (health, work, retirement, gender, new technologies, etc.), which can be found on ORCID.

Sonia de Lucas Santos

Sonia de Lucas Santos is Associate Professor at Department of Applied Economics, Universidad Autónoma de Madrid, having worked there since 2004. She received a PhD in Economics at 2008. She has received recognition from CNEAI during the period 2009–2014for her research work. Her research fields are the environment and emissions; applied economics and public policies in land; circular economy; advanced statistics; business cycles in land; convergence and sustainability. She also has a focus on quantitative methods. Most of her articles have been published in international journals and they can be found on Google Scholar and Web of Science. She has participated in numerous research projects financed by Ministries and Public Institutions. She has also received the prize of Cátedra Telefónia (Universidad de Zaragoza) in 2008 and the prize for younger researchers by Asociación Libre de Economía (AldE) in the Applied Economics Meeting in 2007 in Spain.

Jesús Cuadrado Rojo

Jesus Cuadrado Rojo is an Associate Professor of Construction Engineering in the University of Basque Country (UPV/EHU). He received M.S. (1997) and Ph.D. (2009) degrees in Industrial Engineering, both from UPV/EHU. His research interest and activities include the environment and emissions; circular economy; construction and building in land; forest certification; sustainability and climate change (see his ORCID and other sites).

Preface

This online Special Issue (SI), now in book format, which you hold in your hands or are reading on your device, consists of 16 articles produced between 2021 and 2022. We have had the opportunity to coordinate it or get to know it in depth to produce this preface after its publication.

The proposal for this monograph or SI is based on the evidence and confirms the need to continue the evaluation and research in several key aspects of land, from different approaches and disciplines. Especially making mention of the land, gender, and the older people since it has not worked as much as with other characteristics.

Therefore, the main objective of this SI was from the context of some sustainable development goals (SDGs), to bring together researchers from different fields to analyze different points considering aging and gender. We humbly believe that this ambitious objective set by the co-editors has been met in this monograph, with a great deal of thoroughness.

Regarding the reasons and motivations for writing this Special Issue, as we have previously added, the main motivation resides in the low participation (before this monographic) in research on the topics worked on in these papers. The authors have been motivated to investigate the land considering aging, gender and the environment as intersecting variables that are of great importance and are often forgotten. This monographic has been able to fill, albeit "a grain of sand", this gap that existed in research and evaluation on the mentioned areas.

As we can see, according to the different papers, there are very different but related topics, such as age, gender and environmental discrimination. We can find them in order of publication (from the most recent to the oldest).

These can be divided into different areas such as social sciences and health and technology. A total of 11 papers were directly related to social sciences (e.g., economic issues such as ageing industries and regional economy, sustainability through education, urban ageing, climate change, ecological footprint, evaluation, social images, gender and age studies) and 5 papers were related to technology and health (e.g., design of a citizen participation application, environmental monitoring platforms in smart cities, digitalization, memory problems in nursing homes and territorial and gender differences in home care of relatives with dementia).

Although the two areas are related in the papers, some tend to collect more specific questions referring to sociological and economic studies while others cover social variables from the prism of technological and health sciences. For the latter disciplines, we find studies that can appear in both fields.

Concretely, according to the title, we can divide the articles into three key categories: gender, land, and ageing. There are 7 papers on gender or related to women, 10 on land, spatial or environmental aspects, and 11 on ageing/older people.

A total of 58 authors were involved in the development of these papers, including 40 women and 18 men. Women were analyzed according to sex from the information collected from websites. The idea that women continue to be more interested in the topics of gender and older people is confirmed or, at least in this monograph, it has been reflected.

Of these, 29 belong to the social sciences: 14 to the field of sociology, 9 economists, while the rest are divided into other areas of social sciences such as history, geography, law, statistics, gerontology, and psycho-pedagogy; 10 to technology, divided into different engineering fields such as electronics, agriculture, renewable energy, mechanics, systems and automation, data sciences and biotechnology. Finally, 17 belong to health-related sectors: 10 psychologists, 5 doctors and biomedical doctors, 2 biologists.

It has been observed that of the 16 papers published, 3 were qualitative, 9 were quantitative, 1 was a platform design, and 3 were mixed methods.

The themes and authorships are diverse but all related have been organized around gender, ageing, land/terrain, and environment, as we thought for the title of our proposal (in the year 2020, see "Editorial") to make it as inclusive as possible in terms of all these novel areas and which, together, had hardly been dealt with until before this monograph.

This monograph has been inspired and motivated, in part, by the complex and multidimensional reality of today's world and our societies, which cannot be ignored.

Although the results are unfavorable, and we wish they were not, there is no doubt that we are experiencing an accelerating climate crisis due to human activity. This situation has significant and negative impacts on the quality of life and livelihoods of individuals and communities (especially the older people and women, as emphasized throughout the monograph).

The effects of climate change are being felt in all parts of the world, with marked increases in global warming and sea levels, including floods and storms, landslides, as well as extreme weather events, heat waves and droughts; the impacts of climate change could be irreversible (1), both territorially, socio-spatially, socio-politically, and psychosocially, without doubt.

The latest international data is showing that human activities through greenhouse gases

(GHGs) emissions are the key cause of global warming; surface temperature reached 1.1°C above 1850–1900 in the period 2011–2020. Contributions stem from unsustainable energy use, land use and land-use change, lifestyles and consumption and production patterns in all regions (1), as indicated in some parts of the monograph and expected to be developed further in a future paper in Volume II (now open for publication; see: https://www.mdpi.com/journal/land/special_issues/NJ06IY7FL5).

In turn, the climate crisis is closely linked to human livelihoods and survival, such as food insecurity and human health. These facts are manifested in different ways; on the one hand, as a previous report indicates (2) extreme hunger has more than doubled in the 10 countries most affected by climate change (2). On the other hand, the prevailing global food system was estimated to be responsible for about one third of greenhouse gas (GHG) emissions and is the main source of methane and biodiversity loss (3). Other quantitative indicators related to global warming paint an alarming picture, e.g., it has been visualized those suboptimal temperatures are associated with a substantial mortality burden, which varies spatial-temporally (4).

For all the above reasons, climate change comes from all parts of the world and affects all people, so the issue should be a global priority. Furthermore, its adverse impacts are greatest for the most vulnerable people (1) as mentioned in this monograph: especially the older people and women.

While the pandemic caused by COVID-19 and its social consequences are now behind us, global inequalities have been exacerbated (5,6). It is estimated that rising inflation, as well as other pandemic-related measures, could further increase inequality in the long term (6). Thus, generating knowledge and proposing resilience measures, especially for the most fragile groups, takes on a central role.

These facts are already present; a recent World Bank report, 2023, has shown that domestic food price inflation is running high worldwide (7), indicating alarmingly high inflation in most countries, at least 70% (or more) of them, depending on income levels.

In this sense, it is worth reflecting that climate change as well as the COVID-19 pandemic are the twin crises of the Anthropocene (8). The unsustainable growth of human activities has led to significant changes in the global environment and is leading to a deepening of social inequality (8) without precedents.

The impacts of climate change are an undue burden for the most disadvantaged populations, and especially for women and older people. Despite growing recognition of vulnerabilities and differential efforts to promote environmental sustainability between genders, women, (and older women), still have less economic, political and policy influence and are therefore more exposed to the adverse effects of climate change; however, experiences show that they are powerful agents of change and make significant contributions to sustainable development, despite the structural and socio-cultural barriers they still face (9). In the same vein, we can point out that older people continue to be among the most affected by climate change (10,11).

Importantly, differences in vulnerability arise from non-climatic factors, but from multidimensional inequalities produced by inequitable development, but these differences in turn shape differential risks to the climate crisis. Increased vulnerability is a product of the intersection of different social categories that result in unequal social, economic, and cultural status and exposures to different shocks (9) and which has been alluded to in this monograph and will be further elaborated in the future volume II.

Intersectionality (12–15), both as a concept and as a decades-old approach, recognizes the combined effects of the social categories of race, class, and gender and of other processes such as age and (dis)ability, and that in the published monograph and in the future, special sensitivity and attention are shown to this.

Older people and population ageing are critical issues that should be on the scientific, gender and climate change policy agenda, yet they are often overlooked, both in the design and in the process and evaluation of policies/programmes that have also been addressed, in part, in this monograph. This group is of vital importance, as according to the latest World Population Prospects 2022 report, the proportion of the world's population aged 65 and over is expected to increase from 10% in 2022 to 16% in 2050 (16), which is important both in quantitative and qualitative terms.

While the gender, age and climate crisis construct are a largely unexplored area of knowledge, it is not yet known how to address it (11). It is feared that some groups such as women, people experiencing poverty, minorities and older people will be even more affected by the consequences of the climate crisis along with the exacerbation of intersectional inequalities, as already mentioned and discussed in the Special Issue we are presenting.

For all the above reasons, one of the purposes of this monograph was and is to contribute to a call to action for different stakeholders (decision makers, academics, civil society, and others) to generate a new opportunity to advocate on sustainability issues, but also to focus on inequality and intersectionality to design a new convergent public agenda in the current context of multiple crises.

In addition, there is a need to include and build bridges between intergenerational, sustainable, ecofeminist approaches, with other more contemporary approaches such as artificial intelligence and technology applied to everything human and socio-political.

In the end, we would like to mention the potential of interdisciplinary perspectives (both from the health sciences, social sciences, and technology) and collaborations that have much to contribute, within teams and "shoulder to shoulder", to positively address the complex issues and challenges addressed in this volume.

The Guest Editors of this volume, and co-authors of this preface (María Silveria Agulló-Tomás and Vanesa Zorrilla-Muñoz), thanks the other co-editors (Eduardo Fernández, Blanca Criado, Sonia De Lucas and Jesús Cuadrado) for their collaboration and support. We would also like to thank all the authors of the original articles, from different backgrounds and areas, of each of the 16 papers mentioned.

Thanks to Daniela Luz Moyano and Carolina Marcos Carvajal for their help and special interest in these issues and for their collaboration in the preparation of this preface.

We would also like to thank the journal *Land* for the editing proposal and interest in the topic addressed, and for the support in the process of editing and management of the 16 papers (since their invitation to propose the monograph in 2019, until today), especially Ms. Carol Ma.

Finally, we thank our families and loved ones, all the readers and potential audiences (from different areas close to or linked to the ones we have mentioned) to whom this book is addressed, without whom it would be meaningless.

We welcome any feedback and comments you would like to send us to serve as a dialogue with us (Dr. Vanessa Zorrilla-Muñoz <vzorrilla@umh.es>, Prof. Dr. Maria Silveria Agulló-Tomás <msat@polsoc.uc3m.es>, Prof. Dr. Eduardo Fernandez <e.fernandez@umh.es>, Blanca Criado Quesada <blancacquesada@gmail.com>, Prof. Dr. Sonia de Lucas Santos <sonia.delucas@uam.es>, Prof. Dr. Jesus Cuadrado Rojo <jesus.cuadrado@ehu.es>), with the editors and with the authors.

This Special Issue is not closed, but will continue to live on, and we hope you will find it useful, reflective, and applied, and open to new horizons that are necessary for what has already been researched and evaluated, and for new challenges. We will always be grateful; enjoy the book.

July 2023,

Daniela Luz Moyano, María Silveria Agulló-Tomás, Carolina Marcos Carvajal, and Vanessa Zorrilla-Muñoz.

1. Intergovernmental Panel on Climate Change (IPCC). Summary for Policymakers. In: Climate Change 2023: Synthesis Report. A Report of the Intergovernmental Panel on Climate Change. Contribution of Working Groups I, II and III to the Sixth Assessment Report of the Intergovernmental Panel on Climate Change [Internet]. Lee H, Romero J, editors. Geneva; 2023 [cited 2023 Jul 6]. Available from: https://www.ipcc.ch/report/ar6/syr/downloads/report/IPCC_AR6_SYR_SPM.pdf.

2. Oxfam. Hunger in a heating world: How the climate crisis is fuelling hunger in an already hungry world [Internet]. 2023 [cited 2023 Jun 13]. Available from: https://www.oxfam.org/en/research/hunger-heating-world.

3. The World Bank. What You Need to Know About Food Security and Climate Change. [Internet]. 2022 [cited 2023 Jun 13]. Available from: https://www.worldbank.org/en/news/feature/2022/10/17/what-you-need-to-know-about-food-security-and-climate-change.

4. Zhao Q; et al. Global, regional, and national burden of mortality associated with non-optimal ambient temperatures from 2000 to 2019: a three-stage modelling study. Lancet Planet Health. 2021;5(7): e415–25.

5. The World Bank. January 2022 Global Economic Prospects [Internet]. Washington; 2023 [cited 2023 Jul 6]. Available from: https://www.worldbank.org/en/publication/global-economic-prospects.

6. Narayan A et al. COVID-19 and Economic Inequality: Short-Term Impacts with Long-Term Consequences." Policy Research Working Paper 9902. Washington DC; 2022.

7. The World Bank. Food Security Update. Available from: https://thedocs.worldbank.org/en/doc/40ebbf38f5a6b68bfc11e5273e1405d4-0090012022/related/Food-Security-Update-LXXXVIII-6-29-23.pdf.

8. Asayama S; Emori S; Sugiyama M; Kasuga F; Watanabe C. Are we ignoring a black elephant in the Anthropocene? Climate change and global pandemic as the crisis in health and equality. Sustain Sci. 2021;16(2):695–701.

9. Habtezion S. Overview of linkages between gender and climate change [Internet]. New York; 2016 [cited 2023 Jul 6]. Available from: https://www.undp.org/sites/g/files/zskgke326/files /publications/UNDP%20Linkages%20Gender%20and%20CC%20Policy%20Brief%201-WEB.p df.

10. Fernández-Mayoralas G; Schettini R; Sánchez-Román M; Rojo-Pérez F; Agulló-Tomás MS; Forjaz MJ. El papel del género en el buen envejecer. Una revisión sistemática desde la perspectiva científica. Revista Prisma Social. 2018; 21:149–76. https://revistaprismasocial.es/article/view/2422.

11. Sánchez Millán M; Oliva Rojo E; Rubio Montero M; Moyano D; Zorrilla-Muñoz V; Agulló-Tomás MS. Revisión sistemática de referencias bibliográficas en torno a evaluación, género, envejecimiento y medio ambiente. In: Avances en Educación Superior e Investigación XVI Foro Internacional Sobre la Evaluación de la Calidad de la Investigación y de la Educación Superior (FECIES). Dykinson; 2023. Available from: https://www.forofecies.com/libro-de-resumenes.

12. Bilge S. Recent feminist outlooks on intersectionality. Diogenes. 2010; 57:58–72.

13. Crenshaw K. Mapping the Margins: Intersectionality, Identity, and Violence Against Women of Color. Stanford Law Rev. 1991;43(6):1241–300.

14. Crenshaw K. Demarginalizing the Intersection of Race and Sex: A Black Feminist Critique of Antidiscrimination Doctrine. University of Chicago Legal Forum. 1989;139–68.

15. Gaard G. Women, water, energy: An Ecofeminist Approach. Organ Environ. 2021;14(2):57–72.

16. United Nations Department of Economic and Social Affairs PD. World Population Prospects 2022: Summary of Results. UN DESA/POP/2022/TR/NO. 3 [Internet]. New York; 2022 [cited 2023 Jul 6]. Available from: https://www.un.org/development/desa/pd/sites/www.un.org. development.desa.pd/files/wpp2022_summary_of_results.pdf.

Article

Evaluation of Older People Digital Images: Representations from a Land, Gender and Anti-ageist Perspective

Georgiana Livia Cruceanu [1,*], Susana Clemente-Belmonte [1], Rocío Herrero-Sanz [2], Alba Ayala [3,4], Vanessa Zorrilla-Muñoz [4,5], María Silveria Agulló-Tomás [2,4], Catalina Martínez-Miguelez [2] and Gloria Fernández-Mayoralas [4,6]

[1] Faculty of Political and Social Sciences, Complutense University of Madrid, 28223 Madrid, Spain
[2] Department of Social Analysis, University Carlos III of Madrid, 28903 Madrid, Spain
[3] Statistics Department, University Carlos III of Madrid, 28903 Madrid, Spain
[4] University Institute on Gender Studies, University Carlos III of Madrid, 28903 Madrid, Spain
[5] Fundación Pilares para la Autonomía Personal, 28003 Madrid, Spain
[6] Spanish National Research Council (CSIC), Institute of Economics, Geography and Demography (IEGD), 28037 Madrid, Spain
* Correspondence: georgicr@ucm.es

Abstract: There are numerous sociological and psychosocial studies, both classic and current, that have analysed the images and representations of older people and aging. If gender, intersectional and land perspectives are added, the literature consulted is only a few years old, particularly in Spanish. In addition, research based on fieldwork from virtual image banks is still scarce and recent. The objective of this paper is to evaluate the images from some free access image banks (like Freepik, Canva, Pixabay, or Storyblocks) of older people from a gender, intersectional and socio-spatial and land perspective. Methods: 150 images have been analysed following different selected criteria: 22 variables related to gender, activity, socio-spatial environment, natural space and land, among others, briefly describe the main methods or treatments applied. The key results show a stereotyped and barely diverse image of old age and aging around positive representations, with a notable absence of images related to loneliness as opposed to the presence of social relationships. A feminization has also been observed in the representations, with an imbalance in the activities that are carried out (care in the case of women and leisure in the case of men) and in the visible space (indoor among women and outdoor among men). Older people are still identified with a rural, traditional, and more defined territory and not with more diverse and ecological spaces, which are more frequently attributed to younger profiles. This evaluation contributes to linking this necessary connection of current issues and challenges to ageism, sexism and other exclusions derived from territory and socio-spatial aspects. However, more research is still needed, and, in fact, a second phase of the fieldwork is underway to broaden the sample and to expand further evaluations of images.

Keywords: land; environmental issues; older people; gender; images; evaluation

Citation: Cruceanu, G.L.; Clemente-Belmonte, S.; Herrero-Sanz, R.; Ayala, A.; Zorrilla-Muñoz, V.; Agulló-Tomás, M.S.; Martínez-Miguelez, C.; Fernández-Mayoralas, G. Evaluation of Older People Digital Images: Representations from a Land, Gender and Anti-ageist Perspective. *Land* **2023**, *12*, 18. https://doi.org/10.3390/land12010018

Academic Editors: Tobias Haller and Dingde Xu

Received: 1 September 2022
Revised: 22 November 2022
Accepted: 29 November 2022
Published: 21 December 2022

1. Introduction

Images, which are central to audio-visual culture, arise special interest as qualitative and quantitative material for analysis and social research. However, these digital artifacts [1] give rise to theoretical debates concerning the subjectivity/objectivity of the images, debates from which the mirror metaphor emerges as a critique [2]. From this approach, images are not characterized by reflecting reality itself, i.e., they are not only a window from which one can view the world, but rather a reflection of stereotypes, prejudices, discourses, and cultural codes of the subject that creates the image [2]. In turn, images construct a collective identity of older people and shape a social imaginary about old age [3,4]. Indeed, the analysis of images (and their meanings) provides a glimpse of the social construction of old age [5] or old ages, in plural, although this mosaic of social representations is still not diverse [6].

Unlike social networks, where users are the ones who represent themselves in the virtual space and maintain or create new identities [7], image banks are portals of stock photographs, illustrations, videos, vectors, or icons, among other resources created by professionals, and are used in multiple areas, thus allowing the democratization of the use of images [8]. In the context of the current social and digital transformations, the consumption and production of still images acquires a certain complexity [9] as well as a privileged position by being at the centre of people's lives [10]. Images offer "[individuals and collectives] the possibility of feeling represented, self-defining, capable of interpreting or understanding reality" [10]. In other words, its study "contributes to the understanding of the social world by revealing and updating aspects that are not conscious" [11], such as sexism and ageism, which may go unnoticed, but which, together with racism, are the main forms of discrimination [12].

The perception or social imaginary of old age in Western culture has been constructed negatively, "we associate old age mainly to the decline of all aspects of personal and social life" [13]. Moreover, Simone de Beauvoir classified it as an immutable social construct [14], and it still represents a problem of social and environmental inequalities [15].

Since the production and reproduction of images, values and stereotypes has a great impact on the individual and social levels, there is a need to investigate and evaluate the images that represent old age and older people. The purpose of this article is to answer a series of research questions about the images associated with old age, the way in which older persons are represented and where they are located. In answering these questions, we also consider related aspects such as dependency, the leisure and free time activities in which they participate, the social relationships they maintain, or their emotional state, along with the application of a gender, anti-ageist, intersectional and inclusive perspective.

Background

Studies on the relationship between images, aging, gender and the environment are still scarce but sufficient to show the existence of both an academic and a social interest on the matter. It should be noted that both classical and recent authors, especially from Sociology, Social Psychology or Communication, have dealt with the importance that images (stereotypes, labels, social representations...) have for the identity and better self-perception of the population. However, the references that allude to the images of older people from a gender, socio-spatial and intersectional prism are scarce.

Image banks comprise primary information, the images per se, and secondary information, which includes image description and further information [16], although the latter is not always available. When this happens, there are other ways to obtain information, like with descriptive labels, titles that briefly indicate what the image is intended to represent or with the similar images proposed by the image bank. Thanks to metadata, understood by Mevillet [17] as "the structured set of data describing a resource, book, image, article, video, audio document, etc.", it will be possible to make use of images as an analytical and interpretative tool [4].

The reasons for accessing image banks are diverse; for example, to incorporate an image, to contextualize a newspaper article or to use for an Instagram post. It is necessary to point out that the access and use of these images is not cheap, so not everyone has the economic capacity to acquire them: we are talking about paid image banks such as iStock, Dreamstime or Depositphotos. However, there is a wide range of open access and free-to-download image banks, either with or without the need to register on the platform.

Regarding image evaluation, previous studies have used this audio-visual methodology. One can find, for example, the study conducted by Anne-Vinciane Doucet in 2008 [17] in which she analysed the description of 30 images from environmental image banks. Other studies focus their attention on aspects such as copyright management systems [18], image banks and their characteristics [19] or on the type of images available, their conditions of use and collaborative creation [20]. From another point of view, the evaluation of images has not taken place exclusively in digital spaces; authors such as May Narahara (1998) [21] or Angela M. Gooden and Mark A. Gooden (2001) [22], among other studies in this line, conducted evaluations of images found in textbooks. Others emphasize the image of older people in mass media or in movies [23].

Objective information on old age and aging can be found in statistical institutes. In Spain, the National Institute of Statistics (or Instituto Nacional de Estadística in Spanish), as of 1 January 2021, puts the population aged 65 and over at 9.38 million people, which makes up almost one-fifth of the total population [24]. To this, the phenomena of the feminization of old age (there are more women than men and their life expectancy is higher), of masculinization (overrepresentation of men) in some rural areas, and the aging of aging (increase in the number of people aged 80 and over) are added, which imply the presence of groups of people in more need of attention and care [25].

The most prevalent approaches to old age both in the academic debate and in the social discourse are associated with the most negative aspects that focus on the one hand, on the deterioration of functional capacities and, on the other, on the non-production of goods after retirement. These discourses can negatively affect younger generations in their perceptions and behaviours toward older persons [26], influence professional care for this group [27], and affect self-perceptions about aging itself, which, in turn, can impact health, performance, and longevity [28]. In a meta-analytic review, it was observed that the most negative attitudes occur towards older people and that women are doubly stereotyped for being female and older [29]. Indeed, in the case of women the prejudice is twofold since the sexist stereotype (where the reproductive role of women loses meaning) is added to this negative view [30].

In contrast to the traditional mainstream media discourse on old age as a decadent, unproductive and dependent stage, there is a more recent positive discourse based on the concept and framework for action to promote Active Aging, proposed by the World Health Organization at the Second World Assembly on Aging held in Madrid in 2002 [31–33]. Some authors warn of the risks that an idealized vision of old age may entail, thus becoming a potential source of frustration for older people when certain goals prove unattainable [34].

Sociological studies on images and women, or images and older persons, are numerous. However, research that considers age and gender intersectionality, that is, focused on the study of images and older women, is much scarcer. This assertion is based on the bibliographic approach to studies on images, aging, and gender.

Regarding spaces and the environment, the residential environment can be considered as the closest geographical space where the older people population maintains its social relations and develops its life [35], which is constituted, therefore, by the home and the neighbourhood. In this way, the state of housing, its ownership [36] and a better assessment of the closest residential environment will directly affect the well-being and quality of life of the older persons [35,37]. Furthermore, the characteristics of the physical environment, social and family relations, accessibility and safety are key to the promotion of leisure time activities that help encourage active aging [38].

Indeed, during the aging process, physical and mental capacities tend to diminish, and there is an increase in the time spent in the closest physical environment. Parks, green areas and natural environments, as well as direct contact with nature, are beneficial since they favour the performance of daily life activities and the development of social relationships [39,40]. Thus, the natural environment becomes an extension of the home, creating emotional ties with the environment through experiences and memories. These images in the natural environment are a determining factor in the active aging process [38]. In contrast, the replacement of green areas by urban ones decreases outdoor leisure activities and social relationships and, therefore, has a negative impact on active aging [41]. For this reason, some studies propose social intervention programs for environmental conservation focused on older people [42] and, recently, there have also been some studies that evaluate the few programs aimed at older persons and the environment [43].

It should be added that, despite their importance for older people well-being and quality of life, studies that evaluate the environment through images, whether digital or not, are scarce. Therefore, it is pertinent for research to focus on the way aging is experienced while acknowledging aspects such as gender, the environment and images that reflect old age [6], as well as considering population aging as an enrichment strategy for societies [43] and broadening the scope to environmental gerontology, the geography of aging, physical activity in the natural environment, age-friendly cities, etc., to mention a few of the consulted studies.

2. Materials and Methods

Even if we deal with percentages and figures, a qualitative methodology has been carried out using virtual ethnography, also known as digital ethnography or netnography, as a research technique, which can be defined as a "qualitative research method that adapts ethnographic research techniques to study cultures and communities that emerge through computer-mediated communities" [44]. Through this qualitative digital research technique, whose pioneer is Christine Hine with her work Virtual Ethnography (2000) and in Spain its main reference is Elisenda Ardevol [45–47], a descriptive knowledge of the cultural artifacts [48,49] of interest is obtained; in this case, from digital images retrieved from open access image banks.

An image bank is an online repository and important audio-visual documentation tool [16] of photographs, illustrations and other visual materials that can be accessed, viewed and downloaded for use in a variety of settings. Bearing in mind that there is a wide range of repositories in the virtual space, a series of selection criteria have been established for image banks, which must be: 1. open access, i.e., they must allow images to be viewed and downloaded, and 2. they must have royalty-free images (creative commons) or have the possibility of downloading them for non-commercial use. However, in some cases access is restricted and this may be due to having to register on the platform, attributing authorship by means of a link or subscribing, which implies a financial outlay. Taking these limitations into account, the chosen image banks were Freepik, Storyblocks, Canva, Pixabay and Pexels.

The list of selected concepts, in Spanish and in English, came from more extensive lists. These initial broader glossaries were based on previous studies, both by the authors (their projects, theses, papers) and from other publications and experts on these topics. The final list was also used to substitute words when a concept in an image bank did not show any results. In this way, the following terms were searched in Spanish: "personas mayores" (n = 35), "vejez" (n = 27), "edadismo" (n = 17), "tercera edad" (n = 14) and "anciano/a" (n = 3) and its correlate in English: "elderly" (n = 18), "ageism" (n = 10), "retired people" (n = 26). Other concepts included in the above-mentioned list are: "retired", "elder", "old age", "older people", "older persons", "ageing" and "aging". A few more have recently been incorporated and are being evaluated in a second phase of this research, currently in progress, to broaden the sample. It should be noted that during the virtual fieldwork, the most frequent search suggestions made by people who consult these platforms, which are linked to the introduced concept, were also recorded. In other words, the alternative suggested by the algorithm was included. For example, when searching for "elder people" the virtual image bank recommends searching for "elder people sunset", "older/elder people care", "older/elder people activities", among others. This "spontaneous" pattern offered by the platforms reveals how users search for images that represent older people and, furthermore, it can be considered as an indicator of expressions that are associated with older persons.

The following criteria were defined for the search and selection of images: 1. They must be photographs of people, i.e., other audio-visual materials such as illustrations or videos were not selected, and 2. The filters "most popular" or "most downloaded" were applied in each search, depending on the terminology and options of each image bank.

They did so in order to evaluate the most viewed or downloaded images and were therefore more likely to have been used in personal, academic or work projects, among others.

Once the search has been performed, the selected images are the first ones to have met the above criteria. To avoid any bias linked to the subjectivity of the researchers, the number of images selected was previously determined randomly by an automatic generator. As is often the case in international image banks, some searches do not give back any results, in which case the term is replaced by another from the available list.

The fieldwork was conducted between December 2021 and February 2022. The sample obtained was made up of 150 images of older people that had been extracted from the previously indicated image banks. These audio-visual materials, which reflect situated views [48] on aging, have been stored in our own database as a way to solve the problem of the ephemerality of data in digital environments. In addition to archiving the images, the characteristics observed in the images themselves and in the metadata, together with the link for their identification, have been recorded.

With the contrasted design, each of the selected images was evaluated individually following the structure of 22 variables that describe the people who appear in them, and the environment and the activities they are carrying out (Table 1). Likewise, the following metadata have been recorded to collect additional information to contextualize the content reflected in the images (keywords and title of the images) and to help locate them in the image bank (suggestion box and filter applied, in this case, the most downloaded).

Table 1. Variables used for image evaluation (n = 150 images).

Sex	Type of Group	Skin Colour and Other Physical Traits	Teeth	Hair	Skin	Clothing	Dependency	On Whom It Rests	Disability and/or Use of Devices/Prostheses	Emotional State
Woman	Intergenerational	White	Not visible	Bald	Wrinkled	Not visible	Dependent	Older person	Not visible	Unrecognizable
Man	Intragenerational	Black	In good condition	Long	Stretched	Sport	Independent	Grandson/Granddaughter	Wheelchair	Anger
Nonbinary Man and woman	Does not apply	Asian	In bad condition	Dyed	With make-up on	Casual/informal	Unknown	Other	Cane	Joy
Two men		Oriental	Denture	Short	Aesthetic retouching	Formal			Jaw protheses	Surprise
Two women		Albino		White	Not visible	Mourning attire			Audiphones	Disgust
Group of women		Vitiligo		Grey Transition					Glasses	Sadness
Group of men		Interracial		Not visible					Electric lift	Fear
Mixed group		Mestizo							Blind	Unrecognizable: facemask
		Indigenous							Leg prothesis	
		B/W picture							Arm prothesis	
									Bedridden person	

Image planes	Foreground	Objects	Relational gesture	Natural space	Space	Space description	Activity	Performed activity	Activity description	Literal title of the image in the source
One person	The older woman	Technological	Physical contact	Yes	Residence, residential centres	Open response	Passive	Cultural	Open response	Open response
Several people on the same plane	The older man	Sanitary	Without contact	No	Housing/house		Active	Sport		
Several people, but one stands out	The young person	Sport		Unknown	Collaborative housing		Portrait	Technological		
		Decorations (flowers)			Urbanised street		Other: open response	Domestic		
		Household items			Beach			Care		
		Without objects			Park			Communicative		
		Others: open response			Garden			Portrait		
					Desert			Leisure		
					Sea			Work		
					Mountain			Passive action (sitting down)		
					Countryside					
					Forest					
					No background					
					Other: open response					

3. Results

As a result of the fieldwork, a total of 150 images have been evaluated. However, given that in many cases the images represent couples or mixed groups, more than 200 older people have actually been observed. In addition, it should be clarified that this is a qualitative study that aims to observe, for example, various aspects of daily life and the territorial location where older people are represented have been addressed Thus, the results obtained make it possible to shape and interpret the image of old age that is created and transmitted in different areas, based on the use of images from open access and free downloadable image banks.

3.1. Older People in Digital Images: General Features

Firstly, and in response to the question "How are the older people represented in digital images?", physical and emotional characteristics have been evaluated based on the variables observed in the images themselves and/or in the metadata.

In the 150 images evaluated, older persons appear in the images accompanied by other older people (see Figure 1), family members or health professionals (n = 86). In these cases, the representation of couples composed of a man and a woman of the same age (n = 44), that is, intragenerational couples, stands out. On the other hand, in 64 of the images older people appear alone. More precisely, older women are frequently shown represented alone (n = 35) or accompanied by other women of different ages (n = 16), and older men appear in the images mostly alone (n = 27) or together with other older men (n = 11). As a result of the above, it is worth looking more closely at who is accompanying the older persons. It has been observed that the images show older people together with persons of the same age (n = 55) or they appear in the company of younger people (n = 31).

Figure 1. Couple of older persons. Source, Freepik. https://www.freepik.es/foto-gratis/linda-pareja-ancianos-ventana-busca-alguien-esperando_15203032.htm (accessed on 22 November 2022).

Another aspect of interest that has been noticed in the images is the plane in which the older persons are placed. The plane is understood as the position that a person occupies in the photograph with respect to other people. A person can be found in the foreground, making their presence stand out, or in the background, behind other people. It has been noted that in the case of couples or groups, they are found in the same plane or have greater prominence, either because the focus is on one person or because they are doing something while the others are looking at them. In 61 of the cases, the people who appear in the images are on the same plane while in 25 of them only one person stands out. In the latter, this tends to occur more frequently in couples consisting of a man and a woman, with the older woman being in the foreground.

In terms of physical appearance, out of the approximately 231 older persons who appear in the images evaluated, 181 are white and the rest are distributed among people of African descent (n = 14), Asians (n = 14), Middle Easterners (n = 6) and Latin Americans (n = 2). Out of the remaining 14, 6 are black and white images and could not be identified and 8 images show people of different ethnicities together.

As seen in this image, in 134 of the cases people's teeth are not visible and when they are, 89 are in good condition (3 appear to be false teeth) and only 5 are in poor condition. Another characteristic observed is their hair, with similar results: 76 of the older persons have white hair, followed by people whose hair is in the process of turning grey, which has been called "grey transition" (n = 71), 7 are bald, 8 of the older people have short hair, 12 have long hair and 19 have dyed hair, particularly women. Thirty-eight of the cases could not be identified as any kind due to their hair not being visible.

As with teeth, another aspect that can be observed in the images is the appearance of the skin of the older persons. The results show that 120 have wrinkled skin, 23.8% have stretched skin, 15 have make up on (it should be noted that sometimes these are photographic portraits taken in studios, so that people's appearance is often touched up) and in 9 of the images there seems to be some aesthetic retouching. In the remaining 32 images, the skin is not visible or does not allow a clear classification.

Finally, the clothing of the people appearing in the evaluated images is mostly casual/informal (n = 174), followed by formal (n = 28), not visible (n = 20) (for example, portraits of hands), eight sportswear and only one are mourning clothes.

3.2. Older People: Dependence and Care

When evaluating the images, the focus was also put on the activities the seniors were doing. Sixty-one are photographic portraits, i.e., they are stock photographs created in a studio. Even so, in 51 of the images the older people are performing some active action or activity that requires physical movement (see Figure 2), compared to the 32 that are reflected without performing any activity. The remaining 6 are images that focus on the hands of older people and since there is no significant metadata available, it is not possible to make an accurate classification.

Figure 2. Older woman cycling. Source, Canva. https://pixabay.com/es/photos/parque-la-bicicleta-mayor-solitario-5528190/ (accessed on 20 November 2022).

Based on this distinction, different kinds of activities can be identified and it can be seen more specifically that 43 of them are portraits, in 22 images they are shown to be performing leisure activities, 20 show passive actions, 16 are care activities, 11 are sports activities, 11 show activities related to the use of new technologies, 9 are communicative activities, 7 are domestic, 6 are work-related activities and finally, 5 of the images show cultural activities.

The images also show a difference in terms of the activities carried out by older people according to sex. In this sense and focusing on the cases in which the older persons are alone, regarding men, the activities shown are mainly related to, for example, new technologies (n = 4), leisure (n = 4), work (n = 3) or cultural activities (n = 2). On the other hand, older women are frequently shown to be engaged in passive action activities (n = 8), sports (n = 4), domestic (n = 3), leisure (n = 2) or care activities (n = 1). Depending on the type of group, it is observed that in the images showing a group of men, the activities they carry out are mainly leisure activities (n = 6) while women are represented carrying out care activities (n = 6). In the case of couples made up by a man and a woman, they appear performing care activities (n = 7), passive action activities such as sitting on a bench or sofa or contemplating a natural landscape (n = 7), leisure activities (n = 7) or sports activities (n = 4), such as walking in natural environments or stretching exercises.

Another aspect of interest that the evaluated images allow us to notice is the relationship between care and dependence (see Figure 3). In this sense, it can be observed that the representation of auxiliary technological devices is infrequent: of the people observed in the 150 images, 171 appear without any aid device for mobility and/or prosthesis. Although scarcely, in some images older people use glasses (n = 46), a cane (n = 10), a wheelchair (n = 3), or are in a hospital bed (n = 1). Moreover, in 26 of the 150 images, older people are being cared for by a younger person (n = 20) or another older person (n = 6). Two images show dependent but lonely elderly people and six of the images show older people caring for a minor, which according to the metadata are their grandchildren. However, 116 of all images show independent older people.

Finally, in the images evaluated it was observed whether there was physical contact and the emotional state of older people. Out of the total number of images evaluated, in 57 of them there is visible physical contact, of which in 37 images it is people from the same generation and in 20 it is between people of different ages. The most frequent relational gesture observed is linked to hands and hugs. In terms of their emotional state, older people in the images are mainly shown to be cheerful (n = 124). In 91 of the images, no emotion is shown as they have their backs turned, are wearing masks or because the photograph focuses on their hands. The rest of the emotions represented was sadness (n = 10), surprise (n = 2), anger (n = 2), and fear (n = 1).

3.3. Older People: Environmental Issues and Land

Another objective of this digital research was to observe the spaces in which older people are located in the images evaluated. In 44 of them it is unknown, either because they are portraits without a background or because there is no additional information to contextualize the images. Even so, 68 of the older persons are represented in outdoor spaces, and 38 in indoor spaces, mainly in housing (n = 21) or healthcare environments (n = 6) while only 2 are shown in residences.

As for outdoor spaces, older people appear mainly in urban spaces (n = 43) followed by natural/rural spaces (n = 25) (see Figure 4). Specifically, the spaces in which older people are most often located are parks (n = 22), gardens (n = 11), in the urbanized street (n = 10), in the countryside (n = 8), in the forest (n = 7), at the beach (n = 6), in places near the sea (n = 2), in the mountains (n = 1) or in the desert (n = 1).

Figure 3. Older man with a cane. Source, Pixabay. https://www.canva.com/photos/MADyQ5rAJ-0-photo-of-elderly-man-walking-on-pavement/ (accessed on 20 November 2022).

Figure 4. Older and younger women appear in natural space. Source, Freepik. https://www.freepik.es/foto-gratis/anciana-jardin-joven-nieta_7397387.htm (accessed on 20 November 2022).

In indoor spaces, 22 of the older people appear in couples or groups, 11 women and 5 men. In outdoor spaces, the distribution according to sex is similar, although older women are more represented in urban environments (n = 16) compared to men (n = 12), and in natural/rural environments there are more men (n = 8) than women (n = 7).

As for the activities that older people perform, in the case of natural environments they are sports (n = 6) or leisure activities (n = 3) and, in urban environments, they are mainly leisure activities (n = 15) and passive actions (n = 4).

Finally, regarding the emotional state of older people according to the space in which they are located, the images show older persons are happier outdoors (n = 60), especially in parks (n = 25). In contrast, sadness appears more in images in which the older persons are at home (n = 4) or healthcare environment (n = 1), that is, in indoor spaces. In the other cases the space is unknown since they are portraits.

4. Discussion and Conclusions

This study, by means of virtual ethnography applied to a set of 150 images obtained from open-use image banks, attempts to answer how old age and aging are constructed in the collective imaginary. First of all, attention has been paid to how older people are represented in the analysed images.

It has been observed that there is a feminization of old age [25] since there are many more women portrayed than men, and they usually have more prominence in the images, as they are placed in the foreground.

Another one of the highlights has been the representation of social relationships that older people maintain. Perhaps the most surprising result has been not having found a single image in this sample of a group of women from the same generation. It could be considered as friendship groups of the same age seems to be something exclusive of male old age and not so of female old age. Older women appear alone or in the company of family members (daughters/sons, granddaughters/grandsons) or health professionals, but never in the company of other older women that we could interpret as friends. Unlike women, the men depicted in these images do have a social life, and the leisure activities they engage in are evidence of this gender stereotyped representation.

Then, the image of loneliness, usually attributed to old age and aging [32,33], is relatively absent, and, in fact, in most of the images older people appear accompanied. This representation of older people is predominantly accompanied by an old age idealization that carries certain risks as it may produce frustration in older persons when their experience differs from this representation [34].

The company and the activities they carry out determine the aging of the person, since social and family relationships are key aspects for the promotion of leisure and free time activities that help promote active aging [38]. Therefore, a highly differentiated representation of these aspects between men and women has very relevant implications, both for self-image and for older people's identity as a better valued group, without biased images that are not excluded by gender, age, space, or other key variables [1].

The analysed images portray people who mostly perform active activities, so it can be stated that they show a positive image of aging, an active aging. Leisure and care are the most represented kinds of activities, as well as passive actions which are mainly portraits. However, differences are found between the main activities performed by men and women. The little representation shown in the images of the labour market is starred by men, as is the use of technology or leisure. Women, on the other hand, are portrayed performing mainly passive, sporting, and domestic activities. In addition, they are represented carrying out care activities, but men are hardly seen performing this kind of activity. Moreover, older people are observed doing sports activities (such as cycling, walking, stretching . . .) in natural spaces, which again demonstrates the representation of active aging. It is mainly women who carry out this type of activity, which is a solitary leisure activity. Leisure is also very much determined by the space in which it takes place.

In terms of dependency and care, older people are shown in these images are mostly independent. This is reflected by the infrequent representation of assistive technological devices, although they actually have a very positive impact on the independence of older persons [50], and why older persons are depicted as caregivers but also as recipients of care; they care for and are cared for. In addition to this, the images illustrate active aging in the activities captured, as mentioned above. It is important to highlight the representation of hands in these images, which seems to allude to care, as they are hands that support other hands.

In the analysed sample, a positive representation of old age as a non decadent vital stage has been observed, showing cheerful older people, accompanied by others, with physical contact, and carrying out activities. It is an image close to the objective of active aging [33,34], but also a stereotyped one, as indicated by different studies [3,5], allowing us to glimpse at prejudices associated with this group. In addition, since most of the images are stock photographs, that is, photographic portraits, older people are portrayed with a neat appearance, ranging from their clothing to their physical features (hair, teeth...), which reflects the good health of these people.

Also, most of the people who appear in the images evaluated are white, well-groomed and in couples. It can also be interpreted that, in the cases in which there is also physical contact and/or some object such as a ring, they could be heterosexual couples living together in the same household. With this, it can be concluded that the representation of old age continues to be unvaried [6], as it excludes homosexual or interracial couples, non-binary, and racialized people; rendering invisible numerous minorities in old age and thus corroborating that sexism, ageism and racism continue to be the main forms of discrimination [6,12]. A fundamental question in this digital research is to observe in which spaces older persons are located, and they are more frequently represented in outdoor spaces, but in urban contexts. That is, they are mainly located in parks and gardens, natural spaces transformed and adapted by and for human beings. These spaces, as well as the direct contact with nature, present benefits for older people by favouring the performance of daily life activities and the development of social relationships [39,40]. This is reflected in the images as they show that the leisure activities realised in these spaces consist of board games (dominoes, chess, or cards), dancing, playing with grandchildren, or sightseeing.

On the other hand, older women are mostly represented at home while men are mainly depicted in natural/rural spaces, a phenomenon that is probably due both to the fact that they are doubly stereotyped for being women and for being older [29], and to the masculinization of old age in rural areas [25].In general, there is little representation of older people in residences, and this limitation should be highlighted because it is difficult to distinguish homes from residences.

In the evaluated images older people are not directly shown caring for the environment; it is an absent representation: Older persons are portrayed in natural urban spaces. These images reflect a certain proximity to the natural environment. Nonetheless, older people represented as active actors involved in the care and conservation of the environment is practically non-existent, so it would be promising to promote programs with these objectives focused on older people as proposed by some studies [43].

Finally, the emotional state reflected, in these images, is fundamentally one of great joy, which indicates a positive perspective on old age [51] and shows us, in short, a representation of the well-being and quality of life of this group [35,37]; even though it may also reflect a stereotypical view of old age [34].

It is also noteworthy that in natural/rural environments, older people appear happier than in urban spaces. As for the research itself, it should be added that a second phase of fieldwork is in progress to check whether or not the same trends are being followed, update data and contrast information with a larger sample of images. In addition, the field of study has also been diversified to include other image banks such as Gratisography, Pinterest, Unsplash, SplitShire or OpenPhoto.

As for the specific objectives and design aspects, similar sampling, recording and storage decisions have been made as the ones applied in this first phase of the digital social research. Virtual ethnography also poses specific challenges, such as the problems of access and use of certain cultural artifacts [48], making it difficult to evaluate them even if they are of interest to the study. Another methodological limitation is the problem of data ephemerality, and, for this reason, archival storage decisions have focused on archiving the evaluated images in addition to recording the direct access link. Other study limitations are the need to: (1) analyse other image banks to contrast the representations of older people at an international level, (2) increase the sample of images to be able to, in the future, speak of a greater numerical representativeness, and (3) the trouble that results from the difficulty of different evaluators applying the 22 variables or labels (e.g., about their skin) and agreeing on them. This last methodological limitation was overcome over the course of various debates and reflections that led to the reformulation of some categories and the re-evaluation of some images. Despite all this, the advantages of the study are based on the accessibility and richness of the data and on the fact that it is an ethically non-invasive social investigation.

Author Contributions: Conceptualization, G.L.C., M.S.A.-T. (original idea and coordination of meetings) and G.F.-M.; Methodology, G.L.C., S.C.-B., M.S.A.-T., V.Z.-M. and G.F.-M.; Investigation and process, G.L.C., S.C.-B., A.A. and G.F.-M.; Formal analysis, G.L.C., S.C.-B., R.H.-S. and C.M.-M.; Writing—original draft preparation, G.L.C., S.C.-B., A.A., R.H.-S. and C.M.-M.; Writing—review and editing, G.L.C., S.C.-B., M.S.A.-T., R.H.-S., C.M.-M., V.Z.-M. and G.F.-M.; Visualization, G.L.C., S.C.-B. and V.Z.-M.; Project administration, G.F.-M. and M.S.A.-T.; Funding acquisition, G.F.-M. and M.S.A.-T. All authors have read and agreed to the published version of the manuscript.

Funding: Ministry of Science, Universities, and Innovation, corresponding to the Program of R&D Activities between research groups of the Madrid Region, and co-financed by the European Social Fund. R&D Activities Program ENCAGEn-CM: "Active Ageing, Quality of Life and Gender. Promoting a positive image of old age and ageing, against the ageism", https://encage-cm.csic.es/ (accessed on 20 November 2022) (Ref. H2019/HUM-5698) main research G.F.M, 2020-2023 and Prof. Dr. Agulló-Tomás who has contributed with a personal provision and provisions proceeding of UC3M. This project was granted in 2022 (see more information: http://iegd.csic.es/es/article/programa-actividades-id-encagen-cm-coordinado-csic-recibe-premio-fundacion-pilares) (accessed on 20 November 2022).

Data Availability Statement: The following documents that have been prepared for this work can be consulted: (1) Own image bank that includes the 150 evaluated images https://drive.google.com/drive/folders/1HAYo4Shx_6N-qnk_CRR3OwCJXW07yC1d?usp=sharing (accessed on 20 November 2022), (2) List of some diverse studies on ageing and gender https://docs.google.com/spreadsheets/d/1lkyun1NaM3xH-dqAHIzqmpCtjmtc2m66fAV4w58Nb5Y/edit?usp=sharing (accessed on 20 November 2022) and (3) List of some references on older people and the environment https://docs.google.com/document/d/1fdcWW5-fFcMqpI2qA9RbfslzGSUV5Mv64mQ_wrp3mk0/edit?usp=sharing (accessed on 20 November 2022).

Acknowledgments: The authors appreciate all those who contributed to this work for their support and dedication: to María Loro Rubia for her support during the fieldwork and to Mercedes Sánchez Millán for the translation of the article.

Conflicts of Interest: The authors declare no conflict of interest.

References

1. Rose, G. *Visual Methodologies: An Introduction to Researching with Visual Materials*; SAGE Publications: London, UK, 2016; ISBN 978-1473948907.
2. De Miguel, J.M. *Fotografía. De la Investigación Audiovisual: Fotografía, Cine, Vídeo y Televisión*; Buxó, M.J., De Miguel, J.M., Eds.; Proyecto a Ediciones: Barcelona, Spain, 1999; pp. 23–34. ISBN 84-922438-1-3.
3. Sánchez-Vallduví, M.V. Fotografía, identidad y territorio: La circulación de imágenes sobre indígenas chaqueños y su operación en la configuración de un imaginario social. In Proceedings of the VII Jornadas de Sociología de la UNLP (Argentina), La Plata, Argentina, 5–7 December 2012.
4. García Gil, M. El uso de la imagen como herramienta de investigación. *Campos En Cienc. Soc.* **2013**, *1*, 363–372. [CrossRef]
5. De Andrés Del Campo, S. Hacia un planteamiento semiótico del estereotipo publicitario de género. In *Revista Signa*; UNED: Madrid, Spain, 2006; pp. 255–283. Available online: https://revistas.uned.es/index.php/signa/article/view/6135 (accessed on 31 August 2022).
6. Agulló-Tomás, M.S. Identidad y representaciones sociales: Conceptos, imágenes y construcción psicosocial de las "vejeces". In *Mayores, Actividad y Trabajo en el Proceso de Envejecimiento y Jubilación: Una Aproximación Psico-Sociológica*; Instituto de Migraciones y Servicios Sociales: Madrid, Spain, 2001. Available online: https://e-archivo.uc3m.es/bitstream/handle/10016/30729/mayores_agullo_2001.pdf?sequence=1&isAllowed=y (accessed on 31 August 2022).
7. Bauman, Z. *Identidad*; Losada: Buenos Aires, Argentina, 2005; ISBN 978-8496375208.
8. Barragán-Gómez, R.; Gómez-Moreno, W. Maestros, Imágenes e Imaginarios: Prácticas y saberes en relación con la didáctica de la imagen. *Pedagog. Y Saberes* **2010**, *32*, 23–31. [CrossRef]
9. Martínez-Luna, S. *Cultura visual: La pregunta por la imagen*; Sans Soleil Editions: Vitoria-Gasteiz, Spain, 2019; ISBN 978-84-12-00973 6.
10. Sola-Morales, S. Imaginarios sociales, procesos de identificación y comunicación mediática. *Rev. Prism.* **2014**, *25*, 3–22.

11. Díaz-Herrera, C. Sociología de la imagen en Silvia Rivera Cusicanqui: Conceptualización teórica y metodológica de una disciplina dialéctica, discursiva y rebelde. *Izquierdas* **2020**, *49*, 2021–2049. [CrossRef]
12. Butler, R.N. Ageism: A Foreword. *J. Soc. Issues* **1980**, *36*, 8–11. [CrossRef]
13. Campillay-Campillay, M.; Calle-Carrasco, A.; Rivas-Rivero, E.; Pavéz-Lizarraga, A.; Dubó-Araya, P.; Araya-Galleguillos, F. Ageísmo como fenómeno sociocultural invisible que afecta y excluye el cuidado de personas mayores. *Acta Bioethica* **2021**, *27*, 127–135. [CrossRef]
14. Cano-Acosta, A.L. *Imaginarios estáticos*; Universidad de los Andes (Colombia): Bogotá, Colombia, 2015.
15. Sánchez-González, D.; Rojo-Pérez, F.; Rodríguez-Rodríguez, V.; Fernández-Mayorales, G. Environmental and Psychosocial Interventions in Age-Friendly Communities and Active Ageing: A Systematic Review. *Int. J. Environ. Res. Public Health* **2020**, *17*, 8035. [CrossRef]
16. Codina-Bonilla, L.; Valle-Palma, M. Bancos de imágenes y sonido y motores de indización en la WWW. *Rev. Española De Doc. Científica* **2001**, *24*, 251–274.
17. Doucet, A.V. La descripción de las imágenes en Internet a través del análisis de 30 bancos de imágenes. *Rev. Gen. De Inf. Y Doc.* **2008**, *18*, 81–105.
18. Muñoz-Castaño, J.E. Banco de imágenes: Evaluación y análisis de los mecanismos de recuperación de imágenes. *El Procesional De La Inf.* **2001**, *10*, 4–18.
19. Navarro-Hernando, I. *Análisis de bancos de imágenes: Descripción y evaluación de Corbis, Getty Images, Agesfotostock y Shutterstock*; Universidad de Salamanca: Salamanca, Spain, 2012.
20. Trabadela-Robles, J. Imágenes de acceso abierto y los bancos de imagen: Estudio de casos. *Prism. Soc.* **2017**, *18*, 309–333.
21. Narahara, M.M. *Gender Stereotypes in Children's Picture Books*; University of California: Los Angeles, CA, USA, 1998.
22. Gooden, A.M.; Gooden, M.A. Gender representation in notable children's picture books: 1995–1999. In *Sex Roles*; Springer: Berlin/Heidelberg, Germany, 2001; pp. 89–101.
23. Agulló Tomás, M.S. En tercer plano. Estereotipos, cine y mujeres mayores. In *Medios de Comunicación, Mujeres y Cambio Cultural*; Muñoz, B., Ed.; Dirección General de la Mujer de la Comunidad de Madrid: Madrid, Spain, 2001; pp. 245–276.
24. Instituto Nacional de Estadística. *INEbase-Demografía y Población*; INE Database: Madrid, Spain, 2022.
25. Agulló Tomás, M.S.; Zorrilla Muñoz, V.; Gómez García, M.V.P. Aproximación Socio-Espacial al Envejecimiento y a los Programas Para Cuidadoras/es de Mayores. *Int. J. Dev. Educ. Psychol.* **2019**, *2*, 211–228.
26. Menéndez Álvarez-Dardet, S.; Cuevas-Toro, A.M.; Pérez-Pandilla, J.; Lara, B.L. Evaluación de los estereotipos negativos hacia la vejez en jóvenes y adultos. *Rev. Española De Geriatría Y Gerontol.* **2016**, *51*, 323–328. [CrossRef] [PubMed]
27. Bustillos-López, A.; Fernández-Ballesteros, R. Efecto de los estereotipos acerca de la vejez en la atención a adultos mayores. *Salud Pública De México* **2012**, *54*, 104–105. [CrossRef] [PubMed]
28. Westerhof, G.J.; Miche, M.; Brothers, A.F.; Barrett, A.E.; Diehl, M.; Montepare, J.M.; Wahl, H.-W.; Wurm, S. The Influence of Subjective Aging on Health and Longevity: A Meta-Analysis of Longitudinal Data. *Psychol. Aging* **2014**, *4*, 793–802. [CrossRef] [PubMed]
29. Kite, M.E.; Stockdale, G.D.; Whitley, B.E., Jr.; Johnson, B.T. Attitudes toward Younger and Older Adults: An Updated Meta-Analytic Review. *J. Soc. Issues* **2005**, *61*, 241–266. [CrossRef]
30. De la Mata Agudo, C.; Salas, B.L.; Farré, A.F. Estrategias para la vida en la cuarta edad: Mujeres que viven solas. *Prism. Soc.* **2018**, *21*, 1–27.
31. World Health Organization. *Active Ageing: A Policy Framework*; WHO: Geneva, Switzerland, 2002.
32. Fernández-Ballesteros, R.; Sánchez-Izquierdo, M.; Santacreu, M. Active Aging and Quality of Life. In *Handbook of Active Ageing and Quality of Life: From Concepts to Applications*; Rojo-Pérez, F., Fernández-Mayoralas, G., Eds.; Springer: Berlin/Heidelberg, Germany, 2021; pp. 15–42.
33. Fernández-Mayoralas, G.; Rojo-Pérez, F.; Rodríguez-Rodríguez, V.; Sánchez-Román, M.; Schettini, R.; Rodríguez-Blázquez, C.; Agulló-Tomás, M.S.; Abad, F. Marco teórico y estudio de diseño e implementación de investigación cualitativa en Envejecimiento Activo, Calidad de Vida y Género. In *Envejecimiento Activo, Calidad de Vida y Género. Las Miradas Académica, Institucional y Social*; Fernández-Mayoralas, G., Rojo-Pérez, F., Eds.; Tirant Lo Blanc: Valencia, Spain, 2021; pp. 47–126. ISBN 9788418329142.
34. Bravo-Segal, S. Edadismo en medios de comunicación masiva: Una forma de maltrato discursivo hacia las personas mayores. *Discurso Y Soc.* **2018**, *1*, 1–28.
35. Rojo-Pérez, F.; Fernández-Mayoralas, G.; Forjaz, M.-J.; Prieto-Flores, M.-E.; Martínez-Martín, P. Residential Environment and Health Conditions Among Older-Adults in Community-Dwelling in Spain: What Influences Quality of Life? In *Environmental Gerontology in Europe and Latin America*; Springer: Berlin/Heidelberg, Germany, 2016; pp. 149–174.
36. Zorrilla, V.; Agulló, M.S.; García, T. La vivienda y su entorno social. Análisis cuantitativo desde las personas mayores de 50 años. *Rev. Española De Investig. Sociológicas* **2020**, *170*, 137–154.
37. Rojo-Pérez, F.; Fernández-Mayoralas, G.; Rodríguez-Rodríguez, V.; Rojo-Abuín, J.-M. The Environments of Ageing in the Context of the Global Quality of Life among Older People Living in Family Housing. In *Quality of Life in Old Age*; Springer: Berlin/Heidelberg, Germany, 2007; pp. 123–150.
38. Cano Gutiérrez, E.; Sánchez-González, D. Espacio Público y Sus Implicaciones in El Envejecimiento Activo en el Lugar. *Cuad. De Arquit. Y Asun. Urbanos* **2019**, *9*, 33–44.

39. Peace, S.; Holland, C.; Kellaher, L. *Environment, and Identity in Later Life*; Open University Press: Berkshire, UK, 2005; ISBN 978-0-335-21511-9.
40. Sánchez-González, D.; Adame-Rivera, L.M. Identidad ambiental y envejecimiento en el lugar ante los retos del cambio climático: El caso de Monterrey, México. In Proceedings of the VIII Congreso Internacional de Geografía de América Latina, Antwerp, Belgium, 15–20 September 2014; pp. 69–92.
41. Kaplan, S.; Kaplan, R. Health, Supportive Environments, and the Reasonable Person Model. *Am. J. Public Health* **2003**, *93*, 1484–1489. [CrossRef]
42. Arenas, L.B.M. Trabajo social con adultos mayores en la protección del medioambiente. *Rev. Trab. Soc.* **2019**, *29–30*, 112–129.
43. Rubio, M. La Educación Ambiental y el Envejecimiento Desde una Perspectiva de Género. Análisis de la Concienciación y Participación Ecosocial de las Personas Mayores Desde la Evaluación de Programas. Bachelor's Thesis, Universidad Carlos III de Madrid, Madrid, Spain, 2022.
44. Freixas, A.; Luque, B.; Reina, A. Critical Feminist Gerontology: In the Back Room of Research. *J. Women Aging* **2012**, *24*, 44–58. [CrossRef]
45. Kozinets, R.V. The field behind the screen: Using netnography for marketing research in online communities. *J. Mark. Res.* **2002**, *39*, 61–72. [CrossRef]
46. Ardèvol, E.; Bertrán, M.; Callén, B.; Pérez, C. Etnografía virtualizada: La observación participante y la entrevista semiestructurada en línea. *Athenea Digit.* **2003**, *3*, 72–92. [CrossRef]
47. Estalella, A.; Ardévol, E. Internet: Instrumento de investigación y campo de estudio para la antropología visual. *Rev. Chil. De Antropol. Vis.* **2010**, *15*, 1–21.
48. Hine, C. *Virtual Ethnography*; SAGE Publications: New York, NY, USA, 2000; ISBN 0-7619-5895-9.
49. Serrano-Pascual, A.; Zurdo-Alaguero, A. Investigación social con materiales visuales. In *Metodologías de la Investigación Social: Innovaciones y Aplicaciones*; Arroyo-Menéndez, M., Sádaba-Rodríguez, I., Eds.; Síntesis: Madrid, Spain, 2010; pp. 217–250.
50. Agulló-Tomás, M.S.; Zorrilla-Muñoz, V. Technologies and Images of Older Women. In *Human Aspects of IT for the Aged Population. Technology and Society*; Gao, Q., Zhou, J., Eds.; Springer: Berlin/Heidelberg, Germany, 2020.
51. Freixas, A. *Yo Vieja*; Capitán Swing: Madrid, Spain, 2021; ISBN 9788412390292.

Article

Aging Industries in the Regional Economy: How to Support an Aging China?

Fan Xu [1], Yongming Huang [2,*] and Qiang Wang [3,*]

[1] School of Economic and Management, and Center for Industrial Economics, Wuhan University, Wuhan 430072, China
[2] China Institute of Development Strategy and Planning, and Center for Industrial Economics, Wuhan University, Wuhan 430072, China
[3] School of Business, Shandong Management University, Jinan 250357, China
* Correspondence: hym@whu.edu.cn (Y.H.); wangqiang@caas.cn (Q.W.)

Abstract: This study investigates the law, distribution characteristics, and changing trend of the coordinated development of China's aging industry and regional economy, as well as the factors which influence the degree of coordination between the aging industry and economic development on the provincial level. In doing so, we construct a comprehensive evaluation index system of the aging industry and regional economy development, introduce an entropy weight coupling model, and measure the coupling and coordinated development level of the two systems using data of 31 selected Chinese provinces (municipalities) from 2009 to 2019. The spatial Dubin model is then used to empirically analyze the influencing factors and spatial effect decomposition of the coordinated development of the aging industry and regional economy. We reach the following main results: (1) China's aging industry is developing unevenly, with substantial regional differences, but these differences have narrowed in recent years. (2) China's regional economic disparities have widened. The eastern regions have the highest level of development, while the northeast region's growth rate of GDP has declined since 2014. (3) The coordinated development of the aging industry and regional economy in one region of China has a positive impact on its neighboring regions, and all Chinese regions exhibits high–high, low–low agglomeration characteristics in terms of their degree of coordination. (4) A variety of socioeconomic and demographic factors affect the coordinated development of the aging industry and regional economy. An important implication of these findings is that, China should improve population structure, population quality, and economic development quality in order to achieve a high-level coordinated development of the aging industry and regional economy.

Keywords: aging industry; regional economy; coupling coordination degree; spatial Dubin model (SDM)

Citation: Xu, F.; Huang, Y.; Wang, Q. Aging Industries in the Regional Economy: How to Support an Aging China? *Land* **2022**, *11*, 2096. https://doi.org/10.3390/land11112096

Academic Editors: Vanessa Zorrilla-Muñoz, Eduardo Fernandez, Blanca Criado Quesada, Sonia De Lucas Santos, Jesus Cuadrado Rojo and Maria Silveria Agulló-Tomás

Received: 28 August 2022
Accepted: 12 October 2022
Published: 20 November 2022

Publisher's Note: MDPI stays neutral with regard to jurisdictional claims in published maps and institutional affiliations.

1. Introduction

According to the data of seventh census in 2020, China's population of people aged 65 and above was 190.64 million, accounting for 13.50% of the total population, indicating that the country is rapidly aging [1]. China found itself in a challenging position of "getting old before getting rich" [2]. Dealing with the problem of population aging is critical in determining a country's long-term economic growth momentum, and is also a key factor in great power competition in the 21st century. China is confronted with unprecedented challenges in rural and urban setting [3]. The "14th Five-Year Plan for National Aging Development and Elderly Care Service System Plan" was promulgated in 2021, officially listing active response to population aging as a national strategy [4]. The development of the aging industry will have a significant and far-reaching impact on China's economic and social development during the "14th Five-Year Plan" period, as well as the overall process of building a socialist modernized country.

In recent years, the importance of high-quality development[1] in China has been emphasized [5,6]. There is a significant relationship between regional economy and the aging industry, which promote and develop in tandem. Therefore, the coordinated development of the regional economy and the aging industry becomes an important part of the regional economy's high-quality development. However, what does not match the rapid development of China's economy is that China's aging industry is still in its infancy and is unable to meet the rapidly growing demand for elderly care service, which induced by the rapid population aging [7]. There are regional imbalances and deficiencies in China's economic development and aging industry [8]. For example, the eastern coastal regions of China are economically developed but have a serious problem with aging, while the western regions are underdeveloped despite having a more favorable population structure than the east [9]. At the same time, the elderly care issue is a problem in large cities such as Beijing, Tianjin, and Shanghai [10]. Therefore, studying the status quo and relationship between the regional aging industry and economic development, as well as analyzing the main factors affecting the coordinated development of them, not only solves the aging problem, develops and strengthens the aging industry, but also provides impetus to the high-quality economic development.

To this end, we pursue three objectives in this study. First, we aim to gauge the law of coordinated development of the coupling between the aging industry and regional economy development, as well as the distribution characteristics and variation trend of the coupling coordination degree. Our second objective is to explore the spatial effects of coordinated development between the aging industry and economic development, and to conduct in-depth analysis of related influencing factors. Third, we hope to provide valuable empirical results and theoretical exploration for the formulation of relevant policies.

The remainder of the study is organized as follows. Section 2 provides a review on relevant literatures. Section 3 presents the data sources and the theoretical framework, while Section 4 analyzes the coupling and coordinated development of aging industry and regional economy. Section 5 empirically studies the influencing factors of coupling coordination between aging industry and regional economic development and Section 6 contains our conclusions.

2. Literature Review

Western academic circles were more mature in studying the ageing problem than Chinese academic circles because most European and American countries entered the aging society with various degrees around the 1980s. They generally refer to elderly-related industries, such as the "health industry" and "silver hair industry", but they have no clear concept of the aging industry. While research on the aging industry in China gradually appeared in papers, academic works, policies, and regulations since developing the undertaking of the elderly was clearly proposed in the 1980s [11], but the concept and connotation of the aging industry are not unified. Some scholars proposed that the aging industry is driven by the increase in demand from the elderly consumer market. Tian believed that the "aging industry" is a general term for the non-governmental profit-making activities that provide goods and services to the elderly with the goal of meeting the needs of high-level life and culture for the elderly [12]. In 2020, the National Bureau of Statistics issued the "Statistical Classification of the Aging Industry (2020)", which defined the aging industry as a collection of production activities that ensure and improve the life, health, safety, and participation of the elderly in social development, and provide the public with the production of various elderly care and related products.

China's economy is currently transitioning from a stage of high-speed growth to a stage of high-quality development. Sun Zhijun et al. believed that high-quality development research is still in its early stages, and that high-quality economic development is guided by the concept of coordination and balance to improve economic stability [13]. From the perspective of economic theory, Jin Bei proposed that high-quality development requires implementation of a comprehensive strategy, as well as balance and coordination of various policy objectives to achieve the multi-dimensional and desirable goal of high-quality development [14]. Therefore, coordinated and balanced development is the core content of high-quality economic development. High-quality development is inclusive and a "co-evolution" in nature, achieving economic, social, and natural synchronization and coordination [15]. High-quality development is reflected in the optimization of industrial structure, the synergy, integrity, inclusiveness, and openness of industrial and regional economic development, as well as production efficiency, technological innovation, and green development at the enterprise level [16]. As an industry meeting the needs of specific groups, the aging industry plays a pivotal role in promoting economic development in China. High-quality development of the aging industry is also an important part of high-quality economic development.

The aging of the population stimulated the development of the aging industry, which is closely linked to economic development. Clark Tibbits, known as the *father of gerontology* in the United States, published the book *Handbook of Social Gerontology* in 1960, which revealed the interaction between population aging and social and economic development, and explained the importance of aging issues [17]. With western countries entering the aging society, economists and demographers began to pay attention to the aging industry, pension issues and its economic effects [18]. Western academics carried out research on elderly consumption, aging policies and systems, the development of the aging industry chain, and the relationship between elderly care and society. It is argued that establishing the government welfare planning system on the basis of the market would be more efficient [19]. In the case of Japan, the aging industry is classified into three categories: standard industry, related industry, and derivative industry [20]. The old-to-young dependency ratio was found to be positively correlated with residents' consumption [21]. Furthermore, the rising consumption rate reflects the impact of the aging population on the social economy [22]. The EU has a similar impact on population aging and fiscal solvency policy responses [23].

In China, the aging industry was studied for more than 30 years. Initially, the study was based on foreign (western) research results. The research on aging industry was gradually expanded to include the perspective of development mode, path, development trend, and economic benefits [24]. In comparison to the early days of qualitative analysis and research on aging industry, Chinese scholars have conducted more and more quantitative research in recent years, with the gradual enrichment and improvement of data. Using a panel data model, Huang Qian et al. examined the impact of population aging on innovation [25]. Li Xinguang et al. employed the dual-zone spatial Dubin model (SDM) to analyze the space of China's population aging on economic growth from the perspective of deep aging [26].

Hou and Liao constructed an econometric model using spatial panel data from 30 provinces in China from 2004 to 2018 to examine the relationship between population aging, innovation, and industrial structure transformation [27]. He and Yang used a coupling index system to empirically analyze the degree of coupling coordination between the broadened health industry and the aging industry [28]. In general, research on the aging industry progressed from qualitative to quantitative, and the research content gradually expanded into other fields and industries of the social economy. However, little research has been conducted on the relationship between the aging industry and regional economy development. Based on the coupling coordination degree model, He and Wang took Jiangsu province as an example and analyzed the spatial–temporal evolution of the aging industry, regional economy, and their coupling coordination in Jiangsu Province in 2005, 2010, and 2015, then put forward some policy suggestions [29].

It is therefore imperative to construct a comprehensive evaluation index system of the aging industry and regional economy development to measures the coupling and coordinated development level of the two systems in order to determine the coordinated development level of the aging industry and regional economy in various provinces and cities in China.

Our study contributes to the existing body of literature in two ways. First, to the best of our knowledge, we provide the first comprehensive analysis of the evolution characteristics, influencing factors and spatial effects of the coordinated development of China's provincial aging industry and regional economy on the basis of provincial panel data. Second, we construct the spatial Dubin model to examine the influencing factors and spatial effect decomposition of the coordinated development of the aging industry and regional economy, which enriches the existing research in this topic and enable us to provide targeted policy guidance on how to make a high-quality development of both aging industry and regional economy.

3. Methods and Data

3.1. Data Sources

In this study, data about the aging industry and economic development are collected from *China Civil Affairs Statistical Yearbook* [30], *China Statistical Yearbook* [31], and the official website of the National Bureau of Statistics. The data that cannot be obtained directly are calculated and sorted. Additionally, the proportion and interpolation methods are used to make up for the missing data. For the convenience of discussion, according to the National Bureau of Statistics in 2011, China's economic regions are divided into four regions: the east, the central, the west, and the northeast. The details are as follows[2]: the eastern region includes Beijing, Tianjin, Hebei, Shanghai, Jiangsu, Zhejiang, Fujian, Shandong, Guangdong, and Hainan; the central region includes Shanxi, Anhui, Jiangxi, Henan, Hubei, and Hunan; the western region includes Inner Mongolia, Guangxi, Chongqing, Sichuan, Guizhou, Yunnan, Tibet, Shaanxi, Gansu, Qinghai, Ningxia, and Xinjiang; Northeast China includes Liaoning, Jilin, and Heilongjiang [32].

3.2. Theoretical Framework

3.2.1. Evaluation Index System

Based on the principles of systems, representative, hierarchy, and data availability, and with the reference to previous studies about indicators of the aging industry and economic development [28,29], this study selects different evaluation indicators to construct an evaluation system for the development level of the aging industry and regional economy (See Table 1).

(1)   The aging industry system includes 3 primary and 9 secondary indicators as follows:

A1 = the proportion of the elderly population aged 65 and above, (population over 65/total population) 100%

A2 = the old-age dependency ratio (ODR), (population over 65/population aged 15–64) 100%

A3 = the old-to-young ratio, (population over 65/population aged 0–14) 100%

A1, A2, A3 reflect the demographic structure. Those three indicator are inverse indicators, which means the larger the data, the higher the degree of aging, and the greater the pressure on elderly care.

A4 = the number of pension insurance participants

A5 = the total income of the basic pension insurance fund

These two indicators represent the level of basic pension security.

A6 = the number of elderly care institutions

A7 = the number of employees in elderly care institutions

A8 = the number of elderly care beds per thousand elderly people

A9 = the quality of aging industry practitioners, (the proportion of college and above personnel)

A4–A9 are used to measure the ability of basic elderly care, which are all positive indicators.

(2)   The regional economy system consists of 3 primary and 7 secondary indicators:

B1 = the regional GDP

B2 = the local fiscal revenue

B3 = the whole society's fixed asset investment

B1, B2, B3 are the indicators of economy development scale.

B4 = the per capita GDP
B5 = the per capita disposable income
B4, B5 are used to reflect the economy development quality.
B6 = the tertiary industry added value
B7 = tertiary industry's share of GDP, (B6/GDP)100%
B6, B7 are used to measure the pros and cons of the industrial structure.
B1–B7 are all positive indicators.

Table 1. Evaluation index system of the aging industry and regional economy development.

Name	First-Level Indicator	Secondary Indicator	Unit	Code	Attribute
Aging industry	population structure	proportion of the elderly population aged 65 and above	%	A1	Inverse
		old-age dependency ratio	%	A2	Inverse
		old-to-young ratio	%	A3	Inverse
	basic pension security	pension insurance participants	10 K	A4	Positive
		total income of the basic pension insurance fund	100 Million	A5	Positive
		number of pension institutions	-	A6	Positive
	basic elderly care	number of employees in pension institutions	-	A7	Positive
		number of pension beds per thousand elderly people	-	A8	Positive
		quality of pension practitioners (college and above)	%	A9	Positive
regional economy development	scale	regional GDP	100 Million	B1	Positive
		local fiscal revenue	100 Million	B2	Positive
		whole society's fixed asset investment	100 Million	B3	Positive
	development quality	per capita GDP	Yuan/person	B4	Positive
		per capita disposable income	Yuan/person	B5	Positive
	Industrial structure	tertiary industry added value	100 Million	B6	Positive
		tertiary industry's share of GDP	%	B7	Positive

3.2.2. Standardization and Weights of Indicators

(1) Standardization of indicators

Since each index has a different measurement unit and dimension and cannot be directly compared, the extreme value method needs to be used to normalize the original data of each index. Based on the attributes of indexes, the larger the value of the positive index, the better, whereas the smaller the value of the inverse index, the better. Therefore, we adopt different calculation methods for the normalization of the positive and inverse indexes. Details are as follows:

$$x'_{ijt} = \begin{cases} \dfrac{x_{ijt}-\min(x_{ijt})}{\max(x_{ijt})-\min(x_{ijt})} \times 0.95 + 0.05, & \text{positive} \\[2mm] \dfrac{\max(x_{ijt})-x_{ijt}}{\max(x_{ijt})-\min(x_{ijt})} \times 0.95 + 0.05, & \text{inverse} \end{cases} \quad i=(1,2,\ldots,n) \tag{1}$$

where i represents the province, j represents the evaluation index, x_{ijt} is the value of the j index of the i province in year t. $\max(x_{ijt})$, $\min(x_{ijt})$, respectively, represent the maximum and minimum values of the index; x'_{ijt} is the dimensionless index value after normalization. In order to avoid the occurrence of zero in the data processing, a coefficient of 0.95 and a constant term of 0.05 are added to the formula [33].

(2) Determination of weights

There are several ways to determine indicator weights. In this study, the entropy value method is adopted, and the weight is determined by the principle of information entropy, which can objectively represent the evaluation index of the research object. The entropy method uses the discrete degree of the index to determine the weight of the index. We collect data of 31 selected provinces from 2009 to 2019 and calculate the weight of each index each year utilizing the entropy method.

First, calculate the proportion p_{ij} of the j index value in province i,

$$p_{ij} = x'_{ij} / \sum_i x'_{ij}. \tag{2}$$

Since p_{ij} is calculated respectively by each year, for simplicity, x'_{ij} in Formula (2) is used to represent x'_{ijt} in (1).

Then, calculate the entropy value e_j of the j index

$$e_j = -k \sum_i p_{ij} ln p_{ij} \tag{3}$$

In Formula (3), $k > 0$, $k = 1/ln(n)$, n is the number of provinces.

Next, calculate the difference coefficient (information utility value) g_j of the j indicator

$$g_j = 1 - e_j. \tag{4}$$

Finally, calculate the weight w_j of each indicator

$$w_j = g_j / \sum_j g_j. \tag{5}$$

3.2.3. Calculation of Coupling Coordination Evaluation Index

(1) Aging industry and regional economy development index

$$U_1 = \sum_j w_j x'_{ij}; \; U_2 = \sum_j w'_j y'_{ij} \tag{6}$$

where w_j and w'_j are the weight coefficients of the aging industry and regional economy, respectively. x'_{ij} and y'_{ij} are the dimensionless values of the aging industry and regional economy indicators, respectively, and U_1 and U_2 are the composite scores of the aging industry and the regional economy.

(2) Coupling Index

$$C = 2 \left[\frac{U_1 \times U_2}{(U_1 + U_2)^2} \right]^{\frac{1}{2}} \tag{7}$$

The concept of coupling originated from physics and was later introduced into the field of regional economy. The synergistic influence between regional systems or between elements within the system is usually measured by the degree of coupling, which reflects the strength of the interaction between the two parties.

In the above Formula (7), C represents the coupling index, $C \in [0, 1]$. When C is close to 0, the systems are independent of each other and are in an irrelevant state. The larger the value of C, the stronger the interaction and the greater the correlation. When C is close to 1, it indicates that the degree of coupling between systems is higher, and the dispersion between systems is minimum. However, the coupling index C can only indicate the strength of the interaction between the systems, that is, the degree of correlation between the two. It cannot reveal the level of coordinated development between the regional economy and the aging industry. Therefore, it is necessary to introduce the coupling coordination index that can measure the development degree of the relationship between the two for analysis.

(3)   Coupling coordination index

$$D = \sqrt{C \times T}, \text{ among } T = \alpha U_1 + \beta U_2 \qquad (8)$$

The degree of coupling coordination is mainly used to analyze the level of coordinated development between different systems (or things). The degree of coordination reflects the benign degree of the coupling relationship between the two parties, and can represent whether the systems promote each other at a high level or restrict each other at a low level. A high degree of coupling indicates that there is a strong interaction between the two parties, while a high degree of coordination indicates that there is a benign and high-level mutual promotion between the two.

In the above Formula (8), C is the coupling index, T is the comprehensive index of the aging industry and regional economy development, and α and β are undetermined coefficients, which, respectively, represent the contribution of the two systems in the coordinated operation. In this study, the two systems are regarded as equally important, then $\alpha = \beta = 1/2$. D represents the coordination index, $D \in [0, 1]$, the larger D is, the more coordinated the system is, and vice versa, the smaller D is, the less coordinated. In order to further understand the state of coordinated development of various regions in China, within the value range of the coordination index, with the reference to the classification criteria [29], the coupling coordination degree between the aging industry and the regional economy is divided into 15 categories in five levels, as shown in Table 2.

Table 2. Classification standard of coupling coordination degree between the aging industry and regional economy development.

D	Type	U_1 and U_2	Feature
0.8 < D ≤ 1	Good coordination	$U_1 > U_2$	Regional economy development lags
		$U_1 = U_2$	Well-coordinated development of the two
		$U_1 < U_2$	Aging industry development lags
0.6 < D ≤ 0.8	Moderate coordination	$U_1 > U_2$	Regional economy development lags
		$U_1 = U_2$	Moderate coordinated development of the two
		$U_1 < U_2$	Aging industry development lags
0.4 < D ≤ 0.6	Basic coordination	$U_1 > U_2$	Regional economy development lags
		$U_1 = U_2$	Basic coordinated development of the two
		$U_1 < U_2$	Aging industry development lags
0.2 < D ≤ 0.4	Moderate incoordination	$U_1 > U_2$	Regional economy development lags
		$U_1 = U_2$	Moderately imbalanced development of the two
		$U_1 < U_2$	Aging industry development lags
0 < D ≤ 0.2	Severe incoordination	$U_1 > U_2$	Regional economy development lags
		$U_1 = U_2$	Severely imbalanced development of the two
		$U_1 < U_2$	Aging industry development lags

3.2.4. Kernel Density Estimation Method

As a non-parametric estimation method, kernel density estimation usually fits sample data through a smooth peak function and uses a continuous density curve to describe the distribution of random variables, which has the characteristics of strong robustness and weak model dependence [34], and is a commonly used method to describe the dynamic evolution trend of random variables. The estimation formula of the kernel density function is:

$$f_h(x) = \frac{1}{nh} \sum_{i=1}^{n} K\left(\frac{x - x_i}{h}\right) \tag{9}$$

$K(x)$ is the kernel density function, and h (>0) is the smoothing parameter, which is called bandwidth. In this study, we choose the optimal bandwidth set by Stata software and use Gaussian kernel function to estimate the kernel density function:

$$K(x) = \frac{1}{\sqrt{2\pi}} e^{\left(-\frac{x^2}{2}\right)} \tag{10}$$

4. Coupling and Coordinated Development and Evolution

4.1. Overall Coupling Coordinated Development and Evolution

4.1.1. Space-Time Evolution of Aging Industry Development

Based on the development index of the aging industry, this study used software to draw a spatial-temporal evolution map of the aging industry development of 31 selected Chinese provinces for the years 2009, 2012, 2016 and 2019, as displayed in Figure 1. As shown, the development of the aging industry in different regions of China in 2009 varied significantly. Four provinces had a development index below 0.25, while five provinces had a development index greater than 0.6. By 2019, the aging industry development index of all provinces was greater than 0.25, while only two provinces exceeded 0.6. These findings indicate that the development of the aging industry is improving in most regions of China, especially in less developed western regions where aging industry developed fast, while the aging industry in developed eastern regions (except Guangdong and Jiangsu) deteriorated. Except Liaoning Province in northeast China, the aging industry in the other two provinces in the northeastern made progress.

From a spatial perspective, China's aging industry has large regional differences. During 2009-2019, the higher level of aging industry were in developed provinces and cities in the central and eastern regions (Figure 1). The value of Guangdong Province, ranked first in 2009 with 0.7778, was 3.8 times that of Hainan Province with the smallest value of 0.2074. In 2019, the value of Guangdong Province, ranked first with 0.7812, was 2.8 times that of Gansu Province with the smallest value of 0.2824. The tendency indicates that the regional differences in the development of the aging industry became narrow, and the development of the aging industry in less developed areas accelerated.

From the perspective of sub-regions, as shown in Figure 2, the development of the aging industry in the eastern and central regions is above the national average. There is indication that the overall development trend of the aging industry in the western region gradually improved, the eastern region has a slight decline, and the central and the northeast improved from 2017 after experiencing a large decline.

4.1.2. Space-Time Evolution of Regional Economy Development

China's regional economy developed steadily from 2009 to 2019. The economic development indices of Yunnan and Jiangxi provinces greatly improved. There are three provinces experiencing economic recession, namely Shanxi, Jilin and Liaoning. Among them, the economic development index of Liaoning Province dropped the most, by approximately 40%, from 0.3829 in 2009 to 0.2289 in 2019.

With the development index of the regional economy, this study used software to draw a spatial-temporal evolution map of the regional economy development of 31 selected Chinese provinces for the years 2009 and 2019, as displayed in Figure 3. As shown, the development of China's regional economies varies greatly across the country from a spatial perspective. Over the past decade, the indexes of regional economic development above 0.4 was concentrated in the eastern regions (coastal provinces). In 2009, the value of Guangdong Province, ranked first with 0.6948 in 2009,, was 5.8 times that of Tibet with the smallest value of 0.1193; the value of Guangdong Province, ranked first in 2019 with 0.6836, was 7.3 times that of Tibet with the smallest value of 0.0936. The tendency indicates the disparity in regional economy became larger, and the polarization became more and more serious.

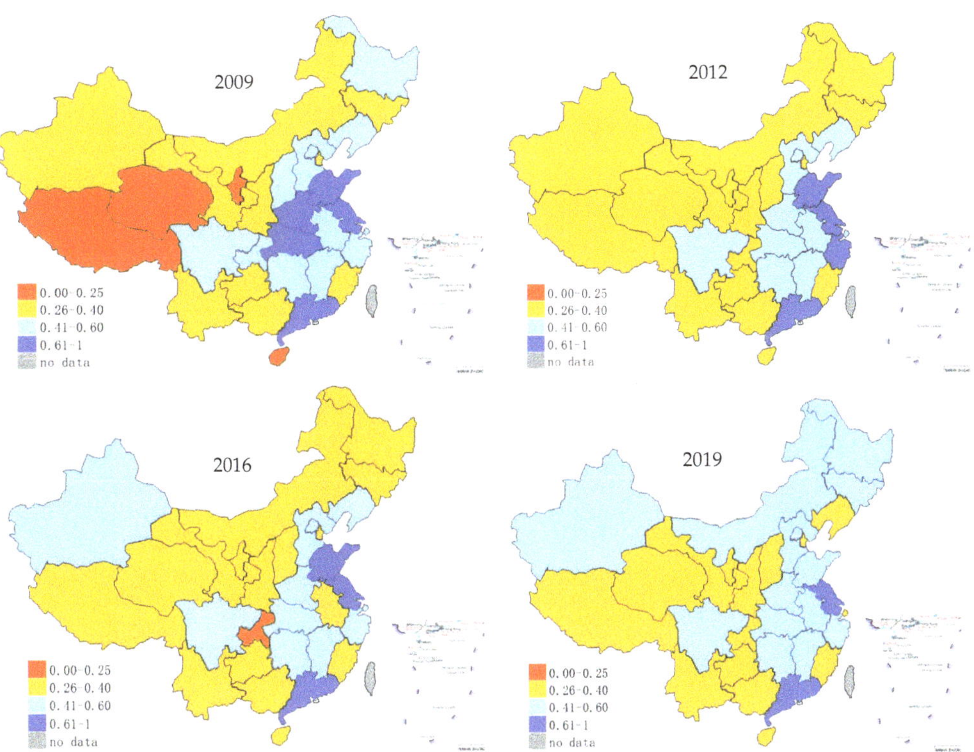

Figure 1. The development level of the aging industry.

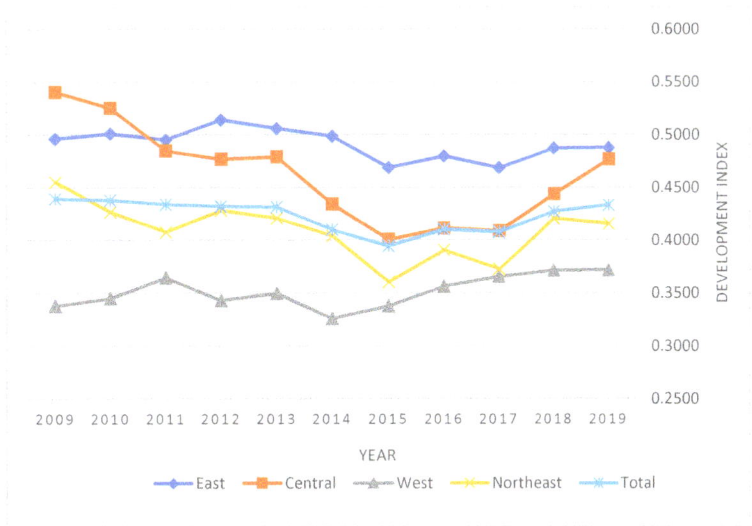

Figure 2. Sub-regional development index of the aging industry.

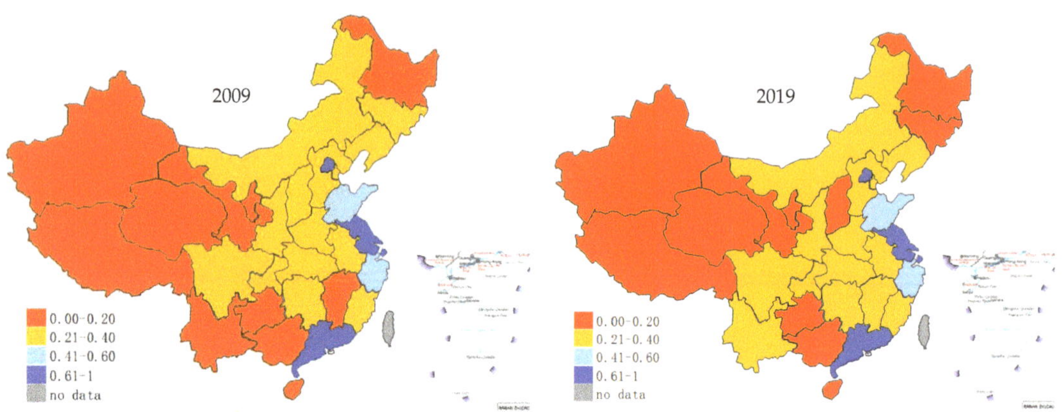

Figure 3. The development level of the regional economy.

From the perspective of the sub-regions, the economic development level of the eastern region is the highest, far exceeding the national average, but it gradually declined in recent years (Figure 4). The development of the central and western regions was relatively stable. While the central region steadily rose and gradually approached the national average level; the western region was stable with a slight decline, and continued to lag far behind the national average. The economy of the northeast region began to decline in 2014, falling down from the same level of the central region to the level of the western region, far below the national average.

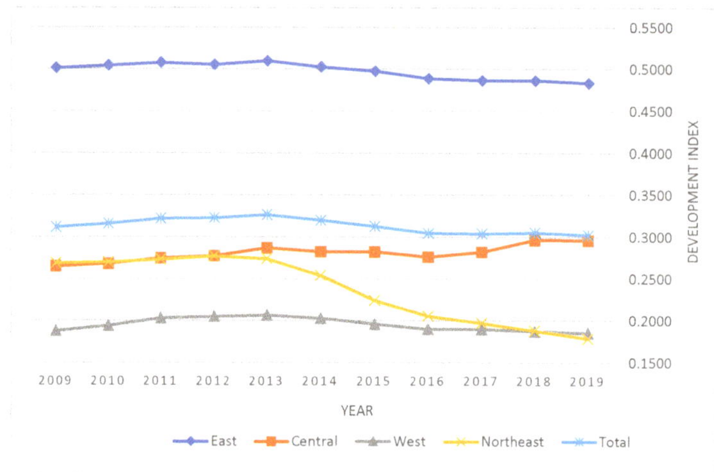

Figure 4. Sub-regional development index of the regional economy.

4.1.3. Space-Time Evolution of the Coordination Index

From 2009 to 2019, the coordination between the aging industry and regional economy development was good, and there was no incoordination. There was little change in recent years, and the quality of coordinated development was not high. Basic coordinated provinces still hold the majority, with 19 in 2009 and 18 in 2019; well-coordinated provinces remained largely unchanged over the 10-year period, with Guangdong first and Jiangsu second; moderately coordinated provinces in 2009 were 10, and 11 in 2019, of which, Liaoning dropped from moderate coordination to basic coordination, while Anhui and Fujian improved from basic coordination to moderate coordination.

According to the division of the coupling coordination type, this study used software to draw a spatial-temporal evolution map of the coordination development between aging industry and regional economy of the 31 selected Chinese provinces for the years 2009 and 2019 as shown in Figure 5. From a spatial perspective, there are regional differences in the coordination between China's aging industry and regional economy development, which is basically positively related to the degree of economic development. As can be found from Figure 5, in the past 10 years, the provinces with better-coordinated development were also the provinces with a higher regional economy level. The coordination index of Guangdong Province, ranked first in 2009 with 0.8574, was 2.1 times that of Tibet with the smallest value of 0.4077. The coordination index of Guangdong Province, ranked first in 2019 with 0.8549, was twice that of Gansu Province with the smallest value of 0.4220. Therefore, the overall coordination level of each region did not change much, and the regional differences in the coordination were less polarized than the regional economy conditions.

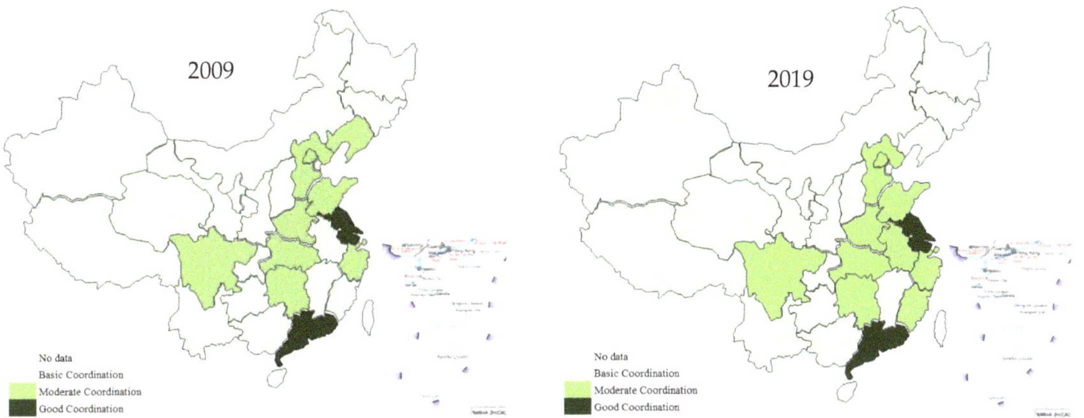

Figure 5. Coordination development level of the aging industry and regional economy.

From the perspective of the sub-regions, as shown in Figure 6, the eastern region had the highest level of coordination, the central region was basically close to the average, and slightly exceeded the average in the last two years. The western region had the lowest level of coordinated development with little improvement, while the coordination status of the northeast region dropped from close to the average in 2009 and approached the level of the western region in 2019. If this tendency continues, the northeast region may become the region with the worst level of coordination.

4.2. Mean Value Analysis

From 2009 to 2019, the mean values of the aging industry development index, regional economy development index, comprehensive development index, coupling index, and coordination index of each province (directly administered) city are shown in Table 3.

Table 4 indicates that the coupling and coordination between the development of the aging industry and economic development in China's 31 provinces was generally acceptable. Guangdong and Jiangsu were in a state of good coordination. Guangdong ranked first and the regional economy development slightly lagged the development of the aging industry. Jiangsu ranked second and the aging industry slightly lagged the development of the regional economy; Shandong, Henan, Sichuan, Hubei, Liaoning, Hebei, Hunan, Zhejiang, Beijing, and Shanghai were moderately coordinated. Among them, Zhejiang manifested the most balanced development. The development of the aging industry in Beijing and Shanghai lagged the economic development, and the economic development of the rest provinces lagged the development of the aging industry. Anhui, Inner Mongolia, Jiangxi, Shaanxi, Chongqing, Jilin, Shanxi, Guangxi, Xinjiang, Yunnan,

Heilongjiang, Guizhou, Hainan, Ningxia, Tibet, Qinghai, Gansu, Fujian, and Tianjin were in a state of basic coordinate. In Fujian and Tianjin, the aging industry development lagged, and economic development in other provinces lagged. In general, although there was no incoordination between the aging industry and economic development, most (over 60%) provinces or municipalities were in a state of basic coordinate. What is more, basic coordinate usually occurred in provinces in the western and northeastern regions

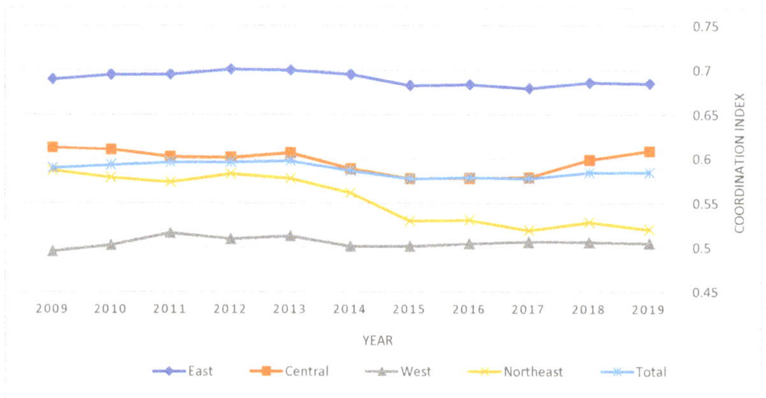

Figure 6. Coordination index of the aging industry and regional economy.

Table 3. Mean analysis table of the coupling coordination index between the aging industry and the regional economy (2009–2019).

Name	Aging Industry Index (U_1)	Regional Economy Index (U_2)	U_1–U_2	Comprehensive Development Index (T)	Coupling Index (C)	Coordination Index (D)	Coordination Level	Rank
Beijing	0.4838	0.6732	−0.1894	0.5785	0.9865	0.7554	Moderate	4
Tianjin	0.3028	0.3758	−0.0731	0.3393	0.9942	0.5808	Basic	15
Hebei	0.4825	0.3120	0.1705	0.3973	0.9767	0.6229	Moderate	11
Shanxi	0.3845	0.1980	0.1866	0.2913	0.9473	0.5253	Basic	21
Inner Mongolia	0.3831	0.2661	0.1169	0.3246	0.9836	0.5651	Basic	16
Liaoning	0.4697	0.3226	0.1471	0.3962	0.9826	0.6239	Moderate	10
Jilin	0.3658	0.2121	0.1537	0.2890	0.9640	0.5278	Basic	20
Heilongjiang	0.3923	0.1784	0.2139	0.2853	0.9271	0.5143	Basic	25
Shanghai	0.4601	0.6446	−0.1845	0.5524	0.9860	0.7380	Moderate	6
Jiangsu	0.6436	0.6994	−0.0558	0.6715	0.9991	0.8191	Good	2
Zhejiang	0.5670	0.5435	0.0236	0.5553	0.9998	0.7451	Moderate	5
Anhui	0.4151	0.2853	0.1298	0.3502	0.9827	0.5866	Basic	13
Fujian	0.3245	0.3554	−0.0309	0.3399	0.9990	0.5827	Basic	14
Jiangxi	0.4654	0.2176	0.2478	0.3415	0.9319	0.5641	Basic	17
Shandong	0.6350	0.5329	0.1021	0.5840	0.9962	0.7627	Moderate	3
Henan	0.5332	0.3411	0.1921	0.4371	0.9756	0.6530	Moderate	7
Hubei	0.5148	0.3338	0.1810	0.4243	0.9770	0.6439	Moderate	9
Hunan	0.4580	0.3099	0.1482	0.3839	0.9812	0.6138	Moderate	12
Guangdong	0.7462	0.6737	0.0725	0.7100	0.9987	0.8420	Good	1
Guangxi	0.3503	0.2034	0.1470	0.2769	0.9641	0.5167	Basic	22
Hainan	0.2659	0.1671	0.0988	0.2165	0.9736	0.4591	Basic	27
Chongqing	0.3091	0.2686	0.0405	0.2888	0.9975	0.5368	Basic	19
Sichuan	0.5462	0.3177	0.2284	0.4319	0.9644	0.6454	Moderate	8
Guizhou	0.3303	0.1575	0.1729	0.2439	0.9351	0.4776	Basic	26
Yunnan	0.3392	0.2067	0.1325	0.2730	0.9701	0.5146	Basic	24
Tibet	0.3195	0.1176	0.2020	0.2185	0.8868	0.4402	Basic	29
Shaanxi	0.3512	0.2424	0.1087	0.2968	0.9831	0.5402	Basic	18
Gansu	0.2863	0.1269	0.1594	0.2066	0.9225	0.4365	Basic	31
Qinghai	0.2906	0.1262	0.1644	0.2084	0.9189	0.4376	Basic	30
Ningxia	0.3074	0.1347	0.1726	0.2211	0.9206	0.4511	Basic	28
Xinjiang	0.4064	0.1743	0.2321	0.2903	0.9166	0.5158	Basic	23

Table 4. Distribution of the mean value of the coupling coordination index between the aging industry and the regional economy in various provinces (2009–2019).

Coordination	Province	Feature
Good coordination	Guangdong	Regional economy development lags
	Jiangsu	Aging industry development lags
Moderate coordination	Shandong, Henan, Sichuan, Hubei, Liaoning, Hebei, Hunan	Regional economy development lags
	Zhejiang	Coordinated development
	Beijing, Shanghai	Aging industry development lags
Basic coordination	Anhui, Inner Mongolia, Jiangxi, Shaanxi, Chongqing, Jilin, Shanxi, Guangxi, Xinjiang, Yunnan, Heilongjiang, Guizhou, Hainan, Ningxia, Tibet, Qinghai, Gansu	Regional economy development lags
	-	Coordinated development
	Fujian, Tianjin	Aging industry development lags
Moderate incoordination	-	
Severe incoordination	-	

4.3. Dynamic Evolution Analysis

In order to objectively reflect the distribution changes of China's aging industry and regional economy development coordination index from 2009 to 2019, this study compared the kernel density maps of the coordination index in 2009, 2014, and 2019 (Figure 7). In general, the kernel curve basically remained unchanged, indicating that the coordination level of aging industry and regional economy development did not change significantly; from the distribution shape, the right tail is longer, indicating that there are obvious regional differences in the coordination index. From 2009 to 2014, the peak value became higher, from single peak to double peak, and the width became larger. Therefore, the difference in the coordinated index of provinces became larger during the period, and polarization was intensified. From 2014 to 2019, the peak value became lower, the double peak turned into a single peak, and the width of the curve narrowed, indicating that the differences and polarization were lessened to some extent.

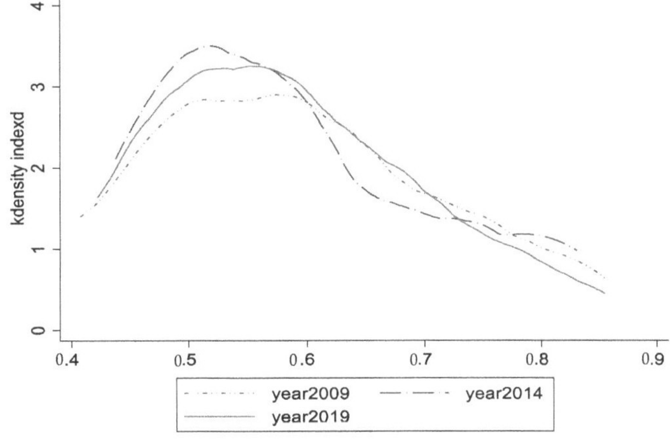

Figure 7. The distribution dynamics of the coordinating development index of the aging industry and the regional economy.

5. Analysis of Influencing Factors of Coupling Coordination

According to the first law of geography, all things are related, and the closer the distance, the greater the connection [35]. Since the Dutch economist Paelinck proposed the term "spatial econometrics" in 1975 [36], after the pioneering work of [37,38], etc., the research on spatial correlation and spatial difference finally formed a framework for qualitative spatial metrology. The branch of econometrics that deals with spatial data is called "spatial econometrics" [39]. Nowadays, spatial econometrics is widely recognized as a mainstream applied econometrics research method and is applied in many fields, such as urban economics, regional economics, real estate economics, and economic geography.

China has a vast territory, and the economic development of the east–west and the north–south is obviously different. The relationship between the aging industry and economic development in a region is also affected by other regions. But traditional econometric methods only consider the linear structure without considering the spatial factors. Utilizing traditional econometric methods to study relationship concerned with spatial factors will easily lead to deviations in the model estimation results. Therefore, it is necessary for us to further explore whether the coordinated development of the aging industry and regional economy in China is affected by neighboring regions. If there is an impact, what are the factors which affect the coordinated development and the extent of the impact? To this end, we introduce spatial correlation analysis to conduct a comprehensive analysis.

5.1. Selection and Description of Variables

The theme of this study is the coordinated development of the aging industry and the regional economy. Therefore, based on the previous indicator system and calculation results, the coordination index D is selected as the explained variable, denoted as (CID); the explanatory variables are selected from the evaluation indicators of the aging industry and regional economy described in the above sections. Above all, among these indicators, indicators with large absolute values not only have relatively large fluctuations, but also have a large degree of dispersion. Therefore, to reduce the possible heteroscedasticity and the difference between the estimated coefficients, the logarithm is used for the index variables with large absolute values; secondly, the variables with severe multicollinearity are eliminated through the VIF test, and finally, nine explanatory variables are obtained:

FROG = proportion of local fiscal revenue of GDP
FAOG = proportion of investment in fixed assets of GDP
LnPCG = logarithm of per capita GDP
ODR = elderly dependency ratio
LnIP = logarithm of the number of people insured in the basic pension insurance
LnINS = logarithm of the number of elderly care institutions
LnBED = logarithm of elderly care beds per one thousand people
EDU = the proportion of elderly care service personnel with college and above
OYR = the ratio of old and young
The descriptive statistics of each variable are shown in Table 5.

5.2. Spatial Correlation Analysis

In order to test the spatial correlation of each variable, we use Stata software to calculate the global Moran's index and Geary's C index for verification. For various spatial weight matrices commonly used at present, the geographic distance (latitude and longitude, inverse of geographic distance) spatial weight matrix w1 is selected for calculation, ** and the results are shown in Table 6.

Table 6 indicates that from 2009 to 2019, the Moran indices of the coordination indices of China's provinces all passed the 1% significance test, and the Geary C index mostly passed the 1% significance test. The Moran index values were all greater than 0, and the Geary C index values were all less than 1, indicating that the overall coordination between the aging industry and regional economy in various provinces manifested positive spatial autocorrelation characteristics.

Table 5. Descriptive statistics of variables.

Variable	Definition	Obs	Mean	Std. Dev.	Min	Max
	explained variable					
CID	coordination index D	341	0.5875	0.1134	0.4077	0.8574
	explanatory variable					
FROG	proportion of local fiscal revenue of GDP	341	0.1149	0.0319	0.0587	0.2452
FAOG	proportion of investment in fixed assets of GDP	341	0.8338	0.2831	0.2109	1.5965
LnPCG	logarithm of per capita GDP	341	10.6376	0.4940	9.2886	11.9940
ODR	elderly dependency ratio	341	0.1353	0.0330	0.0670	0.2380
LnIP	logarithm of the number of people insured in the basic pension insurance	341	6.5892	1.0769	2.2225	8.5928
LnINS	logarithm of the number of elderly care institutions	341	6.6302	1.1172	1.6094	8.1342
LnBED	logarithm of old-age beds per one thousand person	341	3.1317	0.5091	1.3753	4.3780
EDU	proportion of elderly care service personnel with college and above	341	0.2427	0.0866	0.0162	0.4787
OYR	ratio of old and young	341	0.6540	0.2801	0.2072	1.8631

Table 6. Moran's Index and Grary's C Index.

Moran's I				Geary's C			
Year	I	z	*p*-Value	Year	c	z	*p*-Value
2009	0.096 ***	3.718	0.000	2009	0.862 ***	−3.353	0.001
2010	0.102 ***	3.890	0.000	2010	0.862 ***	−3.352	0.001
2011	0.097 ***	3.746	0.000	2011	0.870 ***	−3.097	0.002
2012	0.099 ***	3.784	0.000	2012	0.875 ***	−3.025	0.002
2013	0.095 ***	3.687	0.000	2013	0.879 ***	−2.947	0.003
2014	0.088 ***	3.486	0.000	2014	0.890 ***	−2.663	0.008
2015	0.086 ***	3.449	0.001	2015	0.893 **	−2.466	0.014
2016	0.082 ***	3.332	0.001	2016	0.891 **	−2.531	0.011
2017	0.084 ***	3.402	0.001	2017	0.889 **	−2.536	0.011
2018	0.104 ***	3.969	0.000	2018	0.865 ***	−3.125	0.002
2019	0.103 ***	3.939	0.000	2019	0.866 ***	−3.151	0.002

Note: *** and ** indicate signifcance at the 1% and 5% levels, respectively.

From the Moran scatter plot of the coordination index in 2009 and 2019 (Figure 8), most provinces were concentrated in the first (HH) and third (LL) quadrants, that is, showing high and low aggregate characteristics. From 2009 to 2019, the number of provinces falling into the first and third quadrants increased, indicating that the spatial correlation of the coordination index was gradually significant. Figure 8 also indicates that the developed provinces in the eastern and central regions fell into the first quadrant with high coordination(HH), while most of the western provinces fell into the third quadrant with low coordination (LL) (see Table 7 for the region codes).

5.3. Model Constructing and Testing

5.3.1. Model Constructing

Due to the obvious spatial correlation between coupling and coordination indices of various provinces and municipalities, referring to relevant experience [39], the general expression of the spatial panel model is as follow:

$$Y_{it} = \rho \sum_{j=1}^{n} w_{ij} Y_{jt} + \beta X_{it} + \delta \sum_{j=1}^{n} w_{ij} X_{jt} + u_i + \gamma_t + \varepsilon_{it} \tag{11}$$

$$\varepsilon_{it} = \lambda \sum_{j=1}^{n} w_{ij} \varepsilon_{it} + v_{it} \tag{12}$$

The explained variable Y_{it} is the coupling coordination index between the aging industry and regional economy development in the i region in year t; ρ is the spatial autocorrelation coefficient of the explained variable; X_{it} is the set of explanatory variables in the i region in year t; β is the regression coefficient of the explanatory variable; δ is the spatial autocorrelation coefficient of each explanatory variable; w_{ij} is the spatial weight matrix element; u_i and γ_t are the spatial and temporal fixed effects, respectively; ε_{it} represents the random disturbance termis; λ is the spatial autocorrelation coefficient of each disturbance term; when $\lambda = 0$, the above model is transformed into a spatial Durbin model; when $\lambda = 0$ and $\delta = 0$, the model is transformed into a spatial autoregressive model (SAR); when $\rho = 0$ and $\delta = 0$, the model is transformed into a spatial error model (SEM).

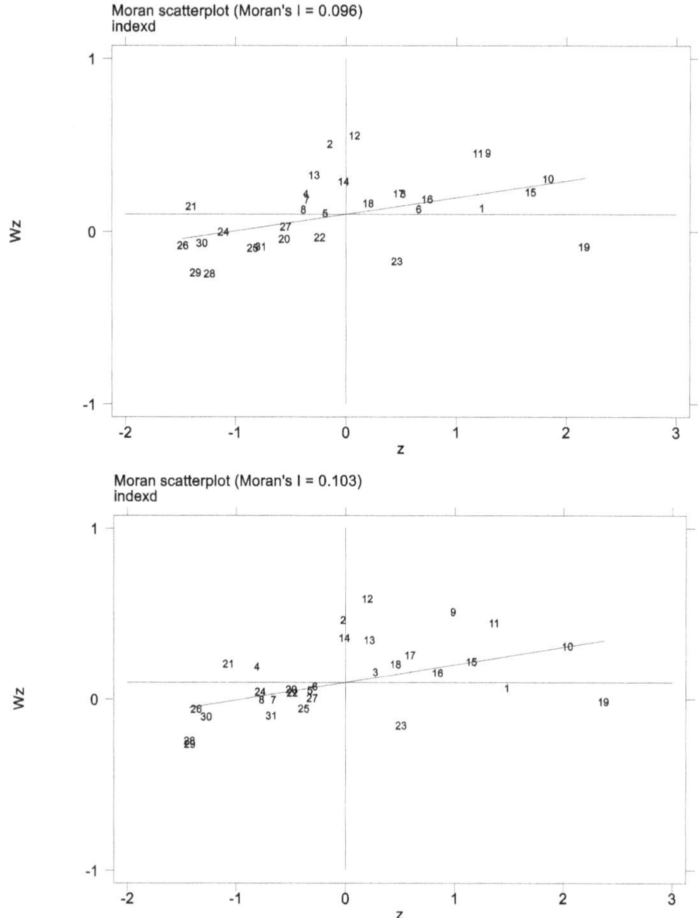

Figure 8. Moran scatter plot of the coordination index (CID) in 2009 (above) and 2019 (below).

Table 7. Provinces and cities code table.

1	Beijing	8	Heilongjiang	15	Shandong	22	Chongqing	29	Qinghai
2	Tianjin	9	Shanghai	16	Henan	23	Sichuan	30	Ningxia
3	Hebei	10	Jiangsu	17	Hubei	24	Guizhou	31	Xinjiang
4	Shanxi	11	Zhejiang	18	Hunan	25	Yunnan		
5	Inner Mongolia	12	Anhui	19	Guangdong	26	Tibet		
6	Liaoning	13	Fujian	20	Guangxi	27	Shaanxi		
7	Jilin	14	Jiangxi	21	Hainan	28	Gansu		

5.3.2. Model Testing

Because the panel data in this study are about 31 selected provinces (municipalities) from 2009 to 2019, which belongs to short panel data, the unit root test is not needed. Through the LM test results, we found that the three tests for spatial error all rejected the null hypothesis of "no spatial autocorrelation", and the two tests for spatial lag one rejected the null hypothesis. These results once again demonstrate that spatial econometric analysis should be utilized [40].

The spatial panel model is generally divided into two types: random effect and fixed effect. According to the results of the Hausman test, the null hypothesis is rejected, indicating that there are significant differences between individual coefficients. Therefore, it is better to choose the fixed effect model. Since there are different types of spatial econometric models, LR and Wald tests are utilized to make choices. The results of LR (*p* value) obviously reject the null hypothesis, indicating that the SDM model cannot degenerate into the SAR and SEM models, and the results of the Wald test also basically support the test results of the LR (see Table 8 for detailed results).

Table 8. Spatial model testing results.

Name		Statistics	*p*-Value	Name		Statistics	*p*-Value
LM test	Spatial error: Moran's I	7.776 ***	0.000	Hausman test LR test	Hausman Fe Re	1754.53 ***	0.000
	Lagrange multiplier	46.831 ***	0.000		SDM-SAR	182.56 ***	0.000
	Robust Lagrange multiplier	67.047 ***	0.000		SDM-SEM	43.55 ***	0.000
	Spatial lag: Lagrange multiplier	0.564	0.453	Wald test	Test For SAR	111.81 ***	0.000
	Robust Lagrange multiplier	20.78 ***	0.000		Test For SEM	14.56 *	0.068

Note: *** and * indicate significance at the 1% and 10% levels, respectively.

Through the above tests, this study finally chooses the fixed effects of the optimal SDM for spatial econometric analysis.

5.4. Empirical Results Analysis

The analysis of fixed effects in spatial measurement includes three forms: spatial fixed, temporal fixed, and spatial and temporal double fixed. According to the estimation results of the SDM model in Table 9, the spatial correlation under the spatial fixed effect passed the significance test, the R-square result is 0.3448, and the log-likelihood value is large, so the model has a high degree of fit with high reliability. This study will conduct a comprehensive analysis based on the estimation results of the spatial fixed effects model.

Table 9. SDM model estimation results.

	Spatial Fixed		Temporal Fixed		Double Fixed	
	CID		CID		CID	
	Main	Wx	Main	Wx	Main	Wx
FROG	0.199 ***	−0.306 *	−0.103	2.410 ***	0.223 ***	0.308
	[0.0488]	[0.1766]	[0.1191]	[0.8399]	[0.0506]	[0.3830]
FAOG	0.0303 ***	−0.0234	−0.0139	0.00259	0.0316 ***	0.0543 *
	[0.0042]	[0.0197]	[0.0139]	[0.0793]	[0.0041]	[0.0311]
LnPCG	0.0749 ***	−0.0972 ***	0.143 ***	0.0509	0.0629 ***	−0.0633
	[0.0104]	[0.0321]	[0.0130]	[0.1144]	[0.0104]	[0.0640]
ODR	−0.420 ***	0.970 ***	0.0890	4.373 ***	−0.455 ***	0.437
	[0.0737]	[0.3226]	[0.1720]	[1.6631]	[0.0775]	[0.6592]
LnIP	0.00318	0.0193	0.0567 ***	0.108 *	0.00476	0.0252
	[0.0059]	[0.0374]	[0.0061]	[0.0635]	[0.0058]	[0.0479]
LnINS	0.00328	−0.0153	0.00559	−0.141 **	0.00472 *	−0.0127
	[0.0023]	[0.0154]	[0.0069]	[0.0673]	[0.0024]	[0.0237]
LnBED	0.0309 ***	0.000508	0.0367 ***	0.0219	0.0294 ***	0.0175
	[0.0029]	[0.0207]	[0.0084]	[0.0791]	[0.0029]	[0.0269]
EDU	0.0424 ***	−0.130 **	−0.104 **	−0.719 **	0.0350 **	0.0928
	[0.0140]	[0.0563]	[0.0419]	[0.3001]	[0.0147]	[0.1095]
OYR	−0.0333 ***	−0.0966 *	−0.0898 ***	−0.410 ***	−0.0240 **	−0.0513
	[0.0095]	[0.0552]	[0.0222]	[0.1350]	[0.0099]	[0.0627]
Spatial rho	0.515 ***		−0.417		−0.246	
	[0.1013]		[0.2681]		[0.2173]	
Variance sigma2_e	0.0000752 ***		0.00151 ***		0.0000687 ***	
	[0.0000]		[0.0001]		[0.0000]	
N	341		341		341	

Standard errors in parentheses; * $p < 0.1$, ** $p < 0.05$, *** $p < 0.01$.

5.4.1. Model Results

In Table 9, the spatial autoregression coefficient of the explained variable (coordination index) $\rho = 0.515$ is a positive value, and the p value is significant at the 1% level, indicating that the explained variable has a positive spatial spillover effect on itself, which is completely consistent with the above-mentioned results verified by the Moran index, that is, the coordinated development of the aging industry and the regional economy in this region promotes the coordinated development of surrounding areas. From the statistical results of column Main (indicating the influence of explanatory variables on the explained variables in the same region), the coefficients of the ratio of old and young (OYR) and the ratio of old-age dependency (ODR) are negative, and reach the 1% level of significance, indicating that the two variables have a negative impact on the coordination index of the aging industry and regional economy in the region, that is, the smaller these two variables are, the more the regional coordination index will increase. Because these two variables are important indicators of population aging, the smaller the indicator's value, the lower the degree of aging in the region, and the less pressure on the aging industry. The result also supports the rationality of the inverse index in the evaluation index system table of regional economy development level (see Table 1). The coefficients, the proportion of local fiscal revenue (FROG), the proportion of fixed asset investment (FAOG), the logarithm of per capita GDP (LnPCG), the logarithm of old-age beds per one thousand person (LnBED), and the proportion of senior college and above pension service personnel (EDU), are positive and passed the 1% significance test, indicating that these variables have a significant positive impact on the coordinated development of the aging industry and the regional economy in the region.

From the statistical results of column Wx (which represents the spatial lag term of the explanatory variable), the logarithm of per capita GDP (LnPCG) and the elderly dependency ratio (ODR) reached the 1% level of significance, the proportion of elderly care service personnel with college and above (EDU) reached the 5% level of significance, the proportion of local fiscal revenue (FROG) and the ratio of old and young (OYR) reached the 10% level of significance, indicating that these variables have obvious spatial effect characteristics. Among them, variables FROG, LnPCG, EDU and OYR have a negative conduction effect on the explained variables in the surrounding areas, while the variable ODR has a positive conduction effect on the surrounding areas. In other words, high local fiscal revenue, high per capita GDP, high quality of elderly care service personnel, and low dependency ratio of the elderly population play a positive role in promoting local economic growth, which will attract various resources (labor population, capital, etc.) from neighbor areas to flow into the discussed area. It will have an adverse impact on the economic development of its neighbor, and then negatively affect the coordinated development of its neighbor.

5.4.2. Spatial Effect Decomposition

In the analysis of spatial econometric models, regression coefficients are unable to fully and accurately reflect the spatial spillover effects of explanatory variables, because the regression coefficients include direct effects and indirect effects, and the coefficients of the spatial lag terms of explanatory variables will also impact the indirect effects [41]. Many scholars regard the estimated coefficient of the spatial lag variable as the "spatial spillover effect" of the variable, which easily causes misjudgment.

Therefore, based on the method of LeSage and Pace, this study reveals the influence of each explanatory variable on the coordination index of the aging industry and regional economy development through direct effects, indirect effects, and total effects [42]. The direct effect represents the average impact of the explanatory variable on the explained variable (coordination index) in the region, the indirect effect represents the average effect of the explanatory variable on the explained variable (coordination index) in other regions, and the total effect represents the average effect of the explanatory variable on the explained variable (coordination index) in all regions. Table 10 shows the estimated results of the direct and indirect effects.

Table 10. Estimation results of spatial effect decomposition.

	Direct Effects	Indirect Effect	Total Effects
FROG	0.192 ***	−0.392	−0.199
	[0.0507]	[0.3629]	[0.3733]
FAOG	0.0297 ***	−0.0177	0.0121
	[0.0043]	[0.0424]	[0.0440]
LnPCG	0.0734 ***	−0.120 *	−0.0468
	[0.0099]	[0.0653]	[0.0653]
ODR	−0.387 ***	1.530 **	1.143
	[0.0773]	[0.7087]	[0.7442]
LnIP	0.00404	0.0438	0.0478
	[0.0059]	[0.0851]	[0.0872]
LnINS	0.00277	−0.0288	−0.0260
	[0.0025]	[0.0337]	[0.0347]
LnBED	0.0316 ***	0.0336	0.0652
	[0.0032]	[0.0454]	[0.0468]
EDU	0.0373 **	−0.219 *	−0.182
	[0.0146]	[0.1178]	[0.1231]
OYR	−0.0381 ***	−0.235 *	−0.274 **
	[0.0099]	[0.1205]	[0.1246]
N	341	341	341

Standard errors in parentheses; * $p < 0.1$, ** $p < 0.05$, *** $p < 0.01$.

From the results in Table 10, the direct effect coefficients of ODR and OYR are negative, and both reach the 1% level of significance, indicating that the population aging has a significant negative impact on the coordinated development of the aging industry and the regional economy in the region. The higher the indexes, the more restrictive the coordinated development of the region. Moreover, the ODR has a significant (5% level) spatial spillover effect, which has a positive impact on the coordinated development of the neighbor regions. The possible explanation is that the increase of dependency ratio of the local elderly population will restrict local economic development, resulting in the outflow of labor and capital into its surrounding areas. Therefore, the level of coordination between the aging industry and economic development in the surrounding areas will be promoted.

The direct effect coefficients of FROG, FAOG, LnPCG, LnBED, and EDU are positive, and all passed the 1% significance test (EDU is 5% significance), indicating that these factors significantly affect the development of the aging industry and regional economy in its own region. Among these coefficients, LnPCG, EDU, and OYR significantly inhibit the coordinated development of the aging industry and regional economy in neighboring areas through indirect effects.

6. Conclusions

This study investigated the law, distribution characteristics, and changing trend of the coordinated development of China's aging industry and regional economy, and the factors which affect the degree of coordination between the aging industry and economic development on the provincial level. In doing so, we constructed a comprehensive evaluation index system of the aging industry and regional economy development, introduced a coupling model based on the entropy weight method, and measured the coupling and coordinated development level of the two systems using data of 31 selected provinces (municipalities) in China from 2009 to 2019. The comprehensive analysis of the study reached the following conclusions:

Firstly, China's aging industry developed unevenly. In the eastern and central regions, the aging industry was more developed than the national average. There were significant regional variations in the aging industry. In recent years, the regional differences narrowed, indicating that the development of the aging industry in underdeveloped regions accelerated.

Secondly, China's eastern region had the highest level of development, far exceeding those of other regions. The economy of the northeast region has declined since 2014, and there was still no sign of improvement. Moreover, the differences in regional economy development widened, and the polarization increased.

Thirdly, the coordination between the aging industry and regional economy development in all provinces is good, and there is no imbalance, but the quality of the overall coordinated development is not high. The western region has the lowest level of coordinated development, and the coordination level of the northeast region gradually decreased.

Fourthly, a variety of socioeconomic and demographic factors affect the coordinated development of the aging industry and the regional economy, including local fiscal revenue, fixed asset investment, per capita GDP, elderly care beds, quality of elderly care service personnel, elderly dependency ratio, and the old-to-young ratio.

Finally, we found that the characteristics of spatial correlation became high–high and low–low agglomerations. Moreover, the coordinated development of the aging industry and the regional economy in a region can promote the coordinated development of its neighbor regions. Per capita GDP, the dependency ratio of the elderly population, the quality of elderly service personnel and the old-to-young ratio have obvious spatial transmission effects, especially the dependency ratio. The increase in the dependency ratio in one region will promote the coordinated development of its neighbor regions.

To support its aging population, China needs take actions in the following aspects:

Above all, to enhance policy and financial support to less developed regions, such as the western region and northeast region, and prevent further polarization of regional economic development.

In addition, in view of the spatial non-equilibrium distribution characteristics of the coordination between the aging industry and economic, especially the low coordination phenomenon in the western region, a differentiated development strategy should be developed to support low coordinated regions.

What is more, to improve the population structure and the quality of the population. To this end, the adjustment of the fertility policy and the implementation of the delayed retirement policy are conducive to the optimization of the age structure of the population (that is, to reduce the old-to-young ratio and the dependency ratio of the elderly), which in turn significantly promotes the coordination of regional aging industry and economic development. The improvement of the national education level, especially the education level of the elderly service personnel, will effectively improve the coordination level. Therefore, on the one hand, the government should increase investment in basic education, and on the other hand, it should increase efforts and policy support in the professional training of elderly service personnel.

Fourthly, the increase in local fiscal revenue and investment in fixed assets will also effectively improve the level of coordination. It is necessary to improve the gaps in public infrastructure and basic public services between regions, strengthen the construction of infrastructure, especially elderly care facilities including elderly beds, and make them match the regional economy development, so as to promote the good coordination between the aging industry and the regional economy.

Author Contributions: Conceptualization, Y.H.; methodology, F.X.; software, F.X.; validation, F.X., Y.H. and Q.W.; formal analysis, Q.W.; investigation, F.X.; resources, Y.H.; data curation, F.X.; writing—original draft preparation, F.X. and Q.W.; writing—review and editing, F.X. and Q.W.; visualization, F.X.; supervision, Y.H.; project administration, F.X.; funding acquisition, Y.H. All authors have read and agreed to the published version of the manuscript.

Funding: This work is supported by The National Social Science Foundation Project of China (No.21BJL008).

Institutional Review Board Statement: This study does not require ethical approval.

Informed Consent Statement: Not applicable.

Data Availability Statement: The data presented in this study are available on request from the corresponding author.

Acknowledgments: The authors would like to thank Jamal Khan, all the reviewers and editors for their insightful comments.

Conflicts of Interest: The authors declare no conflict of interest.

Notes

[1] From a macro perspective, high-quality development includes stable economic growth, balanced regional development, reasonable industrial layout, and green-sustainable development [5,6]

[2] We have to mention, Hongkong, Macao, Taiwan are not included in this study because of the availability of data.

References

1. NBS. *Seventh National Census Communique*; China Statistics Press: Beijing, China, 2021. (In Chinese)
2. Di, X.; Wang, L.; Yang, L.; Dai, X. Impact of Economic Accessibility on Realized Utilization of Home-Based Healthcare Services for the Older Adults in China. *Healthcare* **2021**, *9*, 218. [CrossRef]
3. Wang, Q.; Yang, Y. From Passive to Active: The Paradigm Shift of Straw Collection. *Front. Ecol. Evol.* **2022**, *10*, 3389. [CrossRef]
4. The State Council Issued the "14th Five-Year Plan" National Old-Age Cause Development and Old-Age Service System Plan [N]. Xinhua Daily Telegraph, 2022-02-22(001). (In Chinese)
5. Geng, B.; Zhang, X.; Liang, Y.; Bao, H.; Skitmore, M. Sustainable land financing in a new urbanization context: Theoretical connotations, empirical tests and policy recommendations. *Resour. Conserv. Recycl.* **2018**, *128*, 336–344. [CrossRef]

6. Wang, Q.; Dogot, T.; Yang, Y.; Jiao, J.; Shi, B.; Yin, C. From "Coal to Gas" to "Coal to Biomass": The Strategic Choice of Social Capital in China. *Energies* **2020**, *13*, 4171. [CrossRef]

7. Pan, Y.; Wang, X.; Ryan, C. Chinese seniors holidaying, elderly care, rural tourism and rural poverty alleviation programmes. *J. Hosp. Tour. Manag.* **2021**, *46*, 134–143. [CrossRef]

8. Fang, E.F.; Xie, C.; Schenkel, J.A.; Wu, C.; Long, Q.; Cui, H.; Aman, Y.; Frank, J.; Liao, J.; Zou, H.; et al. A research agenda for ageing in China in the 21st century (2nd edition): Focusing on basic and translational research, long-term care, policy and social networks. *Ageing Res. Rev.* **2020**, *64*, 101174. [CrossRef] [PubMed]

9. Li, Z.; Zheng, X.; Sarwar, S. Spatial Measurements and Influencing Factors of Comprehensive Human Development in China. *Sustainability* **2022**, *14*, 5065. [CrossRef]

10. Gu, H.; Jie, Y.; Lao, X. Health service disparity, push-pull effect, and elderly migration in ageing China. *Habitat Int.* **2022**, *125*, 102581. [CrossRef]

11. Wang, Q.S. *A Humble Opinion on My Country's Aged Work*; Shandong Social Sciences: Jinan, China, 1990; pp. 95–99+90. (In Chinese)

12. Tian, X.L. A comparative study of aging business and aging industry, taking Japan aging business and aging industry as an example. *J. Tianjin Univ. Soc. Sci. Ed.* **2010**, *12*, 29–35.

13. Sun, Z.J.; Chen, M. Xi Jinping's thoughts on high-quality economic development in the new era and its value. *Shanghai Econ. Res.* **2019**, *10*, 25–35.

14. Jin, B. Economic Research on "High-Quality Development". *China Ind. Econ.* **2018**, *4*, 5–18.

15. Feng, Q.B. Five Characteristics and Five Ways of China's High-Quality Economic Development. *China Party Gov. Cadres Forum* (In Chinese). **2018**, *1*, 3.

16. Zhao, D.Q. Reflections and Suggestions on Realizing High-quality Economic Development. *Ref. Econ. Res.* **2018**, *1*, 7–9+48.

17. Clark, T. Handbook of Social Gerontology. *Can. J. Public Health Rev. Can. Sante Publique* **1961**, *52*, 133.

18. Chen, C.; Maung, K.; Rowe, J.W.; Antonucci, T.; Berkman, L.; Börsch-Supan, A.; Carstensen, L.; Goldman, D.P.; Fried, L.; Furstenberg, F.; et al. Gender differences in countries' adaptation to societal ageing: An international cross-sectional comparison. *Lancet Healthy Longev.* **2021**, *2*, e460–e469. [CrossRef]

19. Tamai, T. Economic growth, equilibrium welfare, and public goods provision with intergenerational altruism. *Eur. J. Political Econ.* **2022**, *71*, 102068. [CrossRef]

20. Horioka, C.Y. Japan's public pension system: What's wrong with it and how to fix it. *Jpn. World Econ.* **1999**, *11*, 293–303. [CrossRef]

21. Leff, N.H. Dependency Rates and Savings Rates. *Am. Econ. Rev.* **1969**, *59*, 886–896.

22. Liu, G.; Zhang, G. China's carbon inequality of households: Perspectives of the aging society and urban-rural gaps. *Resour. Conserv. Recycl.* **2022**, *185*, 106449. [CrossRef]

23. Otsu, K.; Shibayama, K. Population aging, government policy and the postwar Japanese economy. *J. Jpn. Int. Econ.* **2022**, *64*, 101191. [CrossRef]

24. Wu, Y. *China Aging Industry Development Report (2014)*; Social Science Literature Press: Beijing, China, 2014. (In Chinese)

25. Huang, Q.; Li, X.; Li, J. The Impact of Population Aging on Innovation: An Empirical Study Based on China's Macro and Micro Data. *Discuss. Mod. Econ.* **2018**, *12*, 25–32.

26. Li, X.; Chen, Y.; Jiang, X.; Chang, J. Analysis of the Spatial Effect of China's Population Aging on Economic Growth—Based on the Spatial Durbin Model of Dual Zone System. *Popul. Dev.* **2020**, *26*, 85–96. (In Chinese)

27. Hou, M.; Liao, T. Population Aging, Innovation Driven and Industrial Structure Transformation and Upgrading—A Study Based on Spatial Durbin Model. *J. Harbin Univ. Commer. Soc. Sci. Ed.* **2021**, *1*, 72–84. (In Chinese)

28. He, Q.; Yang, X. Analysis on the coupling coordination degree of big health industry and elderly care service. *Soft Sci.* **2019**, *33*, 45–49. (In Chinese)

29. He, D.; Wang, Z.W. Research on the temporal and spatial evolution of the coupling coordination degree between the pension industry and the regional economy—Taking Jiangsu Province as an example. *Manag. Mod.* **2019**, *39*, 16–22. (In Chinese)

30. MCA. *China Civil Affairs and Economic and Social Development Statistical Database*; China Statistics Press: Beijing, China, 2022. (In Chinese)

31. NBS. *China Statistical Yearbook*; China Statistics Press: Beijing, China, 2022. (In Chinese)

32. NBS. *Statistical Zoning and Urban-Rural Division Codes*; Beijing, China, 2021. (In Chinese)

33. Ye, S. *Research on the Spatio-temporal Coupling Relationship between China's Basic Public Services and Economic Development*; East China Jiaotong University: Nanchang, China, 2019. (In Chinese)

34. Xin, C.; Chen, Z. Distribution Dynamics, Regional Differences and Convergence of China's Basic Public Service Supply Level. *Quant. Econ. Tech. Econ. Res.* **2019**, *36*, 52–71. (In Chinese)

35. Tobler, W.R. A Computer Movie Simulating Urban Growth in the Detroit Region. *Econ. Geogr.* **1970**, *46*, 234–240. [CrossRef]

36. Paelinck, J. Simple linear planning models: Structure and implications. *Economist* **1975**, *123*, 385–396. [CrossRef]

37. Anselin, L. A test for spatial autocorrelation in seemingly unrelated regressions. *Econ. Lett.* **1988**, *28*, 335–341. [CrossRef]

38. Andrew, C.; Keith, O. Testing for Spatial Autocorrelation Among Regression Residuals. *Geogr. Anal.* **1972**, *4*, 267–284.

39. Zeng, Y.; Jiang, C.; Cui, Z. The impact of digital finance on high-quality economic development: A study based on the spatial Durbin model. *Tech. Econ.* **2022**, *41*, 94–106. (In Chinese)

40. Chen, Q. *Advanced Econometrics and Stata Applications*; Higher Education Press: Beijing, China, 2014. (In Chinese)

41. Liu, C.; Zhao, X. The impact of population aging on economic growth and its spillover effect—Based on the spatial Durbin model. *Explor. Econ. Issues* **2018**, *6*, 21–32. (In Chinese)
42. James, L.; Robert, K.P. *Introduction to Spatial Econometrics*; CRC Press: Boca Raton, FL, USA, 2009.

 *land*

Article

Sustainability through STEM and STEAM Education Creating Links with the Land for the Improvement of the Rural World

Elisa Gavari-Starkie [1], Patricia-Teresa Espinosa-Gutiérrez [2,3] and Cristina Lucini-Baquero [3,*]

[1] Department of History of Education and Comparative Education, National Distance Education University (UNED), 28015 Madrid, Spain
[2] International Doctoral School of the UNED (EIDUNED), National Distance Education University (UNED), 28040 Madrid, Spain
[3] Plant Production and Agrifood Quality Research Group (PROVECAv), St. Teresa of Jesus Catholic University of Avila (UCAV), 05005 Avila, Spain
* Correspondence: cristina.lucini@ucavila.es

Abstract: Rural environment is suffering from serious problems, as reflected in the term "Empty Spain". One of these problems is the pronounced depopulation that rural areas suffer in our time, so creating links with the land thanks to education is of great interest for, among other things, establishing population in rural areas. Interdisciplinary education becomes relevant today as the necessary education in our current world capable of providing answers and solutions to the social demands of our time. Interdisciplinary STEM education had the United States of America as its cradle in the 1990s; later it passed to the acronym STEAM when the Arts were later introduced, this is how you find a true interdisciplinary education. Since 2010, government policies have been developed in the USA, highlighting the Educate to Innovate program and in that same country the STEM4SD Education program, which develops education for sustainability by creating links with the local population. Precisely, this article will collect the educational policies that have been carried out in the USA for the development of this type of education. In this article and thanks to the analysis of certain programs, the importance of interdisciplinary STEM and STEAM education in our days will be exposed for the promotion of sustainability directed towards sustainable development, thereby creating more sustainable societies made up of more sustainable citizens, highlighting the importance of education for sustainability through STEM and STEAM education creating links with the land for the improvement of the rural world, which means establishing population, among other aspects.

Keywords: education for sustainability; STEM; STEAM; land; rural environment; interdisciplinary education; USA; Spain

Citation: Gavari-Starkie, E.; Espinosa-Gutiérrez, P.-T.; Lucini-Baquero, C. Sustainability through STEM and STEAM Education Creating Links with the Land for the Improvement of the Rural World. *Land* **2022**, *11*, 1869. https://doi.org/10.3390/land11101869

Academic Editors: Vanessa Zorrilla-Muñoz, Maria Silveria Agulló-Tomás, Eduardo Fernandez, Blanca Criado Quesada, Sonia De Lucas Santos and Jesus Cuadrado Rojo

Received: 21 September 2022
Accepted: 12 October 2022
Published: 21 October 2022

1. Introduction

If we refer to the current concept of sustainability, we must approach it as a concept with a holistic character capable of providing a sustainable path to human development in the 21st century without future generations losing resources or factors of the planet. When we think about sustainability today we do not lose that holistic characteristic of the concept accepting the foundation of the development of the human being as a sustainable development where all its activities are related within it, considering sustainable development also as a whole, where all actions must go perfectly adjusted to achieve holistic sustainability in all areas.

But we must not forget that the concept of sustainability has undergone an evolution in our time. Each of us will come up with an idea or several ideas when thinking about the concept of sustainability. As Bermejo [1] expounds, several adjectives have been identified with sustainability over time, such as ecological, green and even blue (in relation to the image of the blue planet).

Human responsibility in achieving sustainable development must be present in each of our actions and deeds, as well as in our way of life. But before reaching the point of being responsible citizens, we go through other stages: conscious citizens, strategists and finally responsible citizens. Let us recall in this line the words of Mihalic et al. [2]: The "awareness" phase involves social awareness of all sustainability issues, stimulates sustainable ethics and informs the destination about appropriate and inappropriate behaviors. In the next phase, sustainability challenges are translated into objectives, codified in destination strategies and included in their "agendas" and lists of relevant policy instruments. The last phase is sustainability implementation or responsible "action".

According to the above, for citizens to be first aware, then strategists and finally responsible for working towards the sustainability approach, one of the aspects lies in education. The relationship between education, sustainability and environment, we find in the 2030 Agenda, where it talks about Education for Sustainable Development (ESD) [3,4]

Together with this aspect, we indicate STEM and STEAM education; we must remember the Rocard report [5] of the European Commission (Directorate-General for Research Information and Communication Unit) where the scarcity of vocations in the scientific-technological field, identified as STEM, is pointed out; this scarcity was related to the difficulties (at the level of purchasing power as well as at the gender level) of accessibility to them by the student body. It is important to enhance the knowledge variable in STEM disciplines in the educational and university system and to involve the incorporation of women in work spaces related to these areas of knowledge [6,7]. To reduce gender inequality, countries have had to develop different programs in which the government and academia have combined efforts for gender inclusion in terms of STEM competencies [8].

Bybee [9] is interested in STEM education, whose origin comes from the disciplines Science, Technology, Engineering and Mathematics, and analyzes the different ways in which this concept is used. Bybee [9] puts the focus on STEM education to improve the knowledge of science disciplines in new generations of students, as opposed to the traditional educational approach, although he points out that there is still confusion and debate about the implications of STEM education. However, despite the debate, it is clear that STEM education goes beyond traditional education, as it seeks to provide solutions to current environmental, social, and economic problems by interconnecting science disciplines. Kelley and Knowles [10] note that the loss of interest on the part of students towards science disciplines is related to the scarcity of interrelation between these disciplines as well as lack of real-world applications. As we are seeing, STEM/STEAM education involves factors beyond mere traditional education, so it is essential that, if students are to acquire competencies and skills interrelated with sustainability, the economy and, ultimately, the real world, it is the teachers who must be prepared to transmit such a mission. It is important that teachers perform self-criticism to improve STEM education capacity [11] and know how to integrate STEM education factors so that students can acquire the knowledge and skills of these disciplines in the real world [12]. For this it is necessary that teachers are properly prepared and committed to STEM education [13,14] and that, of course, teachers can develop their activity with job continuity [15] and have adequate teaching tools for their professional activity; in this sense, training programs have been developed in diversified classrooms, in urban areas [16] or in rural areas, although in a very specific way [17].

STEM education can be developed at all stages of education, from pre-school to university. Although STEAM education seems potentially rich to foster engagement and learning of more students, it presents numerous challenges, including the preparation, willingness and confidence of teachers to adopt such a curriculum, as highlighted by Kim and Bolger [18]. In fact, in relation to STEM education, and agreeing with Xu and Ouyang [19] it should be noted that we are facing a complex system, where not only science disciplines interact, but also the educational community, teachers, students and their environment, social, economic and environmental factors.

A relevant aspect in this field of action corresponds to the gender gap. Currently, women are underrepresented in the STEM field and it is important to be able to break this trend by focusing on STEM education at an early age [20,21]. Within this education, it is worth highlighting the gender gap described in national and international [22,23] studies where we can conclude that it is boys versus girls [24–27] who present greater interest in these science disciplines from the early educational stages [28], along with socioeconomic factors and STEM teaching practices. In Germany, Goreth and Vollmer [29] point out that the low participation of women in STEM professions does not correspond to the students' own gender difference but to the "different technical socialization of women and men" so they invite to change the views of the current normative roles for women and men. It is necessary to break the stereotype of the student of STEM disciplines (male, urban, acceptable socioeconomic position) from an early age and open the world of science to the female gender.

On the other hand, within the context of rural education, several studies establish the differences in performance between urban and rural students, in relation to science [30] and mathematics [31,32]. We come across blended learning proposals to improve academic performance in rural areas [33], and school networks to improve rural schools [34]. Another important aspect to improve rural academic success corresponds to the training of rural teachers, especially school principals [35]. However, there are still gaps to be studied in rural schools, as pointed out in the review by Fargas-Malet and Bagley [36], in relation to the promotion of research at early ages, research "with the children", and not "of the children". Likewise, they indicate, as do other authors, that studies have been carried out in relation to the availability of resources, population shortage, lack of funding, which is why many rural schools tend to close. However, they propose strategies to favor these rural schools, from the training of suitable teachers, virtual learning … A very relevant aspect emphasizes the strong links developed between the rural school and the place, so they propose "place-based curricula". The sustainability of rural schools will also depend on the members of the educational community themselves [36].

The rural environment offers a myriad of learning opportunities, both in formal and informal education. As Goodpaster et al. [37] argue, a large part of the rural "way of life" is the management of the social aspects of rural education, both in and out of school, strengthening ties with the land. Studies with implication of education in the rural environment under STEM scenario we can observe it in the one developed by Ihrig et al. [38] where they evaluated, in economically disadvantaged communities, experiences of teachers and high-performing rural students who participated in an extracurricular STEM talent development program, where they supported the use of STEM informal educational environments.

It is in the scenario in which the present work is developed, education for sustainability, but focused on the rural environment, so the main objective of this work is to highlight, through a qualitative approach, rural STEM/STEAM education as a possible solution to the depopulation of rural "empty Spain". Like other educational movements, STEM comes from the political and educational sphere of the USA [9], although in recent years it is promoted internationally [39,40] also in the European Union [8]. The work focuses on the USA, the country in which STEM education emerged and which has developed educational policies and educational programs of interest in this type of education. Subsequently, the situation in Spain has been analyzed in terms of STEM/STEAM, looking for its applicability in rural areas, as a strategy to be taken into account when fighting against depopulation.

1.1. Evolution of International Discourse on the Concept of Sustainability

The idea of sustainability began as a preventive response to the prospect of global or partial collapse of the prevailing model of civilization [41]. Although it is true, the historical journey through the evolution of the concept is of great interest to understand the holistic concept of current sustainability. At this point, let us remember the United Nations Monetary and Financial Conference held in 1944 in Bretton Woods, New Hamshire (USA), with the assistance of 44 countries, which involved a series of agreements to focus monetary and financial development after the near end of the Second World War and there was talk of promoting peace and natural wealth, although everything that was said there was rather aimed at Europe and its recovery after the war situation. Years later the concept of sustainable development would emerge, a concept that was is supported because such a system does not threaten or challenge in any way the neoliberal structures of privilege and reproduction of capital that the capitalist system imposed and spread through the Bretton Woods institutions [42].

In the 1950s, a visible increase in pollution began to take place, leading to an awareness of the situation that produced such pollution in the environment. These facts provided a basis for the development of environmental ideas that forged the future of what would later be known as modern environmentalism. Some ideas began to be forged in this decade that would be the basis for modern environmentalism or ecomodernism [43] that would serve as a beginning to face the future of the development of modern society.

A decade later, in the 1960s, is when we can certainly locate the origins of the aforementioned modern environmentalism, due to the environmental detriment that was already latent at that time and the appearance of a growing concern for the then-existing environmental deterioration. As Bermejo [1] states, in the 1960s there began to be awareness of the proliferation of serious environmental problems that occur especially in OECD countries. With such a situation we remember in 1968 the celebration in Rome of the so-called Club of Rome where great connoisseurs of the situation met to discuss the concern for ecology and the increase in environmental concern of the moment.

In 1972 the Stockholm Summit is held, an international conference in which dangerous levels of pollution and environmental problems are considered. We must emphasize that in this Summit the importance of education in environmental issues is mentioned. As would be collected 30 years later at the Johannesburg Summit at the Stockholm Summit they agreed on the urgent need to address the problem of environmental deterioration [44].

In the eighties, reference should be made to the fact that the United Nations published "Our Common Future", the so-called Brundtland Report, published exactly in the year 1887, a report of great interest due precisely to the issues addressed for the first time on the concept of sustainable development. It defines the concept of sustainable, lasting development, that is, ensuring that it meets the needs of the present without compromising the ability of future generations to meet their own (Brundtland Report, 1987). Precisely in the Preface of the President of the Brundtland Report explicit reference is made to responsibility, *"We live in a time in the history of nations in which coordination of political action and responsibility is needed more than ever"* [45]. Responsibility is a value that, when put into practice to achieve sustainable development, will mean that each action or activity of the human being has that connotation of help and social improvement for the future in common, including the different stages of social awareness [2].

Also in the Preface of the President of the Brundtland Report we find the demand of necessity that was made *"What is needed now is a new era of economic growth, growth that is powerful as well as socially and environmentally sustainable"* [45]. We see in these words the not only environmental focus of sustainability, but also social, something of great importance in the magnitude of the concept.

In 1992, the Rio Summit [46] takes place, in which the three components of sustainability are already observed and which was collected in the Johannesburg Summit as Nogales [47] cites *"the integration itself of the environment and development, with the three components of sustainability included: social, economic and environmental"*. At this Summit, a global strategy is agreed to achieve sustainable development based on global cooperation, establishing 27 principles on the rights and responsibilities of nations in the progress and well-being of humanity [48]. If we refer to Principle 7 *"Developed countries recognize their responsibility in the international pursuit of sustainable development . . . "* and Principle 13 States should develop national legislation regarding liability and compensation for victims of pollution and other environmental damage" we summarize that the concept of responsibility of the human being is exposed as an essential value at this time for the scope of sustainable development. In Principle 10 *"States should facilitate and encourage public awareness and participation by making information available to all"* something very important appears in all this and it is the concept of awareness: raising awareness among the population by States will be crucial to deal with all environmental issues. Principle 11 *"States shall enact effective laws on the environment"*. This principle already makes us reflect on the laws that can be enacted and that are effective in protecting the environment, living beings, their habitats, the land they inhabit, etc. And here we could expand this idea and take it to the educational field, to the educational laws of each country. In Principle 22 of such a document, an explicit reference is made to local communities *"playing a fundamental role in environmental management and development due to their traditional knowledge and practices"*. The local already takes on achieving sustainable development of great importance for all that it implies in terms of culture and knowledge.

In 2002, the Johannesburg Summit is held, we must highlight two documents: the Johannesburg "Declaration on Sustainable Development" and the "Action Plan". The Declaration once again reflects the importance of responsibility and of local communities *"we assume the collective responsibility to promote and strengthen, at the local, national, regional and global levels, economic development, social development and environmental protection, pillars interdependent and synergistic of sustainable development"*. As noted by Nogales [47], this Summit in Johannesburg is a meeting on sustainability, in which they are going to talk about how to achieve greater development while respecting the environment, for developed countries and those that are developing.

In 2015, an important event took place when world leaders adopted goals aimed at achieving sustainable development, the so-called SDGs (inheritance of the old Millennium Development Goals, MDGs). The 17 SDGs consist of 169 goals to achieve a better planet for all by 2030. The 2030 Agenda for Sustainable Development proposes action in all areas and at all levels, from the international to the local.

The words sustainability and sustainable development have been appearing more and more repeatedly in academic-scientific studies, on the agenda of the main political parties and in all the normative proposals that have to do with public policies [41]. The concept of sustainability in our days implicitly carries a value of responsibility at all levels of action of the human being, whether on an international, national, state or local scale.

1.2. Education as a Promotion of Sustainability in Rural Areas Creating Links with the Land

The 2030 Sustainable Development Agenda with its 17 goals aims to achieve a more sustainable planet by the year 2030. All the SDGs to be met are of great relevance, highlighting among all of them SDG 4, the quality education that we will provide to young generations and children, which will be our future societies. We have currently suffered a pandemic with marked characteristics, the COVID-19 pandemic, which also led to an educational change. As Bosch et al. [49] cite, it is required that young people acquire an integrated and interdisciplinary preparation for science and particularly to understand complex problems of engineering, biology, environment, spread of diseases and epidemics, among other problems.

The need to provide quality education to the greatest possible number of people is present in all the proposed objectives [50]. SDG 4 is the protagonist as it implicitly refers to quality education in our time. If we ask ourselves what we mean by a true quality education, we must answer that it is an education capable of responding to the needs of today's society and the society of school-age generations. A quality education is an education capable of forming people who can solve current problems, people who are responsible in their acts and actions towards themselves, their environment and future generations, people capable of achieving sustainable development and capable of moving in different scenarios of holistic sustainability. With a quality education, students must be prepared for their demands and needs of the current age, adulthood and for future generations.

As Cantu-Martínez [51] cites, environmental deterioration can be observed both locally and globally. Sustainability must be understood as a whole, there are no environmental borders on the planet and the deterioration that the planet has suffered in recent decades, due to pollution and the mismanagement of its resources by human beings, is present and observable at a local and global level and therefore we must think of local and global sustainable development. The damage produced in nature, the breakdown of biodiversity and the loss of natural resources occur at a local and global level. Therefore, since this deterioration is present at a local and global level, it can be observed from local spaces. And for all this, solutions to these facts can be found from the local level and local quality education plays a fundamental role in such achievement.

Land in rural areas is of great importance to its inhabitants. The land shows a space to live but also provides food and social welfare, it is more than the physical space where you live. The territory is not an "objectively existing" physical space, but a social construction, that is, a set of social relationships that give rise to and at the same time express an identity and a sense of purpose [52]. The territory in the rural environment offers endless opportunities and relationships to its inhabitants.

The school plays a fundamental role for the improvement of individuals and is capable of providing solutions to the problems of today's societies. The rural environment also suffers from problems of great magnitude but also offers solutions thanks to the resources that it possesses so valuable to offer quality education, fulfilling the aforementioned SDG 4 of the 2030 Agenda. In this scenario, the rural school has a role of great relevance for promote sustainability by being able to create links with the land. Attachment to the land in a sustainable way can be worked on in the rural environment by creating those social bonds with an emotional base in the rural space.

The school institution can be the space from which social capital is generated, aimed at strengthening citizenship and participation in the construction of the territory and its development [53]. Rural territorial development must be worked from a sustainable approach, observing and working on the opportunities that rural areas offer for it. The school is a necessary means for rural development that can only be sustainable or it will not be. And for all this to take place, socio-emotional-spatial links must be created between the individuals of the rural environment and the rural environment itself. As stated [53] the challenge is to be able to implement policies and actions … linked to school spaces, and build and expand citizenship while simultaneously contributing to the competitive insertion of the territory that allows the construction of a strategy for its development. Educational policies and actions in schools should lead us to an improvement in the quality of life in rural areas and in their rural communities, allowing rural sustainable development in all aspects and with all the meanings that the word sustainability implies, so rural development in the rural territory must be a sustainable social development, a sustainable economic development and a sustainable environmental development.

As UNESCO [50] states, sustainable development begins with education. With all this, we must see in education and the rural school a bridge to follow to create more sustainable societies capable of giving holistic responses to existing problems, including existing problems in rural areas and their territory. Education for the sustainable development, the so-called ESD, is very necessary in our time, presenting some characteristics [54]:

> *"Is concerned with the well-being of all four dimensions of sustainability: the environment, society, culture and the economy, is locally relevant and culturally appropriate, is based on local needs, perceptions and conditions but recognizes that the meeting local needs often has international impacts and consequences. It concerns formal, non-formal and informal education, accepts the constantly evolving nature of the concept of sustainability, approaches content taking into account the context, international issues and local priorities and is interdisciplinary"*.

Analyzing all the above characteristics, we observe endless possibilities in education to achieve the holistic meaning of sustainability, which currently has four dimensions and not three, as we discovered at the Rio Summit, that interdisciplinary education is of great importance at the local level for meet your local demands. The actions that are carried out locally and in its territory reach national and international dimensions, let us observe the planet as a whole, without environmental borders, where the actions that take place in any land of any local community have an impact on the common planet for all.

1.3. Objetives

Our main objective of study is to consider rural education as a possible solution to rural depopulation in "empty Spain", creating links with the land between the rural population and its location in the territory. We have analyzed the problems of schools in rural areas and we have reviewed the STEM/STEAM proposals that are being developed in the USA and Spain, in order to advance in the defense of rural STEM/STEAM education as a solution to rural depopulation. The general objective is to highlight the importance of education in rural areas in order to bring about changes that will lead to the permanence of rural areas.

To achieve this objective, the specific actions developed in this study were as follows: initially, the situation in the USA was analyzed in relation to rural education and STEM and STEAM education, reviewing the educational programs and laws that have been developed in the USA, since it is the country where STEM education emerged. Subsequently, the situation of rural and STEM education in Spain has been analyzed. Finally, a diagnosis of this type of education in the Spanish rural environment has been made, trying to find, through STEM and STEAM education, a solution to the problem of depopulation in rural areas, getting young people to stay and live in rural areas, and consequently, that the villages are not as aged. Likewise, through STEM and STEAM education, a solution is sought so that rural women remain in rural areas because they have the same opportunities as men. Finally, it should be noted that carrying out an interdisciplinary STEM and STEAM education using the resources of the rural environment will create an awakening of sensitivity in the students towards their environment and the land that surrounds them, developing links between the land and the students.

To accomplish this objective, we have developed a qualitative methodology, initially identifying the concepts of interest, their relationships through scientific literature in the USA and Spain. After reviewing the articles, we conducted an analysis and synthesis of the key factors that should be taken into account to develop rural STEM education. The results are shown below.

2. Rural STEM/STEAM in USA

2.1. Rural Situation in USA

Although there is no single definition of "rural" [55], we understand rural regions as areas with less populated settlements than urban regions, where the population is fundamentally dedicated to developing an economy based on the primary sector, mainly agriculture and livestock.

According to the National Center for Education Statistics, 9.3 million children are in public rural schools. In the report by Provasnik et al. [56] the situation of rural education in the USA is shown. The new classification system of the National Center for Education Statistics (NCES), classifies twelve categories of location of school districts. As far as rural location is concerned, he distinguishes between three types of rural areas:

- rural areas that are on the periphery of an urban area.
- rural areas that are at a certain distance from the urban.
- remote rural areas.

The rural population in the United States is declining over the last decades (Figure 1) despite the fact that the rural area is greater than the urban area in the US, according to the US Census Bureau.

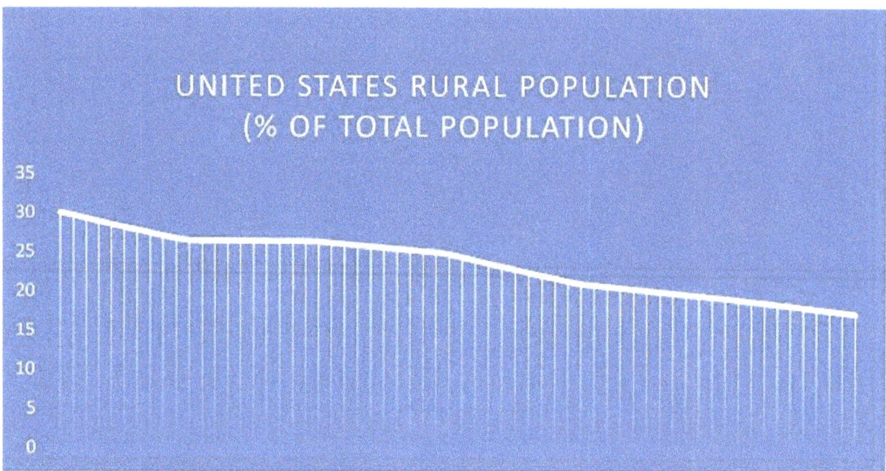

Figure 1. World Bank Rural Population Estimates (1960–2021). Rural population refers to people living in rural areas as defined by the national statistics office. It is calculated as the difference between the total population and the urban population. Source: World Bank.

In 2016, rural areas covered 97% of the land area of the country, but contained 19.3% of the population (about 60 million people). Population growth has slowed in recent years due to lower birth rates and declining net international migration, while death rates are rising due to the country's aging population.

In 2020, there were approximately 57.47 million people living in rural areas in the United States, compared to approximately 274.03 million people living in urban areas.

The decrease that the general rural population has suffered during the last decade (2010–2020) has not been carried out homogeneously throughout the country, since there are rural areas with a developing population (those with greater purchasing power) compared to rural areas in decline (those with the highest level of poverty). This change in the population does not only respond to the level of wealth, since the type of developed local economy has favored, in some cases, the increase in population. This is the case of an economy based on tourism and leisure, both for young people and retirees [57].

In relation to education in the rural environment, we must point out that rural America does not present homogeneity neither in the geographical aspect nor in the populations, since it is made up of diverse communities, where different races and cultures coexist. Rural education is under threat due to racial and class inequality [55]. Differences in graduation rates based on race (white vs. Latino and African American) remain, with rural public schools reflecting the class and racial segregation of the surrounding area.

One of the main problems to overcome this inequality is closely linked to public funding, it is essential to increase the budget dedicated to these public rural schools. Nugent et al. [58] points out the importance of being able to meet the budgetary needs of almost ten million children who attend schools in rural communities across the country, which represents about 20% of the nation's student population and more than 23% of the state. Public funds continue to be claimed to achieve competitiveness in rural schools, since there are situations where access to advanced courses is limited for students from rural areas. An example is found with Advanced Mathematics, whose offer in rural areas is half that in urban areas. To this situation, we must add the shortage of rural teachers in key areas, such as STEM subjects and English, as well as the high turnover of rural teachers due to the fact that their salary is usually lower than that of urban teachers [59]. This situation is also reflected in Canada, in the study conducted by Barter [60]. In 2006, research was initiated to evaluate this problem in rural regions and to extrapolate it to different rural localities. From the point of view of young graduate students and future teachers, problems related to the curriculum, lack of consolidation in the position, heterogeneity in teaching assignments … were identified. These issues should be taken into account to improve the competitiveness of the rural school.

In order to improve the activity of the rural teacher, the study by White and Kline [61] provided a series of resources for rural teachers, adapted to the rural school, in order to help students in education at school and in the community. This proposal is in line with the suggestions of Corbett [62], who pointed out that the preparation of rural teachers should go beyond professional training to "support ways of thinking about teaching in rural contexts that are not standard and directly address problems persistent and pressing rural issues such as: population loss, resource industry restructuring, resource depletion, environmental and habitat degradation, and political land use.

Throughout this section, we can highlight the importance between school and location. Thus, when we talk about localized education, we must work with the proposals of Corbett [63], since the particularities of the school's location must be part of the school's curriculum, in such a way that it is crucial to be able to connect the community in the rural school, in order to reach the academic and educational success of the students. However, according to Corbett, it is not about nationalistic policies, but about adapting and taking advantage of the resources of the community itself.

Thus, in rural areas, the school takes on a broader meaning than in urban regions, as it implies a greater commitment to the community in which it is based, improving and increasing the services necessary for its proper functioning, and offering new opportunities [55] so that strong links are established between the classroom, the school and the rural community.

2.2. Sustainability in Educational Laws and Use of STEM/STEAM Programs Promoting Sustainability in the USA

Interdisciplinary education allows us to work as a whole on sustainable development. ESD is interdisciplinary because no discipline can appropriate ESD for itself; all disciplines can contribute to ESD [54].

STEM education, an acronym for Science, Technology, Engineering, and Mathematics began to gain strength in the US in the 1990s. The subsequent STEAM education also introduced the Arts, making it more complete. STEM/STEAM education is an education capable of solving current problems. As cited by Sanders [64], design and scientific inquiry are routinely employed concurrently in the engineering solutions to real-word problems. According to Toma and Greca [65] the meaning of STEM education is reflected in its main objectives: (a) respond to economic challenges … , (b) identify the needs of workers … to adjust to current labor and social requirements, and (c) emphasize the need to solve technological and environmental problems. A major environmental problem in our time is climate change and its work in schools is of great importance in our time. As Watson [66] exposes with climate change and other environmental concerns becoming evermore present and cumbersome, there is a necessity for increased education on these issues, especially in the United States.

In 2010 is when this concept of STEM education gains strength as it is present in the government policies of that country. As stated by Bosch et al. [49] President Obama announced the "Educate to innovate" Program convening key community leaders and a series of associations, companies, foundations, non-profit organizations and science and engineering societies [67]. In this section we must highlight the publications Improving K-12 STEM education and Improving STEM higher education [68–72]. In addition, the Administration took steps to further bring STEM education to certain groups by highlighting girls and young women Through innovative arrangements such as the NASA/Girl Scouts of the USA partnership, the Department of Energy's Women in STEM mentoring program, and Numerous other commitments agencies across the Administration and the private sector are creating opportunities for students [67,73]. In the USA, a STEM Education Plan is drawn up every five years, the plan is a mandatory component of the America COMPETES

Act of 2010, which requires that the Committee on STEM (CoSTEM) [74] and the White House Office of Science and Technology Policy (OSTP) draw up a plan every five years related to STEM education in that country.

Also launched the STEM plan, highlighting in this section Progress Report on the Implementation of the Federal STEM Education Strategic Plan (17 December 2020) [75], Charting a Course for Success: America's Strategy for STEM Education (December 2018), Summary of the White House Release Event for the 2018 STEM Education Strategic Plan (4 December 2018) [76], Summary of the 2018 White House State-Federal STEM Education Summit (28 June 2018) [77], Progress Report on the Federal Implementation of the STEM Education Plan (16 October 2019) [78].

Recently STEM education is also gaining in importance. In June 2022 the Departments of State and Homeland Security are announcing new actions to advance predictability and clarity for pathways for international STEM scholars, students, researchers, and experts to contribute to innovation and job creation efforts across America [79]. And more recently, in September 2022, it has been published Interagency Roadmap to Support Space-Related STEM Education and Workforce, a report by the National Science and Technology Council of the Office of Science and Technology Policy.

Practically all American Universities have organized STEM Institutes or Centers [49]. An institution to highlight in this section is Arizona State University, as it has the first School of Sustainability in the world. We emphasize at this point that the state of Georgia was the first to include STEM education in the Curriculum. In the 13-year pre-university basic formal education (K-12), STEM education is worked on. The Washington State K-12Integrated Environmental and Sustainability Education Learning Standards describe what all students should know and be able to do in the area of Environmental and Sustainability Education [80,81]. New Jersey became the first state to put climate change in K-12 education standards, teaching climate change topics in classrooms from kindergarten through high school.

In the USA, a STEM education is developed for sustainable development (STEM4SD), promoting an idea of STEM education in a transdisciplinary framework, acknowledging the complex context of global challenges and the need for integrating values, ethics, and world views towards the development of sustainability mindsets and using science to do social well [82]. STEM4SD Education is an education for today's world where Interdisciplinary STEM education is developed with a focus on sustainable development. One of the centers that works with STEM4SD Education is the Smithsonian Center for Science Education (SSEC) who promote STEM education for sustainable development.

According to Pahnke et al. [82] the offers of STEM Education for Sustainable Development (STEM4SD Education) must be guided by some guiding principles, let's see the following:

> *"Empower present and future generations to use science, technology, engineering, and math (STEM) skills and reflective reasoning to solve complex sustainability problems".*

This makes us reflect on the importance of a STEM education for the future that works on all the Goals of the 2030 Agenda, to achieve a fairer and healthier planet by 2030, where sustainable development is not a utopia, but rather a constant work of each one of us. STEM education is an opportunity for today's world to use all the knowledge of Science, Technology, Engineering and Mathematics for the service of the human being to make an education that creates more responsible citizens capable of responding to existing problems in the current world, always from a sustainable point of view, whatever the problem, recalling at this point the current concept of sustainability as a holistic concept that includes all the sustainabilities that can occur in our current world.

2.3. Sustainability through STEM and STEAM Education Creating Links with the Land for the Improvement of the Rural World

If we continue with what Pahnke et al. [82] expose about the offers of STEM Education for Sustainable Development (STEM4SD Education) and that some rectors should be guided, we will now remember the following:

"Encourage autonomous thinking and responsible action that takes place in the context of the student and involves the social and natural environment of the institution, providing the opportunity to implement and experience real changes in the community of students, even on a small scale, which then strengthens their agency capacity".

This brings to mind what we discussed in the previous section on the concept of sustainability in our days and that it implicitly carries an action of responsibility in the human being in each of his acts and actions. By promoting responsible action through STEM4SD Education, that autonomous thought, which involves the student and the environment (understanding here the local, the rural if it is present, the land of the rural then) makes us understand that there will be real changes, that they will be present in the acts and actions of the students reflecting in their activities and their behavior with the environment. With STEM4SD Education, skills are developed so that students can make connections between the personal and their local world, and the personal and their global world.

USA has launched policies and laws to improve rural education. In 2021, the bipartisan Rural STEM Education Research Act was passed, a law that supports research to address the challenges facing rural communities. Laws like this allow to emphasize learning in the rural world with educational foundations in place, giving students the opportunity to work through a STEM education in their communities and in their rural environment.

We must also mention the program (REAP) that was created to help rural schools. We must also mention the Rural and Ready STEM initiative, which is an initiative to improve STEM education in rural areas. According to Buffington in an interview conducted for the Education Development Center in 2019, Rural and Ready STEM has three goals: to help rural teachers improve their science teaching and learning, to help rural districts support and maintain high-quality science learning and use technology to support STEM practice.

Many articles suggest that parents' views of STEM education in rural areas are inconsistent with the realistic need for students to learn principles and skills from the field in order to compete in today's economy [83]. From this point of view, one wonders if it is not important that students develop attachment to place through STEM education for the development of place (local/global) in a sustainable way. It is important to learn the principles, the skills to work the land, but let's not forget how important it is to create links with that land if we hope that future generations want to stay on it or want to return to the land they grew up on to work it. As Williams and Nierengarten [83] maintain, this issue needs to be addressed because not only does parental attitude influence student interest and achievement in STEM, but, in rural areas, parents often play an important role in school councils and school curriculum development.

STEM education in rural areas is extremely important for rural communities, their territory and students. It is important that rural areas *"allow students to find meaning and challenging work in the communities they love"* [84]. The rural environment has something that the urban environment does not and endless resources for a quality education to work on sustainability and to create relationship links between students and their environments, between the student and the local, between the student (who will one day be an adult and will be part of a sustainable society) and the land that surrounds him.

Let us highlight at this point the interview conducted with Pam Buffington, Shona Vitelli and Jill Neumayer DePiper of Education Development Center EDC [85], who live and work in rural communities, and who believe that these areas have many assets when it comes to learning and STEM teaching. When asked if there are advantages to teaching STEM in rural areas, DePiper answered that *"Certainly. There is an amazing amount of opportunity here for place-based education"*. As Vitelli puts it, *"The advantage of teaching science in a rural area is that you can use the natural environment as a teaching tool"*. Buffington comments that *"All students have a right to a high-quality STEM education, and we want to harness the natural and human resources available to us in these rural communities to make that happen"*. From what we can see from her words, the rural, the local, the land has endless resources for quality STEM education. As DePiper quotes *"Rural children have a connection, a passion for the place where they grow up"*. This comment from DePiper leads us to think about what that connection can be translated into, what it can lead to as an adult for that rural child. As DePiper continues, *"Rural kids may become the best advocates we have to protect our coastlines, our forests, our mountains, because they'll think, "That's where I grew up""* STEM education, and even better STEAM, an opportunity to create links between students and their environment, the land in which they live with its living beings and their habitats, links that can be produced at an early age and that will translate into a series of actions and acts also in adulthood.

3. Rural STEM/STEAM in Spain

3.1. Rural Situation in Spain

Data from the Ministry of Agriculture, Fisheries and Food (MAPA) [86] show that the rural population in Spain continues to decline in recent decades; According to the 2020 Census, there are 7,538,929 inhabitants, which represents 15.9% of the total population in rural areas, occupying an area of the country with 84% of the total. Within the rural population, a notable difference has been detected in relation to the proportions between men and women within the rural environment compared to the urban environment: in this rural environment, compared to the urban environment, there are 9.2% more men than women. In the case of young people, the Ministry highlights a lower percentage (35.2%) than the urban one, with the worrying figure of the decrease in said rate since 2011 (9.1%), which provides a clear image of the population aged in rural areas.

These data give us an idea of the small population that lives in rural areas, so the term used of "Empty Spain" [87] fits perfectly with the small population located in large rural areas. As in the USA, the typology of rural areas is classified according to the proximity or distance from the urban nucleus, since it is related to the distribution of the population. We find areas to be revitalized (remote), intermediate or peri-urban, according to RD 752/2010 [88].

Almonte and García [89] point out *"a prolonged loss and aging of the population that puts its environmental, demographic and social sustainability at risk"* in the Iberian Peninsula. We are facing an aging and masculinized rural population, so we must focus on finding solutions that increase the female and youth population in rural areas, as Gallego [90] highlights. This fact means that we must ensure the survival of the rural environment by rejuvenating the population, providing basic and necessary services, including a rural school linked to the territory. Some recent studies have been identified where different educational stages and STEM are analyzed [91,92] exclusively STEM/STEAM programs focused on the gender gap have been developed in recent years [93–96].

3.2. Education in Rural Areas and STEM/STEAM Education

At this point we must point out that the new Education Law, Organic Law 3/2020, of 29 December, (by which Organic Law 2/2006 is modified), integrates, among other issues, education in sustainability, and the Sustainable Development Goals of the 2030 Agenda, very relevant aspects in rural STEM/STEAM education.

If we analyze how education has evolved in rural areas, we find, as indicated in the Spain 2020 Report [97], very few data and few studies on rural schools to be able to draw a priori conclusions on the characteristics at the national level. The disparity of the rural environment implies that we are facing a heterogeneous rural school, with low population density and early school dropout [98]. A relevant issue to deal with is early dropout in rural school. At the level of VET, Salvá-Mut et al. [99] detects discrepancies between territories at the national level, obtaining better results in areas of northern Spain compared to some regions of the Mediterranean and the south with lower results, with greater abandonment of the school, however, it is not specified in rural or urban area.

Domínguez and Sánchez [100] analyze the PISA 2015 report on Spain, from the point of view of rural versus urban education. It is evident that the population, outside the cities, has been drastically reducing, and along with it, public services in the rural community, including education, have been reduced. Over the last few decades, rural schools have been closed, affecting the correct educational development of students located in rural areas. This fact is not exclusive to Spain, since, in Europe, the situation of rural society is also affected by depopulation and reduction of goods and services, presenting lower academic performance than that obtained by students from urban centers. This academic performance, an essential factor in characterizing the quality of education, is determined not only by the social and purchasing power of the student's family, but also by geographical location: it is in this scenario that the rural school should play an important role creating a link between students and the land. It should be remembered that, according to the PISA Report [101], not all communities present the same results, this demonstrates the heterogeneity of education in Spain, depending on the location of the school. The first positions are occupied by Navarra, Castilla y León, Madrid and La Rioja, compared to the communities that occupy the last positions, such as Andalusia, Extremadura and the Canary Islands.

As has been indicated, by the end of the 20th century many schools located in rural areas had already closed due to emigration to the cities. In addition, the teaching staff was reduced, they became itinerant, so that job and educational stability was seriously threatened. It should be added that the rural population is dispersed, as well as the concentration of schoolchildren is reduced, therefore, the design of the classroom in a rural school is quite different from that of an urban classroom, since, in the first case, the same teacher and the same classroom accommodated students of different ages, known as "multigrade classrooms", as indicated by Olivares et al. [102], and abilities, which increased the difficulty for students to acquire the skills required for their age correspondent, Santero-Sánchez and Macías [103] point out that this drastic reduction in the number of teachers has an impact on the lack of specialization in the teaching of subjects as opposed to the urban teacher, who does not have such a large and diverse teaching load and can devote himself to teaching the subjects of his specialty. In the case of mathematics, Santamaria [104] points out differences in the student body in this competence due to the location of the center, that is, the rural school versus the urban school. In this work, Santamaria [104] analyzes the rural issue in the PISA 2015 report. He raises positive questions that may be essential in rural centers to increase the assessment of subjects such as mathematics or science, and integrates proposals for the future National Plan for Rural School, however, currently we only find some regional initiative focused on diagnosing the situation of rural education, as is the case of Aragon, Report of the School Council of Aragon [105], or to diagnose specific cases such as multigrade classrooms in rural environment. This is the case of González et al. [106] in Andalusia; where they analyze one of the characteristic aspects of rural education: multigrade classrooms, identifying management difficulties of this type of classroom by teachers, are relevant issues that must be taken into account, in order to achieve a competitive rural education through these multigrade classrooms: *"training initial and continuous teaching, the provision of material resources and infrastructures and a greater assignment of teachers"* according to González et al. [106].

Martínez and Bustos [107] refered to the general problem of the rural school under the "invisibility" scenario, a concept that accommodates the multitude of factors that affect the loss of education in rural regions, those that range from educational policies to the administrations unaware of the casuistry of the rural world, including the lack of coordination and management between the responsible entities, as well as the lack of digitization, school dropout. Faced with this fact, Olivares et al. [102] propose the rural school in the 21st century *"open to the digital territory from the local territory"*.

However, despite the difficulties presented by the rural school, it is clear that, in the 21st century, this form of education, developed in regions with low population density, is capable of adapting to new technologies, albeit very slowly. This is indicated by Pérez and Martínez [17] in Asturias. The new rural classrooms must be able to adapt to new technologies without losing the essence of local education. With the new tools we will be able to increase the link between the land and rural students, favoring "new learning scenarios" [108]. Thus, the rural classroom is presented with great potential as a source of wealth in innovation, motivation and adaptation of new methodologies, thanks to the integration between the students, the school, the community and the land. This strong bond favors the ability to develop STEM/STEAM projects.

Tamargo et al. [109] in the Autonomous Community of the Principality of Asturias, analyze the interest of Secondary school students in STEM/STEAM subjects, at an urban and rural level. We are interested in this study since a gender difference is detected, although with certain considerations: among the students who choose technical sciences, it is the boys who outperform the girls by 60% in this branch; while in the case of Health, Natural and Arts Sciences, it is the female students who outperform males by more than 50%, both in urban and rural settings. The gender difference between STEAM subjects does not show large differences in rural and non-rural environments, except in Arts and Mathematics. The study of Arts stands out, where in rural areas female students are more interested than male students, while in urban areas the opposite is true. Continuing with Mathematics and Exact Sciences, we find an inverted scenario: it is the female students in urban areas who show a greater interest (although less than that of the male students) than the female students in rural areas, who show little interest in this subject. On the other hand, an interesting issue identified in the article is the weakness related to the educational system: they detect the *"inexistence of educational policies specifically related to the integral STEAM model"*.

Spain still needs to continue deepening research and integration of the STEM/STEAM model, especially in rural areas, as it lacks sufficient research to date. We corroborate this with recent bibliometric studies [110–112] where there is hardly any Spanish representation, although none of the articles identified correspond to rural areas. Ferrada et al. [110] analyzes publications between 2010 and 2018, identifying a single article from Spain on STEM in SCOPUS, although without referencing rural areas. On the other hand, in the bibliometric analysis of Sánchez and Martínez [111] conducted between 2010 and 2020 on STEM education, identifies the United States as the largest producer of published articles (33), compared to Spain (4). However, none of the publications referred to STEAM disciplines in rural areas. The results show that STEM studies have continued uninterrupted until today, although the interest generated in the scientific community has been irregular.

An important aspect to evaluate is the role that this type of education plays in the future of students. Analyzing the demand for graduates with STEM/STEAM training, it has become evident that this type of education generates professionals who are currently in demand. In a European scenario, it should be noted that an increase in the demand for professionals trained in the STEM/STEAM education system has been estimated, focused on the Vocational Training level, trying to find an outlet for occupations related to STEM sectors, among them, the IT sector. It is estimated a higher demand, with greater skills and abilities provided by the STEM study, such as creativity, cooperation, problem solving ... able to innovate thanks to multidisciplinary teamwork, CEDEFOP [113] indicates that *"around 48% of STEM-related occupations require middle-level (upper secondary) qualifications, many of which are acquired through initial Vocational Education and Training (VET) at the upper secondary level"*, however, according to CEDEFOP (2015), Spain has a level of STEAM graduates below the European average [114].

According to this report, within Europe there is no homogeneity in relation to VET graduates in STEM subjects. It is worth noting that Bulgaria, Estonia and Cyprus account for more than 40% compared to Belgium, Denmark and the Netherlands for less than 20%. Spain also does not reach the European average during the years 2006, 2001 and 2015

(Figures 2 and 3). Outside of Europe, the United States, along with Canada and Australia, do have a higher percentage of STEM graduates. At the national levelMoso-Diez, et al. [115] analyzed the differences between autonomous communities, in relation to VT and STEM subjects, evidencing the interrelation between VT specialization and supply and demand.

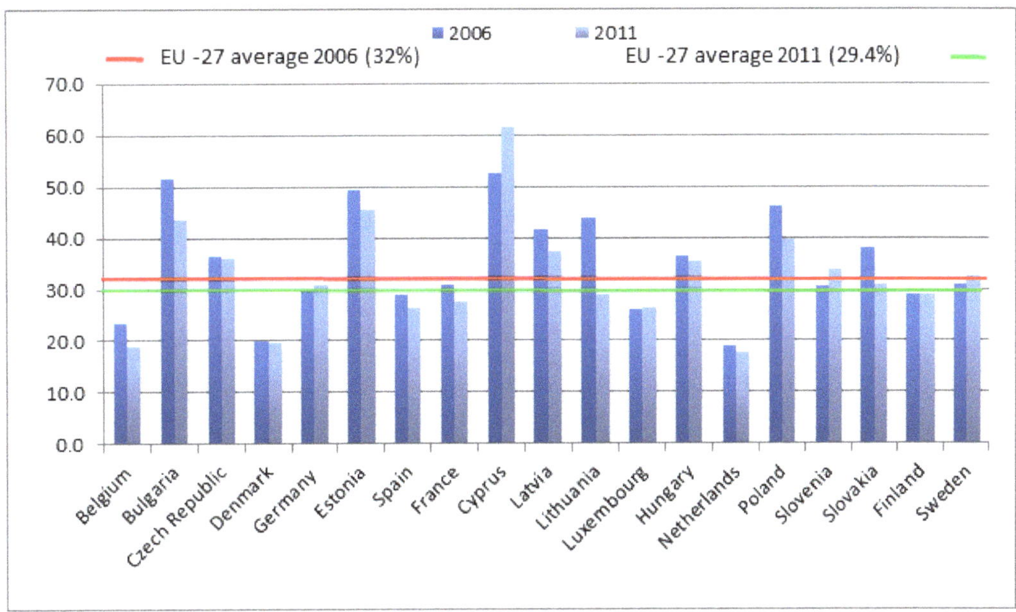

Figure 2. Percentage of VET graduates in STEM subjects in 2006 and 2011. Source: CEDEFOP 2014.

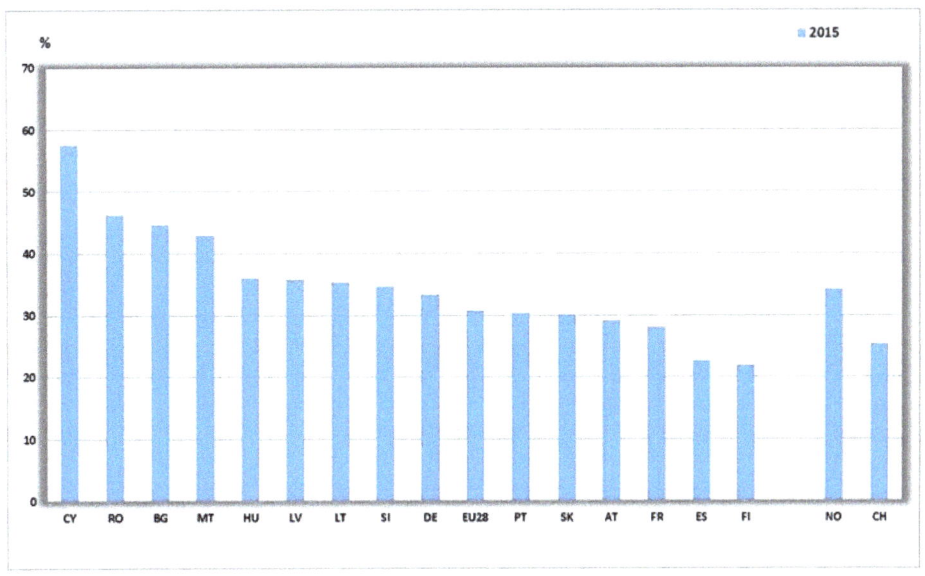

Figure 3. Percentage of VET STEM (science, technology, engineering and mathematics) graduates in 2015. Source: CEDEFOP 2015. Source: CEDEFOP calculations based on Eurostat data/UOE data collection on education.

In these graphs we can see that in 2015, although the European average has risen slightly (30.8%), Cyprus has the highest percentage (57.6%) while Spain remains below 25%, without reaching the European average of VET graduates in STEM subjects (Indicator 2050: STEM graduates from Upper Secondary VET).

According to this report, the supply of ICT and STEM graduates from upper secondary education is insufficient to meet the demand. Very few young people are studying STEM subjects. Entry requirements and dropout rates are high and women's participation is low. It is clear that, as the title of Usart et al. [27]: *"The field of STEM does not attract female talent"*.

Sancha [116] proposes competencies in STEM disciplines in Vocational Education, betting on a green and digitized education. More people, especially women, are now being encouraged to study ICT and STEM subjects. In Spain, trying to promote these STEM careers among women, several studies have been published where they analyze this situation [26,117] to give visibility to women and overcome educational barriers.

One of the example can be found in the scarce participation of women in scientific dissemination channels, as is the case of YOUTUBE [118]. Tomassini [119] analyzes the gender gap in scientific-technological disciplines, identifying some gaps in STEM training; coincides with the differences detected in López Capdevilla [120], which proposes incorporating a greater number of female scientific references in the syllabus of ESO subjects to achieve a greater impact on students and a greater familiarity of women and science, not only use it occasionally. In addition, Moso-Diez et al. [121] points out that the "loss" of women and girls in STEM disciplines is detected from the beginning of the choice of related subjects, even those who choose this path have greater difficulties in leading STEM careers. This gender gap is detectable in STEM VET.

An integrated STEM proposal is proposed by Thibaut et al. [122] to increase motivation for STEM subjects and careers, however, there is no consensus in the development of instructions for this approach, so it is necessary to continue investigating to stimulate students in these subjects, providing new tools and methodologies as pedagogical alternatives, in an integral way for students [123].

Within the university world, the promising results of Martín-Sánchez et al. [124] at the University of Extremadura, to train future teachers, point to a strategy where they establish the connection between ESD and STEM through a Service Learning (SL) methodology, providing more tools to future STEM teachers. Martínez-Campillo et al. [125] confirmed the effectiveness of this methodology with undergraduate students of Business Administration at the University of León, providing comprehensive training and higher academic performance.

4. Discussion and Conclusions

One of the biggest problems in rural Spain is depopulation. In addition, the existing population in rural Spain is very old, male-dominated and with hardly any generational replacement. The rural environment is an area in demographic decline. This depopulation is a consequence of some facts, it is not a cause and as such we must face it for the improvement of the situation. We must emphasize that people have the right to be able to remain in rural areas with stability in their lives, be it economic, social, emotional and, of course, environmental. If we want to bring about change, we have to create the conditions for development in rural areas, a development that must be sustainable or it will not be. There are four fundamental factors to fix the population in rural areas: education, health, technology and employment. Education is the protagonist and the other three factors revolve around it; let us remember then the 17 SDGs of the 2030 Agenda that revolve around SDG 4. If we want changes for today and tomorrow, we must think about the education we give to our current generations of children and young people. For the rural environment to continue to be well educated, it will be necessary for rural people to have sustainable actions and deeds that speak of sustainability in its four dimensions.

Education in the rural environment presents a scarcity of research articles, not only in Spain, but also at the international level. There are common issues such as the scarcity of resources, which cause a loss in the effectiveness of education, either due to the lack of interest of the students, or due to deficiencies of the teachers, gender, socioeconomic or political issues. Both in the USA and in Spain, problems are detected in the academic success of students in rural areas compared to urban areas, such as lack of infrastructure, low population density, lack of funding, lack of management in the administrations ...

On the other hand, STEM and STEAM research has not been homogeneous over time; currently, in Spain it has focused on the visibility of women in scientific fields, integrating gender differences. It is important that teachers become involved in the new strategies proposed in STEM/STEAM education to reach the largest number of students from an early age. Within the STEAM disciplines, we have identified mathematics as a subject with difficulties in rural areas. The Mathematics White Paper [126] presents the STEAM approach as a source of empowerment of educational competences related to mathematics, proposing 64 measures to promote science in Spain, from different areas, including gender equality, dissemination and research. Most of the research has been carried out in urban areas and rural education is invisible to STEAM programs. It is proposed to correct this situation, as the intrinsic characteristics of rural education favor the successful development of STEAM programs in gender equality. It is necessary to provide funding, resources and a stable staff of motivated teachers to promote STEAM careers in rural areas. Conducting interdisciplinary STEM and STEAM education using the resources of the rural environment, using their land, among other things, will create an awakening of sensitivity in students to the land around them that will have short and long term consequences for the care of the land itself in rural areas.

As we have discussed, internationally, there is little research related to STEM education in rural areas. Rural education suffers from lack of resources, lack of teachers and low school performance. In the USA, STEAM and STEM programs are being developed in a more advanced way than in Spain, both in urban and rural areas. The point is that we are facing a possible solution to the Spanish rural depopulation, making use of STEM/STEAM disciplines from an early age, to foster the link between the land, its resources and the educational community as an essential part of the new generations of rural professionals. We found few success stories in rural schools, as an example we point out the work in Victoria, Australia [127] under the broad integrative concept of STEM education described by Bybee [9]. According to Murphy [127], the success of STEM disciplines in these rural schools was located in the strong link of the school with the land, with the community, with its local resources and with a motivated and committed educational community, as well as the promotion of these disciplines in the real world.

In relation to the gender gap evident in STEM/STEAM disciplines at the international level, it has been confirmed that, in Spain, the presence of women in scientific-technological disciplines is still scarce. This situation is not exclusive to Spain, as it is a global problem, as described in the European W-STEM project [128] where strategies are sought to improve the incorporation of women in science. Likewise, the 2018 UNESCO report [129] identifies obstacles in female STEM careers, where only 30% of the positions linked to STEM disciplines correspond to women. It is paramount to be able to generate interest in science from an early age, to facilitate access and motivation in scientific-technological careers. It should be pointed out that, as Echazarreta et al. [130] the economic inclusion of women in the professional world continues to be a barrier for this gender, so that we find an inequality at the international level in terms of economy and professional women, so that, if we focus on the rural scenario that concerns us, we start from the same situation of imbalance. In Spain, the Ministry of Education and Vocational Training has recently created the figure of the STEAM Alliance, from whose portal the contribution of girls and women in STEAM disciplines is encouraged from a very early age, coinciding with the work developed in Cáceres by Sandoval et al. [131] which is committed to the training of women scientists from the time they start school.

Joining the concepts of rural Spain, STEM/STEAM and women, it is important to highlight the efforts to advance in the field of women and rural STEM/STEAM, despite the little research developed to date. We highlight the program *"Conéctateen el medio rural. Digital skills in rural Spain. Breaking the gap"*, proposed in April 2022 by the *"Association of families and women in rural areas"* (AFAMMER), whose objectives are mainly focused on girls and women, seeking to stimulate their curiosity and skills in STEM careers, ensuring their entry into the digital economy. A recent example is that of the Canary Islands, where the rural STEAM Lab (https://laboratoriosteamrural.com/ accessed on 15 September 2022) has been presented, a project in which 24 rural schools have been trained in learning techniques focused on STEAM topics, and which has generated great acceptance by teachers and students. We also highlight the work of the bMaker project (https://bmaker.es/ accessed on 16 September 2022), with STEAM methodology, to increase STEAM vocations in rural areas.

Finally, it should be noted that in Spain, several authors are calling for a National Rural School Plan that integrates the new methodologies, since there is still a lack of greater involvement on the part of the institutions to enhance the value of the educational system based on STEM/STEAM strategies; localized educational plans could be planned, depending on the locality where the rural school is located, incorporating all the necessary disciplines in each of the educational stages.

We should look to US government programs and policies, specifically STEM4SD Education, to train children from an early age towards environmental sensitivity and sustainable development. The rural classroom has great potential to promote the ability to develop STEM/STEAM projects linked to the territory, and with gender equality, from the beginning of the students' educational stage, fixing population in "empty Spain". The rural environment and its territory need to move towards a social transformation born from the development of eco-sustainable competence in the local rural school. The rural environment, its inhabitants and its territory need solutions that allow a sustainable local and rural development and the interdisciplinary STEM/STEAM education for sustainability achieves such purposes.

Author Contributions: Conceptualization, E.G.-S., P.-T.E.-G. and C.L.-B.; methodology, E.G.-S., P.-T.E.-G. and C.L.-B.; software, E.G.-S., P.-T.E.-G. and C.L.-B.; validation, E.G.-S., P.-T.E.-G. and C.L.-B.; formal analysis, E.G.-S., P.-T.E.-G. and C.L.-B.; investigation, E.G.-S., P.-T.E.-G. and C.L.-B.; resources, E.G.-S., P.-T.E.-G. and C.L.-B.; data curation, E.G.-S., P.-T.E.-G. and C.L.-B.; writing—original draft preparation, E.G.-S., P.-T.E.-G. and C.L.-B.; writing—review and editing, E.G.-S., P.-T.E.-G. and C.L.-B.; visualization, E.G.-S., P.-T.E.-G. and C.L.-B.; supervision, E.G.-S., P.-T.E.-G. and C.L.-B.; project administration, E.G.-S., P.-T.E.-G. and C.L.-B.; funding acquisition, E.G.-S. All authors have read and agreed to the published version of the manuscript.

Funding: This research was funded by funding entity Vice-rectorate for Researchgrant number 096-034305 Grants for UNED Research Projects. ESEJERD (Study of the Japanese educational system as an international reference in Disaster Risk Reduction Education: Educational System, Civil Associations and School Architecture).

Institutional Review Board Statement: Not applicable.

Informed Consent Statement: Not applicable.

Conflicts of Interest: The authors declare no conflict of interest.

References

1. de Segura, B.G.R. From sustainable development according to Brundtland to sustainability as biomimicry. *Bilbao Hegoa* **2014**, *1*, 59.
2. Mihalic, T.; Mohamadi, S.; Abbasi, A.; Dávid, L.D. Mapping a Sustainable and Responsible Tourism Paradigm: A Bibliometric and Citation Network Analysis. *Sustainability* **2021**, *13*, 853. [CrossRef]
3. Agbedahin, A.V. Sustainable development, Education for Sustainable Development, and the 2030 Agenda for Sustainable Development: Emergence, efficacy, eminence, and future. *Sustain. Dev.* **2019**, *27*, 669–680. [CrossRef]
4. Shulla, K.; Filho, W.L.; Lardjane, S.; Sommer, J.H.; Borgemeister, C. Sustainable development education in the context of the 2030 Agenda for sustainable development. *Int. J. Sustain. Dev. World Ecol.* **2020**, *27*, 458–468. [CrossRef]
5. Rocard, M.; Csermely, P.; Jorde, D.; Lenzen, D.; Henriksson, H.W.; Hemmo, V. *Science Education Now: A New Pedagogy for the Future of Europe*; European Commission Directorate Generalfor Research Informationand Communication Unit: Brussels, Belgium, 2007.

6. Arredondo, F.G.; Vázquez, J.C.; Velázquez, L.M. STEM y brecha de géneroen Latinoamérica. *Rev. El Col. San Luis.* **2019**, *9*, 137–158.

7. Arredondo, P.; Miville, M.L.; Capodilupo, C.M.; Vera, T. *Women and the Challenge of STEM Professions*; Springer: Berlin/Heidelberg, Germany, 2022.

8. Bogdan Toma, R.; García-Carmona, A. De STEM nos gusta todo menos STEM. Análisis crítico de una tendencia educativa de moda. *Enseñanza Cienc* **2021**, *39*, 65–80.

9. Bybee, R.W. The Case for Education Challenges and Opportunities. In *National Science Teachers Association*; NSTA Press: Arlington, VA, USA, 2013.

10. Kelley, T.R.; Knowles, J.G. Aconceptual framework for integrated STEM education. *Int. J. STEM Educ.* **2016**, *3*, 11. [CrossRef]

11. Basile, V.; Azevedo, F.S. Ideology in the mirror: A loving (self) critique of our equity and social justice efforts in STEM education. *Sci. Educ.* **2022**, *106*, 1084–1096. [CrossRef]

12. Ku, C.J.; Hsu, Y.S.; Chang, M.C.; Lin, K.Y. A Model examining middle school students' STEM integration behavior in anational technology competition. *Int. J. STEM Educ.* **2022**, *3*, 3. [CrossRef]

13. Sulaeman, N.; Efwinda, S.; Putra, P.D.A. Teacher readiness in STEM education: Voices of Indonesian Physicsteachers. *J. Technol. Sci. Educ.* **2022**, *12*, 68–82. [CrossRef]

14. Slater, T.F.; Kilty, T.J.; Burrows, A.C. Improving the Integration of STEAM Concepts for Teacher Preparation and Practice. In *Literacy in Teacher Preparation and Practice: Enabling Individuals to Negotiate Meaning, Chapter 9*; Patrick, M., Jenlink, S.F., Eds.; Information Age Publishing Inc: Charlotte, NC, USA, 2022; pp. 203–218.

15. Goldhaber, D.; Krieg, J.; Theobald, R.; Goggins, M. Front end to backend: Teacher preparation, workforce entry, and attrition. *J. Teach. Educ.* **2022**, *73*, 253–270. [CrossRef]

16. Ogodo, J.A. Developing STEM Teachers' Cultural Competence through an Urban Teaching Curriculum: A Cultural Border Crossing Experience. *J. Sci. Teach. Educ.* **2022**. [CrossRef]

17. Perez, M.; Martinez, L. Digitization of Asturian rural schools: Rural teachers 2.0 and local development Teachers. *Profr. Rev. Currículum Form. Profr.* **2011**, *15*, 109–123.

18. Kim, D.; Bolger, M. Analysis of Korean Elementary Pre-Service Teachers' Changing Attitudes about Integrated STEAM Pedagogy. *Int. J. Sci. Math. Educ.* **2017**, *15*, 587–605. [CrossRef]

19. Xu, W.; Ouyang, F. The application of AI technologies in STEM education: A systematic review from 2011 to 2021. *Int. J. STEM Educ.* **2022**, *9*, 59. [CrossRef]

20. Alam, A. Psychological, Sociocultural, and Biological Elucidations for Gender Gap in STEM Education: A Call for Translation of Research into Evidence-Based Interventions. In *Proceedings of the 2nd International Conference on Sustainability and Equity (ICSE-2021). Atlantis Highlights in Social*; World Leadership Academy and Kalinga Institute of Industrial Technology (KIIT): Odisha, India, 2021.

21. Kenneth, A. Gap in STEM Education: Why is there a Decline in Women Participation? *Int. J. Res. STEM Educ.* **2022**, *4*, 55–63. [CrossRef]

22. Cheryan, S.; Master, A.; Meltzoff, A.N. Cultural stereotypes as gatekeepers: Increasing girls' interest in computer science and engineering by diversifying stereotypes. *Front. Psychol.* **2015**, *6*, 49. [CrossRef]

23. Denessen, E.; Vos, N.; Hasselman, F.; Louws, M. The relationship between primary school teacher and student attitudes towards science and technology. *Educ. Res. Int.* **2015**, *2015*, 534690. [CrossRef]

24. Kim, A.Y.; Sinatra, G.M.; Seyranian, V. Developing a STEM Identity among Young Women: A Social Identity Perspective. *Rev. Educ. Res.* **2018**, *88*, 589–625. [CrossRef]

25. Oon, P.T.; Cheng, M.M.W.; Wong, A.S.L. Gender differences in attitude towards science: Methodology for prioritising contributing factors. *Int. J. Sci. Educ.* **2020**, *42*, 89–112. [CrossRef]

26. Perez, S.G. Impact of role models on the influence of determining factors in girls' career choice in the STEM field. Ph.D. Thesis, Universidad CEU San Pablo, Madrid, Spain, 2021.

27. Usart, M.; Sánchez-Canut, S.; Lores, B. The field of STEM does not attract female talent. *Fund. Obra Soc. La Caixa* **2022**. Available online: https://elobservatoriosocial.fundacionlacaixa.org/-/el-ambito-de-las-stem-no-atrae-el-talento-femenino# (accessed on 20 September 2022).

28. Carrasquilla, O.M.; Pascual, E.S.; San Roque, I.M. La brecha de género en la Educación STEM. *Revis. Educ.* **2022**, *396*, 151–175.

29. Goreth, S.; Vollmer, C. Gender does not make the difference: Interest in STEM by gender is fully mediated by technical socialization and degree program. *Int. J. Technol. Des. Educ.* **2022**, 1–23. [CrossRef] [PubMed]

30. Echazarra, A.; Radinger, T. Learning in rural schools: Insights from PISA, TALI Sand the literatura. In *OECD Education Working Papers*; OECD Publishing: Paris, France, 2019.

31. Bazán-Ramírez, A.; Hernández-Padilla, E.; Castellanos-Simons, D. Educación y apoyo familiar, y logro en matemáticas en dos contextos sociodemográficos diferentes. *Propósitos Represent.* **2022**, *10*, e1354. [CrossRef]

32. Hillier, C.; Zarifa, D.; Hango, D. Mind the Gaps: Examining Youth's Reading, Mathand Science Skills across Northernand Rural Canada. *Rural Sociol.* **2022**, *87*, 264–302. [CrossRef]

33. Ghimire, B. Blended Learning in Ruraland Remote Schools: Challenges and Opportunities. *Int. J. Technol. Educ.* **2022**, *5*, 88–96. [CrossRef]

34. OECD. Adapting the school network to changing needs in urban, ruraland remote areas. In *Responsive School Systems: Connecting Facilities, Sectors and Programmes for Student Success*; OECD Publishing: Paris, France, 2018.

35. Ishii, Y.; Meke, E. An Analysis of Rural-Urban Learning Performance Inequalityin Malawi Primary Education. *East Afr. J. Educ. Stud.* **2022**, *5*, 194–211. [CrossRef]

36. Fargas-Malet, M.; Bagley, C. Issmall beautiful? Ascoping review of 21st-century research on small rural schools in Europe. *Eur. Educ. Res. J.* **2022**, *21*, 822–844. [CrossRef]

37. Goodpaster, K.P.; Adedokun, O.A.; Weaver, G.C. Teachers' perceptions of rural STEM teaching: Implications for rural teacherretention. *Rural. Educ.* **2012**, *33*, 9–22. [CrossRef]

38. Ihrig, L.M.; Lane, E.; Mahatmya, D.; Assouline, S.G. STEM excellence and leadership program: Increasing the level of STEM challenge and engagement for high-achieving students in economically disadvantaged rural communities. *J. Educ. Gift* **2018**, *41*, 24–42. [CrossRef]

39. Irwanto, I.; Saputro, A.D.; Ramadhan, M.F.; Lukman, I.R. Research Trends in STEM Education from 2011 to 2020: A Systematic Review of Publications in Selected Journals. *Int. J. Interact. Mob. Technol.* **2022**, *16*, 5. [CrossRef]

40. Kayan-Fadlelmula, F.; Sellami, A.; Abdelkader, N.; Umer, S. A systematic review of STEM education research in the GCC countries: Trends, gapsand barriers. *Int. J. STEM Educ.* **2022**, *9*, 2. [CrossRef]

41. Fernández Buey, F. Sustainability: Word and concept. *Mag. Gen. Subdirectorate State Mus.* **2011**, *8*, 16–25.

42. Cervantes Dueñas, J.O. The Bretton Woods Institutions: (Neoliberally) Sustainable Development. *Environ. Obs.* **2014**, *17*, 23–43.

43. Rossi, E. A future to build: The proposal of modern environmentalism. *Ibero-Am. J. Bioeth.* **2017**, *4*, 1–17.

44. United Nations. *Johannesburg Declaration on Sustainable Development. Third United Nations World Summit on Environment and Development*; United Nations: New York, NY, USA, 2002.

45. United Nations. *Report of the World Commission on Environment and Development "Our Common Future" Brundtland Report*; United Nations: New York, NY, USA, 1987.

46. United Nations. *Rio Declaration on Environment and Development*; United Nations: New York, NY, USA, 1992.

47. Nogales Naharro, M.A. Rural development and sustainable development. Ethical sustainability CIRIEC-Spain. *J. Public Soc. Coop. Econ.* **2006**, *55*, 7–42.

48. Saura Calixto, P.; Hernández Prados, M.A. The evolution of the concept of sustainability and its effect on environmental education. *Teor. Educ.* **2008**, *20*, 179–204.

49. Bosch, H.E.; Di Blasi, M.A.; Pelem, M.E.; Bergero, M.S.; Carvajal, L.; Geromini, N.S. New pedagogical paradigm for Sciences an Mathematics teaching. *Av. En Cienc. Ing.* **2011**, *2*, 131–140.

50. UNESCO. Sustainable Development Begins with Education. 2021. Available online: https://es.unesco.org/news/desarrollo-sostenible-comienza-educacion (accessed on 25 August 2022).

51. Cantu-Martinez, P.C. Environmental education and the school as an educational space for the promotion of sustainability. *Educ. Electron. J.* **2014**, *18*, 39–52.

52. Schejtman, A.; Berdegué, J. *Rural Territorial Development. RIMISP*; Latin American Center for Rural Development: Brighton, UK, 2004.

53. Riella, A.; Vitelli, R. Territorial development, citizen ship and rural schools: Are flection for the Uruguay an case. *Pampa Interuniv. J. Territ. Stud.* **2005**, *1*, 131–146.

54. CONAMA. *Challenges of Education for Sustainability in the XXI Century. Environmental Education Working Group of the National Congress of the Environment*; CONAMA: Madrid, Spain, 2017.

55. Tieken, M.C.; Montgomery, M.K. Challenges Facing Schools in Rural America. *State Educ. Stand.* **2021**, *21*, 6–11.

56. Provasnik, S.; KewalRamani, A.; Coleman, M.M.; Gilbertson, L.; Herring, W.; Xie, Q. *Status of Education in Rural America*; US Department of Education, National Center for Education Statistics, Institute of Education Sciences: Washington, DC, USA, 2007.

57. World Bank. Available online: https://datos.bancomundial.org (accessed on 1 September 2022).

58. Nugent, G.C.; Kunz, G.M.; Sheridan, S.M.; Glover, T.A.; Knoche, L.L. *Rural Education Research in the United States*; Springer: Berlin/Heidelberg, Germany, 2017.

59. Player, D. *The Supply and Demand for Rural Teachers*; Roci-Rural Opportunities Consortium of Idaho: Boise, ID, USA, 2015; pp. 1–30.

60. Barter, B. Rural education: Learning to be rural teachers. *Workplace Learn.* **2008**, *20*, 468–479. [CrossRef]

61. White, S.; Kline, J. Developing arural teacher education curriculum package. *Rural. Educ.* **2012**, *33*, 36–43.

62. Corbett, M. Reading Lefebvre from the periphery: Thinking globally about the rural. In *Self-Studies in Rural Teacher Education*; Springer: Cham, Switzerland, 2016; pp. 141–156.

63. Corbett, M. Place-based education: A critical appraisal from a rural perspective. In *Rural Teachereducation*; Springer: Singapore, 2020; pp. 279–298.

64. Sanders, M. STEM, STEM education, STEM mania. *Technol. Teach.* **2009**, *68*, 20–26.

65. Toma, R.B.; Greca, I.M. *Interdisciplinary Model of STEM Education for the Primary Education Stage*; University of Burgos: Burgos, Spain, 2016; pp. 1–5.

66. Watson, A. Sustainability Education in Primary and Secondary Schools: Great Needs and Possible Solutions. Chancellor's Honors Program Projects. 2017. Available online: https://trace.tennessee.edu/utk_chanhonoproj/2026 (accessed on 2 May 2022).

67. The White House. Educate to Innovate. 2010. Available online: http://www.whitehouse.gov/issues/education/educateinnovate (accessed on 20 June 2022).

68. Honey, M.; Pearson, G.; Schweingruber, H. (Eds.) *STEM Integration in K-12 Education: Status, Prospects, and an Agenda for Research*; National Academies Press: Washington, DC, USA, 2014.

69. English, L.D. Reconceptualising statistical learning in the early years. In *Reconceptualising Early Mathematics Learning*; English, L.D., Mulligan, J., Eds.; Springer: Dordrecht, The Netherlands, 2013; pp. 67–82.

70. English, L.D.; Mousoulides, N. Bridging STEM in a real-world problem. *Math. Teach. Middle Sch.* **2015**, *20*, 532–539. [CrossRef]

71. English, L.D. STEM education K-12: Perspectives on integration. *Int. J. STEM Educ.* **2016**, *3*, 3. [CrossRef]

72. English, L.D. Advancing elementary and middle school STEM education. *Int. J. Sci. Math. Educ.* **2017**, *15*, 5–24. [CrossRef]

73. Adams, E.C.; Oduor, P.; Wahome, A.; Tondapu, G. Reflections on two years teaching earth science at the women in science (Wisci) steam camp. *J. Women Minor. Sci. Eng.* **2022**, *28*, 23–39. [CrossRef]

74. Committee on STEM Education of the National Science & Technology Council. *Charting a Course for Success: America's Strategy for STEM Education*; Committee on STEM Education of the National Science & Technology Council: Washington, DC, USA, 2018.

75. The Office of Science and Technology Policy. *Progress Report on the Implementation of the STEM Education Strategic Plan*; Office of Science and Technology Policy: Washington, DC, USA, 2020.

76. The White House Office of Science and Technology Policy. *Summary of the White House Release Event for the 2018 STEM Education Strategic Plan*; The White House Office of Science and Technology Policy: Washington, DC, USA, 2018.

77. The White House Office of Science and Technology Policy. *Summary of the 2018 White House State-Federal STEM Education Summit*; The White House Office of Science and Technology Policy: Washington, DC, USA, 2018.

78. The Office of Science and Technology Policy. *Progress Report on the Federal Implementation of the STEM Education Plan*; Office of Science and Technology Policy: Washington, DC, USA, 2019.

79. The White House. Fact Sheet: Biden-Harris Administration Actions to Attract STEM Talent and Strengthen our Economy and Competitiveness. 2022. Available online: https://www.whitehouse.gov/briefing-room/statements-releases/2022/01/21/fact-sheet-biden-harris-administration-actions-to-attract-stem-talent-and-strengthen-our-economy-and-competitiveness/ (accessed on 19 September 2022).

80. Dorn, R.; Kanikeberg, K.; Burke, A. *K-12 Integrated Environmental and Sustainability Education Learning Standard*; OSPI: Washington, DC, USA, 2014.

81. Bryan, L.A.; Moore, T.J.; Johnson, C.C.; Roehrig, G.H. Integrated STEM education. In *STEM Road Map: A Framework for Integrated STEM Education*; Jonhson, C.C., Peters_Burton, E.E., Moore, T.J., Eds.; Routledge: New York, NY, USA, 2015; pp. 23–37.

82. Pahnke, J.; O'Donnell, C.; Bascopé, M. 2019. Using Scienceto Do Social Good: STEM Education for Sustainable Development. Position Paper Developed in Preparation for the Second "International Dialogueon STEM Education "(IDoS) in Berlin. Available online: https://www.haus-der-kleinen-forscher.de (accessed on 6 December 2019).

83. Williams, J.M.; Nierengarten, G. Recommendations from the North Star State. *Rural. Educ.* **2011**, *33*, 15–24. [CrossRef]

84. Peterson, B.; Bornemann, G.; Lydon, C.; West, K. Ruralstudentsin Washington State: STEM asstrategy for building rigor, postsecondary aspirations, and relevant career opportunities. *Peabody J. Educ.* **2015**, *90*, 280–293. [CrossRef]

85. Education Development Center EDC. 2019. Igniting STEM Education in Rural Communities. There are Rich Learning Opportunities for Students and Teachers Living in Rural Areas. Available online: https://www.edc.org/igniting-rural-stem (accessed on 25 June 2021).

86. Ministry of Agriculture, Fisheries and Food (MAPA). Agrinfo no 31. *Demography of the Rural Population in 2020*. Available online: https://www.mapa.gob.es/ (accessed on 25 August 2022).

87. Redondo de Sa, M.; Postigo Mota, S. The emptied Spain. *Nurs. Role Mag.* **2021**, *44*, 21–32.

88. RD 752/2010:49692. Royal Decree 752/2010, of June 4, Which Approves the First Sustainable Rural Development Program for the Period 2010–2014 in Application of Law 45/2007, of December 13, for the Sustainable Development of the Rural Environment. Available online: https://www.boe.es/buscar/doc.php?id=BOE-A-2010-9237 (accessed on 20 May 2022).

89. Almonte, J.M.J.; Garcia, F.J.P. Population and territorial sustainability of rural areas in Spain and Portugal. *Geog. Noteb.* **2022**, *61*, 61–87.

90. Gallego, R.S. Género y repoblación rural. Mujeres autóctonas e inmigrantes en la España interior. In *La España rural: Retos y oportunidades de futuro*; Cajamar Caja Rural: Almería, Spain, 2022; pp. 181–197.

91. Castro Rodríguez, E.; Montoro Medina, A.B. Educación STEM y formación del profesorado de Primaria en España. *Rev. Educ.* **2021**, *393*, 341–364.

92. García-Jiménez, L.; Torrado-Morales, S.; Díaz Tomás, J.M. El rol de la mujer en la ciencia y la docencia en comunicación: Análisis a partir de los programas universitarios en España. *Rev. Comun.* **2022**, *21*, 91–113. [CrossRef]

93. García-Peñalvo, F.J.; García-Holgado, A.; Dominguez, A.; Pascual, J. *Women in STEM in Higher Education: Good Practices of Attraction, Access and Retainment in Higher Education*; Springer Nature: Singapore, 2022.

94. Sempere, I.H., Vidal, C.E. Estudios STEM y la brecha digital de género en bachillerato. *Edutec. Rev. Electrónica Tecnol. Educativa* **2022**, *81*, 55–71.

95. Verdugo-Castro, S. La Brecha de Género en los Estudios Universitarios del Sector STEM en el Espacio Español de Educación. Ph.D. Thesis, Programa de Formación en la Sociedad del Conocimiento, Universidad de Salamanca, Salamanca, Spain, 2022.

96. Verdugo-Castro, S.; Sánchez-Gómez, M.C.; García-Holgado, A. Opiniones y percepciones sobre los estudios superiores STEM: Un estudio de caso exploratorio en España. *Educ. Knowl. Soc.* **2022**, *23*, 1–15. [CrossRef]

97. Mora, S.; López-Ruiz, J.A.; Sánchez, A.M.C. *Spain Report 2020*; Comillas Pontifical University: Madrid, Spain, 2020; Volume 27.

98. Santamaria Luna, R. Early school leaving in rural areas of Europe and Spain. *Adv. Educ. Supervision.* **2015**, *24*, 1–44.

99. Salvá-Mut, F.; Ruiz-Pérez, M.; Psifidou, I.; Oliver-Trobat, M.F. Intermediate vocational training and early dropout from educatio-nand training in Spain: A territorial approach. *Bordón, J. Pedagog.* **2020**, *72*, 95–116.

100. Domínguez, C.M.; Sánchez, R. The educational system in the Spanish rural areas. A comparative analysiswitheducation in thecity in Spain. *Rev. Univ. Eur.* **2022**, *37*, 215–248.

101. PISA. Program for the Comprehensive Assessment of Students. 2018. Organization for Economic Cooperation and Development (OECD). Available online: https://www.oecd.org/pisa/pisa-es/ (accessed on 1 September 2022).

102. Olivares, P.A.; Tomas, R.B.; Peñafiel, L.D.; Lacruz, J.L.; Terrado, P.R. *The Challenge of the Rural School: Making the Invisible Visible*; Critica y Fundamentos Collection: Barcelona, Spain, 2021; Volume 54, 220p.

103. Santero-Sanchez, R.; Macías Domínguez, C. The educational system in the Spanish rural area. *A Comp. Anal. Educ. City Spain* **2022**, *37*, 215–248.

104. Santamaria, R. The rural school and external evaluations in Spain. *PISA as anexample. Tempsd'Educació* **2020**, *59*, 57–90.

105. School Council of Aragon. *The Educational System in the Aragonese Rural Territory: Rural Schools, Grouped Schools and Other Educational Centers*; School Council of Aragon: Aragon, Spain, 2022.

106. González Alba, B.; Cortés González, P.; Leite Méndez, A. Multigrade classroom sin rural Andalusia. Teaching visions. *IE J. Educ. Res. Rediech* **2020**, *11*, e860. [CrossRef]

107. Martínez, J.B.; Bustos, A. Globalization, new ruralities and schools. Faculty. *J. Curric. Teach. Train.* **2011**, *15*, 3–12.

108. Del Moral Perez, M.E.; Villalustre Martinez, L. Cooperative learning through ICT in rural schools. In *Information and Communication Technologies in Educational Contexts. New Learning Scenarios*; Sandoval, Y., Sands, A., Lopez, E., Cabero, J., Aguaded, J.I., Eds.; Santiago de Cali University Publishing House: Santiago de Cali, Colombia, 2012; pp. 91–104. ISBN 978-958-8303-85-7.

109. Tamargo-Pedregal, L.Á.; Agudo-Prado, S.; Fombona, J. STEM/STEAM interests of secondary school students in rural and urban areas in Spain. *Educ. E Pesqui.* **2022**, *48*, 1–22.

110. Ferrada, C.; Díaz-Levicoy, D.; Salga do Orellana, N.; Puraivan, E. Bibliom etric analysis on STEM education. *Spaces* **2019**, *40*, 2.

111. Sánchez, D.; Martinez, A. STEM education, an emerging research field: Biliometric analysis between 2010–2020. *Investig. De Ciências* **2021**, *26*, 195. [CrossRef]

112. Vargas Sánchez, D.L.; García-Martínez, Á. STEM education, an emerging research field: Bibliometric analysis between 2010–2020. *Investig. De Ciências* **2021**, *26*, 195–219. [CrossRef]

113. CEDEFOP. Rising STEMs. Available online: https://www.cedefop.europa.eu/en/data-insights/rising-stems (accessed on 21 January 2022).

114. CEDEFOP. European Center for the Development of Vocational Training. *Spain: Skills forecasts to 2025. Thessaloniki: Cedefop.* 2015. Available online: https://www.cedefop.europa.eu/en/publications-and-resources/country-reports/spain-skills-forecasts-2025 (accessed on 5 February 2020).

115. Moso-Diez, M.; Mondaca-Soto, A.; Gamboa, J.P.; Albizu-Echevarría, M. A Quantitative Cross-Regional Analysis of the Spanish VET Systems From a Systemic Approach: From a Regional Comparative VET Research Perspective. *Int. J. Res. Vocat. Educ. Train.* **2022**, *9*, 120–145. [CrossRef]

116. Sancha Gonzalo, I. 2022. Teachers and Trainers in a Changing World—Spain: Building up Competences for Inclusive, Green and Digitalised Vocational Education and Training (VET). CedefopReferNet Thematic Perspectives Series. Available online: http://libserver.cedefop.europa.eu/vetelib/2022/teachers_and_trainers_in_a_changing_world_Spain_Cedefop_ReferNet.pdf (accessed on 15 October 2022).

117. Ramos Rodriguez, A.M. *Analysis of the Presence of Women Scientists in Secondary Education Textbooks and Their Impact on STEM Careers*; End of Degree Project; Polytechnic University of Cartagena: Cartagena, Colombia, 2021.

118. Cambronero-Saiz, B.; Segarra-Saavedra, J.; Cristófol-Rodríguez, C. Analysis of the engagement of the main popular science youtubers from a gender perspective. *Gend. Issues: Equal. Differ.* **2021**, *16*, 511–525.

119. Tomassini, C. Gender gaps in science: Systematic review of the main explanations and research agenda. *Educ. Knowl. Soc.* **2021**, *22*, 1–13. [CrossRef]

120. López Capdevilla, S. Referents in science teaching, do gender differences persist? *Empirical study of knowledge of men and women scientists and its influence on the scientific vocation in an institute in Madrid.* **2020**, *8*, 4. Available online: https://hdl.handle.net/11162/216323 (accessed on 20 September 2022).

121. Moso-Diez, M.; Mondaca-Soto, A.; Gamboa, J.P.; Albizu-Echevarría, M. A quantitative analysis of the underrepresentation of women in science, technology, engineering, and mathematics (STEM) programs within vocational education and training in Spain. In Proceedings of the European Conference on Educational Research (ECER), Geneva, Switzerland, 2–3 September 2021; Volume IV, pp. 177–185.

122. Thibaut, L.; Ceuppens, S.; De Loof, H.; De Meester, J.; Goovaerts, L.; Struyf, A.; Boeve-de Pauw, J.; Dehaene, W.; Deprez, J.; De Cock, M.; et al. IntegratedSTEMEducation:ASystematicReviewofInstructionalPracticesinSecondaryEducation. *Eur. J. STEM Educ.* **2018**, *3*, 2.

123. Marin-Marin, J.A.; Moreno-Guerrero, A.J.; Dúo-Terrón, P.; López-Belmonte, J. STEAM in education: A bibliometric analysis of performance and co-words in Web of Science. *Int. J. STEM Educ.* **2021**, *8*, 41. [CrossRef] [PubMed]

124. Martín-Sánchez, A.; González-Gómez, D.; Jeong, J.S. Service Learning as an Education for Sustainable Development (ESD) Teaching Strategy: Design, Implementation, and Evaluation in a STEM University Course. *Sustainability* **2022**, *14*, 6965. [CrossRef]

125. Martínez-Campillo, A.; Sierra-Fernández, M.d.P.; Fernández-Santos, Y. Service-learning for sustainability entrepreneurship in rural areas: What is its global impact on business university students? *Sustainability* **2019**, *11*, 5296. [CrossRef]

126. de Diego, M.D.; Chacón Rebollo, T.; CurberaCostello, G.; Marcellán Español, F.; Siles Molina, M. *Mathematics White Book*; Ramón Areces Foundation and Royal Spanish Mathematical Society: Madrid, Spain, 2020; p. 573.

127. Murphy, S. Leadership practices contributing to STEM education success at three rural Australian schools. *Aust. Educ. Res.* **2022**, 1–19. [CrossRef]

128. Garcia-Holgado, A.; Garcia-Peñalvo, F. The W-STEM Projectand Womenin Science. 10.5281/zenodo.5675000. EuropeanUnion. ERASMUS + Capacity-building in Higher Education Call for Proposas EAC/A05/2017—Ref.598923-EPP-1-2018-1-ES-EPPKA2-CBHE-JP. 2021.

129. UNESCO. Women in science: The Gender Gap in Science. *UNESCO Stat.* **2018**, *51*, 1–4. Available online: http://uis.unesco.org (accessed on 20 April 2022).

130. EchazarretaSoler, C.; Costa, A.; Montenegro Gálvez, D.I.; Juárez Postigo, L.; Álvarez Gómez, F. LaeconomíadobleX: Ladesigualdadeconómicadelasmujeresenlosdiferentesentornosglobales, uncambiodeparadigmanecesario. *Commun. Pap.* **2022**, *11*, 134–143.

131. Sandoval Grados, R.; Neck of Gold Ramos, M.J.; Arroyo Higuero, M.J. *Miss Thebes STEAM. Cáparra: Magazine of Innovation and Educational Experiences of the Centers for Teachers and Resources of the Province of Cáceres*; Teachers and Resources Centers of the province of Cáceres, Ed.; Covadonga Rodríguez Hernández: Cáceres, Spain, 2021; Volume 20, pp. 137–143.

Article

Design and Application of a Citizen Participation Tool to Improve Public Management of Drought Situations

Juan Francisco Casero-Cepeda [1,*], Dani Catalá-Pérez [2] and Antonio Cano-Orellana [1]

1   Department of Applied Economics II, University of Seville, 41018 Sevilla, Spain
2   Department of Business Organisation, Polytechnic University of Valencia, 46022 Valencia, Spain
*   Correspondence: jcasero@us.es

Abstract: The growing scarcity of water for human consumption in southern Europe is driving today's public administrations to search for new ways of optimising its availability. Within this context, the purpose of this paper is to analyse whether citizen participation is an appropriate way of improving the management of available water, as several international organisations suggest. This study is part of a research project carried out by the University of Seville in Spain on behalf of the city of Seville's metropolitan water supply company, hereinafter EMASESA. A qualitative method is applied in this research using pre-mortem testing techniques, enabling a specific participation tool to be designed, called the EMASESA Water Observatory, which this article describes in detail. The tool produced specific measures aimed at better addressing drought situations. In view of the practical application of this newly designed tool, we conclude that citizen participation is indeed useful in identifying solutions to improve public water policies and drought management. It is also concluded that the tool's design calling for active participation is a positive factor in its application. Finally, the tool has also demonstrated that it generates knowledge that can be used to address other water-related issues and challenges, beyond those related to water availability.

Keywords: water; drought; climate change; environment; sustainability; citizen participation; programme assessment; evaluation

Citation: Casero-Cepeda, J.F.; Catalá-Pérez, D.; Cano-Orellana, A. Design and Application of a Citizen Participation Tool to Improve Public Management of Drought Situations. *Land* 2022, *11*, 1802. https://doi.org/10.3390/land11101802

Academic Editors: Vanessa Zorrilla-Muñoz, Maria Silveria Agulló-Tomás, Eduardo Fernandez, Blanca Criado Quesada, Sonia De Lucas Santos and Jesus Cuadrado Rojo

Received: 26 August 2022
Accepted: 12 October 2022
Published: 14 October 2022

Publisher's Note: MDPI stays neutral with regard to jurisdictional claims in published maps and institutional affiliations.

1. Introduction

Water, a natural resource that is essential to life, is a human right recognised by the United Nations since 2010 [1]. This right implies sufficient, safe, acceptable, physically accessible, and affordable water for personal and domestic use [1]. Two billion people, equivalent to 25% of the world's population, still suffer from lack of access to safe drinking water in the world [2]. This circumstance underlies the recognition of this right and its fulfilment [3,4] being established in Goal 6 of the 2030 Agenda, dedicated to "Clean Water and Sanitation", the first target of which calls for universal access to safe drinking water for the entire world population by 2030 [5].

In Europe, most countries have 100% of their population connected to the public water supply [6]. However, around 20% of Europe's land cover and 30% of its population is affected by water stress [7], i.e., when water demand exceeds water availability. Situations of water scarcity are becoming increasingly frequent in southern Europe due to lower rainfall and advancing desertification [8], aggravated by climate change [9,10].

Over the last three decades, the surface area and population affected by the number and intensity of droughts in Europe has increased by 20% [11].

It is true that another influence on lower water availability is the growth in the European population over the last 50 years, which has led to an increased demand for water on the continent and a reduction in Europe's renewable water resources [7].

In cities with high average temperatures throughout the year, such as Seville in southern Spain, there have been numerous droughts, the most serious and recent being those that occurred between 1991 and 1995 and in 2017. This reality is common to southern Europe and to cities and territories in this region. The droughts have stimulated increased efficiency in human water consumption. Consumption per inhabitant per day in Europe and in countries such as Spain has fallen in recent years to 128 [12] and 133 litres [13], respectively. These data show society's degree of awareness with regard to water, a factor that has a positive impact in situations of drought.

Apart from human consumption, there are many other factors that affect water availability, especially external ones such as the volume of inflow. Thus, rivers in general, for example the Guadalquivir in Seville, are estimated to experience reduced inflowing water volume between 8 and 20% by the year 2030 [14], due to higher average temperatures and lower rainfall.

Precisely because of its acute vulnerability in this matter [14,15], and in line with a concerted effort towards sustainable urban transformation [16], cities such as Seville, Paris, and Amsterdam declared a climate emergency in 2019, before the same declaration was made at the European level with the approval of the European Green Pact [17].

Faced with this problem, leading international organisations, such as the United Nations through UNESCO [18] and the Organisation for Economic Co-operation and Development, hereinafter OECD [19], are calling for more inclusive and democratic water governance [20], as well as a more proactive change in water consumption management. Through this, they aim to identify solutions and establish measures under a rights-based approach [21], so that different measures can be developed [22].

Within this context of searching for alternative ways to address the problem via new public actions, those international organisations have recently opted for citizen participation [18–20]. However, they do not articulate what this participation means and put it into practice to improve the management of water availability and drought situations.

The United Nations, in particular, acknowledges the importance that citizen participation can have on keener management of water resources, on the formulation of public water policies, and as a means of ensuring those policies meet their objectives [18].

Likewise, the OECD [20] establishes participation, along with effectiveness and efficiency, as essential dimensions for improving public water policies. By creating areas or circles of trust with the general public and guaranteeing citizen inclusion, government agencies can legitimize democratic actions through greater social equity, and thus, strengthening their decision-making processes [23].

Consequently, in view of this problem and the recommendations from international organisations, the purpose of this paper was to analyse whether citizen participation was an appropriate way to improve management of available water and drought situations in cities such as Seville.

Thus, our first hypothesis is that establishing a specific, properly articulated citizen participation tool that offers information and specialised knowledge would be an effective method for finding new solutions to the problem under consideration.

Our second hypothesis proposes that such a tool for citizen participation could, in turn, generate solutions for other water problems, beyond those related to water availability.

Therefore, the main objective of this research was to identify how to articulate citizen participation, develop a specific citizen participation tool for the specific case in question, and determine how to implement participation to address the challenges brought about by drought in cities such as Seville. In this study, we reviewed the literature to see if any instrument for citizen participation had already been developed or applied to other settings and, if not, design a specific tool for our purposes.

2. Materials and Methods

This study used three main sources for its research: information analysis, focus groups, and a questionnaire.

First, information analysis was carried out [24] to ascertain whether any tools for citizen participation in the field of water had already been applied, with the intention of replicating them for cities in Europe that are particularly affected by drought situations, such as Seville. A literature search was run using 'participation', 'water', and 'drought' as keywords, and 'Europe' as the geographical reference area. The search was run on major international multi-disciplinary databases (WOS and Scopus) and another search was conducted using Google as the general search engine.

As a result of this exploratory phase, the Paris Water Observatory [25] was identified as a reference at the international level, whereas, within Spain, we discovered the Terrassa Water Observatory [26] and the Observatory of Water Prices in Catalonia [27].

The next stage was to study their replicability by assessing the origin, purpose, objectives, and practical articulation of the three reference models found.

The Observatory of Water Prices in Catalonia was excluded because its aim, objective, and articulation [27] solely focused on the price of water, which did not match the purpose of our research.

As for the Paris and Terrassa Water Observatories, created in 2010 and 2018, respectively, the reason they were created was from a demand by citizens to have the water service in their respective territories municipalized [25,26]. The underlying objective, in both cases, was to guarantee a public high-quality water supply service with citizen control once water has been brought back under public management. These two observatories are articulated as extra-governmental bodies [25,26] in which citizens play a proactive role.

Unlike these cases, cities such as Seville already have a public water supply service. Therefore, the demand to municipalize a service that is already public was not consistent with our objective in starting processes of citizen participation. This situation occurs in many cities in Spain and southern Europe.

Furthermore, the initiative to set up these citizen participation processes did not come from the public, but rather from the water operating company itself, to identify solutions and alternatives that better deal with drought situations and manage the entire water cycle in general at the local level, in response to suggestions from international organisations [18–20]. Given that the initiative comes from the water company itself, the organisation and articulation of the tool has a governmental character, with citizens having, at least initially, a reactive role.

Therefore, the reference models found in the exploratory phase differed from our initiative in terms of motivation, goals, and organisational and citizen roles (Table 1); thus, we chose not to replicate these tools.

Table 1. Comparative analysis of citizen experiences aimed at better water management and the Paris and Terrassa Water Observatories. (Source: authors).

Comparison Variable	Paris and Terrassa Water Observatories	Other Citizen Experiences
Origin and motivation	To municipalize the water utility servcice	To apply recommendations of international organisations on implementing citizen participation processes
Initiators of participation	Citizens	Public water utilities
Main goal	To guarantee the water service is public in character	To identify solutions for drought situations and, in general, to improve the entire water cycle
Organisation	Extra-governmental, independent of public administration	Governmental
Citizen role	Proactive	Reactive

Consequently, the difference between the two studied initiatives and our case, where cities already have a municipal public management service, meant that these experiences were not replicable. This led to the decision to specifically design a tool for citizen participation.

We chose a qualitative study design using focus groups, following Krueger [28], who proposed that meetings of a limited group of people with homogenous characteristics (in this case, people with expertise in water management) to discuss a researched topic, in a horizontal way, was an effective method. We also used pre-mortem analysis in the focus groups [29,30], a technique that imagines that a strategy has failed; in the first phase, the causes of failure are identified, and in the second phase, possible solutions to avoid failure are defined.

The pre-mortem analysis technique was employed because of its value in deciphering and anticipating possible risks when undertaking projects that are subject to significant uncertainties and areas of ignorance during implementation, such as this unprecedented participation tool. Thus, possible risks were drawn up for each focus group and, once identified, the focus groups suggested solutions to be implemented to achieve optimal adaptation of the participation tool.

These focus group meetings lasted two hours, during working hours, and involved 24 key informants, who made up the total population of managers in the company. The 24 key informants participated in all meetings freely and voluntarily, without receiving any kind of incentive.

The purpose of the tool had already been defined, providing an answer to the 'why' and 'what for' of the tool and which matched the purpose of our research. Thus, the objective was to identify new strategies for better management of available water. This purpose is applicable to water operators that also have public management and are looking for new solutions to the common challenge of drought.

The choice of our key informants is justified, first, because they are the people that are going to be using the participation tool, so their participation in designing the tool helps ensure maximum suitability. Second, they are the people with the keenest understanding of the public water management company's interests

A total of four focus groups meetings were held, with the same participants. In the first meeting, the name of the tool was discussed to facilitate its design under an already established name, and it was agreed that it should be called Water Observatory, following the example of the Paris and Terrassa Water Observatories, in order to make the tool more recognisable and identifiable within the water sector.

The main topics of interest to the participants when defining the Water Observatory were identified. Questions about the design of the participation tool included:

- What are the key points that should determine the participation tool?
- How should the tool be articulated, i.e., what architecture should it have?
- Who should participate, and why?

The second meeting designed the key ideas of the participation tool, thus answering the first of the questions posed in the previous meeting. The participants in this session agreed on the following as key ideas of the tool:

- Horizontal nature, so that all participants in the Water Observatory have equal opportunity to contribute ideas, regardless of its format and dynamics, thus ensuring a truly participatory nature.
- Balanced nature, so that there are an equal number of internal members or members belonging to the public company as external members in the established participation areas. This was important given that the central goal of the tool was to generate knowledge shared between the company and society.
- Flexible nature, with both permanent and visiting members, depending on their profile and the specific topic to be assessed.
- Advisory nature, offering solutions to problems in this subject, but without displacing the company's decision-makers, given that the public company has a Board of Direc-

tors that is responsible for decision-making, and who, legally, cannot be replaced by any other body.

- Geared towards results and impacts, given the Observatory's objective of prioritising the search for measures to cope with drought situations in Seville.

At the third meeting, in response to the question on how the Observatory should be organised, it was decided to structure the participation tool in advisory panels, as opposed to using other techniques. This opened up space for reflection, geared towards the underlying objective. It was agreed to establish as many advisory boards as there are management areas in the public company. An Observatory structure based on the company's own structure would provide the dual advantage of making its organisation more easily identifiable from the outset and, most importantly, facilitate the assignment of tasks and responsibilities when running the Observatory.

The work in the fourth and final meeting was aimed at identifying who should participate in and form the Water Observatory. It was decided to establish criteria for participation, as opposed to an assembly, to guarantee the legitimacy of each participant in the Observatory by establishing institutional legitimacy, legitimacy via expertise and knowledge, legitimacy as sufferers of the problem, and legitimacy via representativity. This way, in the nature of pre-mortem analysis, the company could establish a participation system that would prevent any key agent being deemed to lack legitimacy to participate in the advisory panels, or any lack of legitimacy in the proposals reached.

All the participants were recruited by direct invitation from the company, based on the above-mentioned participation criteria. Thus, all neighbourhood organisations, environmental associations, consumer associations, public administrations, and professional associations in the area were invited. In addition, all university experts from the nearest universities, whose areas of specialisation included water management and drought, were invited to participate. As indicated in the results, eight advisory roundtables were held with the people who were invited to participate.

Finally, we implemented a tool for citizen participation, called the Water Observatory. Setting up the Observatory made it possible to test its validity as a method for applying participatory processes in questions of water management, as it is described in the following sections.

Finally, as a third method of assessment, a questionnaire was used to evaluate the effectiveness of the Observatory once it had been implemented. This questionnaire was addressed to a total of 36 participants in the Observatory's advisory panels after they had been set up, who therefore represented the targeted sample. The questionnaire was divided into a total of 31 items, in two blocks, including: assessment of participation on the advisory panels and uses and usefulness of the results achieved through the advisory panels. The results of the questionnaire are discussed in the following sections. The questionnaire was online, using a Google form. The questionnaire included multiple choice questions, using a Likert scale.

3. Results

Implementation of the Water Observatory tool enabled the proposed hypotheses to be tested.

From a quantitative point of view, it should be noted that eight advisory committees of the Water Observatory had been held to seek proposals for dealing with drought situations. As shown in Table 2, the solution adopted to articulate the Water Observatory was through advisory boards. The average number of participants in these advisory boards was 36 people, 52.7% of whom were external participants, or those not belonging to the Seville public water company, and 47.3% were internal participants, or company personnel. This complied with the parity of the tool, an aspect that is also included in Table 2 as a solution to the risk of the establishment of non-feasible proposals.

Table 2. Summary of the Water Observatory design through pre-mortem analysis: risks to be avoided and solutions taken.

Questions Concerning the Design of the Water Observatory	Solution Taken	Risks to Be Avoided
What are the main key points that should determine participation tool?	Horizontal nature	Inequality between participants
	Balanced nature	Setting unfeasible proposals
	Flexible nature	Setting general rather than specific proposals
	Advisory nature	Exceeding the Observatory's sphere of competence.
	Geared towards results and impacts	Participation tool becoming an end in itself
How should the tool be articulated, i.e., what architecture should it have?	Advisory panels, one per division in the firm organisation.	Restrictions to air views openly
		Inability to manage the Observatory
		Difficulty to assign responsibilities for suitable implementation of the Observatory
Who should participate, and why?	Channels that legitimise participation: institutional, expertise, and knowledge, sufferers of problem, representativity.	Lack of legitimacy of people forming part of the advisory panels
		Lack of legitimacy of the proposals made

Source: authors, from focus group meetings held.

In terms of results, the advisory panels enabled the first of our hypotheses to be accepted, given that a direct result of the Water Observatory was the specialist information and knowledge on solutions and alternatives to the problems of water availability in Seville. We produced an Emergency Plan for drought situations for the city of Seville developed through the Observatory's advisory panels.

Likewise, the second of the established hypotheses was accepted, given that solutions have been generated for other water-related problems beyond those that refer to the availability of water in the area, as a result of the advisory panels held by the Water Observatory. These other problems related to the quality, accessibility, and affordability of water. In response to these other problems, the Water Observatory developed the Seville Emergency Climate Plan. The questionnaire also demonstrated the validity and usefulness of the Water Observatory as a tool for citizen participation in improving the management of water availability for the city of Seville.

In terms of its usefulness, as shown in Table 3, the Observatory was perceived by participants as a tool that favoured transparency, generated circles of trust, and ensured the proposals produced were both appropriate and legitimate. It was also seen as a tool that improved information, communication, accountability, and decision-making. Additionally, it prevented conflicts and encouraged public–private collaboration and co-operation between institutions.

Table 3. Evaluation of usefulness of the Water Observatory. External participant % rating and employee % rating.

Participants in the Water Observatory from outside the company		Low consideration (25–50%)	Medium consideration (50–75%)	High consideration (75–100%)
	High consideration (75–100%)		Information channel (72%, 97%) Communication channel (69%, 97%)	Transparency (100%, 100%) Generates areas of trust (100%, 89%) Suitability of agreed measures (100%, 89%) Legitimacy of agreed measures (89%, 89%)
	Medium consideration (50–75%)	Prevent conflicts (42%, 61%)	Accountability (53%, 64%)	Decision-making (94%, 78%)
	Low consideration (25–50%)		Public-private co-operation (61%, 42%)	Collaboration between institutions (89%,
		Low consideration (25–50%)	Medium consideration (50–75%)	High consideration (75–100%)
		Participants in the Water Observatory that are company employees		

Source: Prepared by the authors based on questionnaire to participants in the Water Observatory.

Regarding its validity as an instrument, the results of the questionnaire also showed that the Water Observatory was positively viewed in terms of its dynamics and organisation. As shown in the table below (Table 4), more than 80% of the Observatory's participants considered that the mechanism for proposing Observatory discussion topics, participation, dynamics used, people participating, and calendar of advisory panel meetings were suitable.

Table 4. Evaluation of the dynamics and organisation of the Water Observatory.

Indicator	Assessment	
	Percentage of Positive Responses	**Remark**
1. Degree of suitability of Observatory discussion topics.	100.0%	Suitable discussion topics.
2. Mechanism for proposing Observatory discussion topics.	88.9%	Topics for discussion proposed.
3. Assessment of the length of time and participation.	81.5%	Suitable length of discussion panel meetings.
	85.2%	Availability of sufficient time to intervene at advisory panel meetings.
	88.9%	Availability of sufficient opportunities to voice an opinion at advisory panel meetings.
4. Assessment of the dynamics used.	88.9%	Assessment of the dynamics used.
5. Assessment of number and profile of the people participating.	85.2%	Assessment of the number of participants on advisory panels.
	88.9%	Assessment of the profiles of the members of the Observatory.
6. Assessment of the schedule and calendar of advisory panel meetings.	85.2%	Assessment of the regularity with which advisory panel meetings were held.
	81.3%	Assessment for morning rather than afternoon/evening sessions.

Table 4. *Cont.*

Indicator	Assessment	
	Percentage of Positive Responses	**Remark**
7. Assessment of the information and documentation received before advisory panel meetings.	96.3%	Information received in good time.
	96.3%	Ample information and documentation.
	92.6%	Documentation open to the possibility of including changes.
	91.1%	Information received is clear and precise.
8. Assessment of feedback process.	96.3%	Outcome of the panel meeting received after it was held.
	96.3%	Documentation received after panel meeting was sufficient.
	96.3%	Was able to make contributions after participating on the advisory panel.
	96.3%	Participation outcomes and results were suitably collected in reporting.
9. Assessment of how results are published.	93.0%	Internal communications suitable.
	91.1%	External communications suitable.
10. Interest in continuing participation in the Water Observatory.	92.6%	Interest in continuing to participate in the Water Observatory.
11. Overall assessment of the Water Observatory.	37.0%	Excellent
	63.0%	Positive

Source: Prepared by the authors based on the results of questionnaire to participants in the EMASESA Water Observatory.

Furthermore, over 90% considered the information and communication received, discussion topics and feedback process, together with the issues addressed, to be adequate and indicated their interest in continuing participation in the Observatory, which represents an indicator of interest in the newly designed participation tool, rated as excellent or good.

4. Discussion

According to Aguilar [31], today's new public governance acquires the responsibility of solving society's problems, needs, and challenges in ways that should be shared between governments and citizens, so that public administrations are, as McBride et al. [32] indicate, articulators of a networked society and acquire a new role as collaborative agents.

In view of this approach, consistent with that of Denhardt and Denhardt [33] on public services being a co-creation between governments and citizens, this research explored the articulation of participatory processes in the specific case of public water policies and the management of drought situations in Seville, in response to recommendations from international organisations [18–20] to apply participatory strategies in this area.

Taking participation to be an instrumental activity [34] in improving public water interventions and policies, the Water Observatory was designed within the framework of this study as a space for participation and co-creation [33].

It is common for citizen participation to follow a proactive model, aimed at a specific demand. However, reactive models can also be observed, in which citizens are not particularly interested in participating, a priori. Then, when faced with an initiative from the water management company itself, the citizen participates as a reaction to it. In line with Bekkers et al. [35], instead of diminishing its role and ceding its responsibility and decision-making capacity, the government benefits from citizen participation by reducing its regulatory and managerial space for a more open and collaborative one. This boosts government institutional learning and the continuous improvement of its public policies and programmes. This research puts such an approach to test, using the questionnaire to determine whether the people in charge of the public water management company in Seville acknowledge that the Water Observatory is a valuable instrument that effectively improves their decision-making processes.

The incorporation of citizen participation in public action is still incipient and lacks abundant empirical evidence, both generally, as pointed out by Wirtz et al. [36], and in the specific area of climate change, as indicated by Mees et al. [37].

Hence, this was a further challenge for the current study, given the lack of earlier experiments, as demonstrated in the literature review. The few research papers that existed [25,26] had very different purposes and objectives to the case of Seville.

The main contribution of this work is, therefore, an experience of citizen participation at the local level, applied to the case of public water policies. Furthermore, it did not stem from public demand within the framework of a water municipalisation process [25,26], but rather as a way of improving decision-making processes initiated by the actual public administration.

Given the way this tool was applied, the results it achieved, and its usefulness, it may be helpful to replicate this method in other territories.

The tool designed for this research led to both a specific proposal for improving drought management, the Drought Emergency Plan, and additional proposals on other issues linked to public water policy beyond water availability, e.g., the Climate Emergency Plan.

These results confirm the usefulness of the new tool, not only for issues relating to drought (first hypothesis), but also for other issues (second hypothesis).

It is important to highlight the collaboration of the people and entities taking part in the Observatory's advisory panels, in line with reports from earlier research papers [36].

The use of a basic questionnaire that provides information on the development of the participatory tool, such as the Water Observatory, is a useful qualitative tool. In other cases, other tools were used, such as interviews, etc. In this study, the questionnaire was conceived as a complementary element to assess the quality of the participation process. In other situations, the interest may be different and, consequently, another tool may be more suitable.

As for limitations to the research, the impact of the COVID-19 pandemic resulted in a large portion of Observatory's advisory panels having to be held online. Furthermore, the number of participating citizens was small. Other similar experiences prior to the pandemic had shown higher participation, with assembly meetings that brought together more than a hundred people; however, for our case, the average participation was 36 people.

Another limitation of this research is that the Water Observatory does not respond to any international, national, and/or local statutory requirement, but has been designed and implemented at the initiative of the current management team at the public water management company in Seville. In other words, it is voluntarily instituted. This circumstance may pose a risk to the continuity of this citizen participation tool when the company's management team changes, as it occurs every five years.

For future research directions, it would be beneficial to establish a system of indicators and a system for evaluating the proposals resulting from the Water Observatory. To this end, there needs to be more investigation of evaluation proposals, such as those of Agulló-Tomás et al. [38] or Pérez et al. [39], which would make it possible to measure and analyse the effects the solutions proposed by this Observatory have on the problem in question. Also, impact evaluations from both a contribution analysis and an attribution analysis should be carried out, in line with the World Bank's approach, given that this is the international reference for evaluating the impact of public policies [40]. In the real context, the results of the Observatory have already been applied through an improvement plan that the public company is developing, with the proposals reaching the advisory roundtables. It would also be important to incorporate gender perspectives and consider age and ageing variables.

5. Conclusions

Citizen participation, when properly articulated, provides a valuable instrument in the management of the full water cycle, with which we can identify solutions and alternatives to drought situations, as well as other problems, needs, and challenges that concern public water policy.

Cities such as Seville are particularly vulnerable to drought situations and the effects of climate change. The design and application of a tool for citizen participation, such as the Water Observatory, helps the city be more resilient and successfully mitigate these effects. Moreover, the Water Observatory provides a useful resource for governance, making public water policy more transparent, and offering spaces for reflection and shared knowledge between public administrations and society. Thus, this tool can strengthen decision-making processes by defining new and more legitimate public actions.

Author Contributions: All authors have contributed equally to the final version of this article. All authors have read and agreed to the published version of the manuscript.

Funding: This research has been funded by Empresa Metropolitana de Abastecimiento y Saneamiento de Aguas de Sevilla S.A, EMASESA, within the framework of the project entitled "Technical assistance for the monitoring, updating and revision of the strategic plan, sustainable public management (SPM), and development and implementation of a participatory assessment model of EMASESA's public policies", carried out by the Clave Europa Research Group at the University of Seville's Department of Applied Economics II.

Conflicts of Interest: The authors declare no conflict of interest.

References

1. UN. *Resolution Adopted by the General Assembly on 28 July 2010*; United Nations: New York, NY, USA, 2010.
2. WHO; UNICEF. Progress on Household Drinking Water, Sanitation and Hygiene 2000–2020. 2021. Available online: https://data.unicef.org/resources/progress-on-household-drinking-water-sanitation-and-hygiene-2000-2020/ (accessed on 10 August 2022).
3. Anand, P.B. Right to water and access to water: An assessment. *J. Int. Dev.* **2007**, *19*, 511–526. [CrossRef]
4. Fernández Aller, C.; Luis Romero, E. El derecho humano al agua y al saneamiento: Un tema clave en la intersección Ecología-Derechos Humano. *Ambienta* **2015**, *113*, 1–13.
5. UN. Sustainable Development Goals. Goal 6: Ensure Access to Water and Sanitation for All. 2015. Available online: https://www.un.org/sustainabledevelopment/water-and-sanitation/ (accessed on 10 August 2022).
6. Eurostat. Population Connected to Public Water Supply (% of Total Population). 2022. Available online: https://appsso.eurostat.ec.europa.eu/nui/show.do?dataset=env_wat_pop&lang=en (accessed on 10 August 2022).
7. European Environment Agency. Water Resources Across Europe—Confronting Water Stress: An Updated Assessment. 2021. Available online: https://www.eea.europa.eu/publications/water-resources-across-europe-confronting (accessed on 10 August 2022).
8. Abraham, E. Tierras secas, desertificación y recursos hídricos. *Ecosistemas* **2008**, *17*, 1–4.
9. Forzieri, G.; Feyen, L.; Russo, S.; Vousdoukas, M.; Alfieri, L.; Outten, S.; Migliavacca, M.; Bianchi, A.; Rojas, R.; Cid, A. Multi-hazard assessment in Europe under climate change. *Clim. Chang.* **2016**, *137*, 105–119. [CrossRef]
10. Mukherjee, S.; Mishra, A.; Trenberth, K.E. Climate Change and Drought: A Perspective on Drought Indices. *Curr. Clim. Chang. Rep.* **2018**, *4*, 145–163. [CrossRef]
11. Comisión Europea. Agua. Available online: https://ec.europa.eu/environment/pubs/pdf/factsheets/water_scarcity/es.pdf (accessed on 10 August 2022).
12. INE. La Situación del Agua en España y en el Mundo, en Gráficos. 2022. Available online: https://www.epdata.es/datos/graficos-situacion-agua-mundo-espana/333 (accessed on 10 August 2022).
13. Aquae Fundación. ¿Cuánto es el Consumo de Agua por Persona en Europa? Available online: https://www.fundacionaquae.org/wiki/europa-consumimos-una-media-128-litros-agua-persona-dia (accessed on 10 August 2022).
14. Ley de Andalucía 8/2018, de 8 de Octubre, de Medidas Frente al Cambio Climático y para la Transición Hacia un Nuevo Modelo Energético. 2018. Available online: https://www.boe.es/diario_boe/txt.php?id=BOE-A-2018-15238 (accessed on 10 August 2022).
15. Ministerio de Medio Ambiente. Libro Blanco del Agua en España. 2000. Available online: http://www.cedex.es/CEDEX/LANG_CASTELLANO/ORGANISMO/CENTYLAB/CEH/Documentos_Descargas/LB_LibroBlancoAgua.htm (accessed on 10 August 2022).
16. Coq-Huelva, D.; Asián-Chaves, R. Urban Sprawl and Sustainable Urban Policies. A Review of the Cases of Lima, Mexico City and Santiago de Chile. *Sustainability* **2019**, *11*, 5835. [CrossRef]

17. European Commission. A European Green Deal. Available online: https://ec.europa.eu/info/strategy/priorities-2019-2024/european-green-deal_en (accessed on 25 August 2022).
18. UNESCO. The United Nations World Water Development Report 2019: Leaving no One Behind. 2019. Available online: https://en.unesco.org/themes/water-security/wwap/wwdr/2019 (accessed on 10 August 2022).
19. OECD. Recomendación del Consejo de la OCDE Sobre el Agua. 2016. Available online: https://www.oecd.org/water/Recomendacion-del-Consejo-sobre-el-agua.pdf (accessed on 10 August 2022).
20. OECD. Principles on Water Governance. 2015. Available online: https://www.oecd.org/governance/oecd-principles-on-water-governance.htm (accessed on 10 August 2022).
21. Abramovich, V. Una aproximación al enfoque de derechos en las estrategias y políticas de desarrollo. *Rev. Cepal.* **2006**, *88*, 35–50. [CrossRef]
22. Paneque, P.; Vargas, J. Drought, social agents and the construction of discourse in Andalusia. *Environ. Hazards* **2015**, *14*, 224–235. [CrossRef]
23. Brugué, Q. Recuperar la política desde la deliberación. *Rev. Int. Organ. Int. J. Organ.* **2011**, *7*, 157–174. [CrossRef]
24. Millas, M.; Huberman, A.M.; Saldaña, J. *Qualitative Data Analysis: A Methods Sourcebook*; SAGE: Phoenix, AZ, USA, 2014.
25. Observatoire Parisien de l'eau. Une Nouvelle Politique Parisienne des Eaux? 2015. Available online: http://www.observatoireparisiendeleau.fr/articles/147 (accessed on 10 August 2022).
26. Observatorio del Agua de Terrassa. Publicaciones OAT. 2018. Available online: https://www.oat.cat/es/comunicacion/ (accessed on 10 August 2022).
27. Observatorio del Precio del Agua en Cataluña. Available online: https://aca.gencat.cat/es/laca/observatori-del-preu-de-laigua/ (accessed on 10 August 2022).
28. Krueger, R.A. *Focus groups: A Practical Guide for Applied Research*; SAGE Publications, Inc.: Thousand Oaks, CA, USA, 2015.
29. Klein, G. Performing a Project Premortem. *IEEE Eng. Manag. Rev.* **2008**, *36*, 103–104. [CrossRef]
30. Serrat, O. The Premortem Technique. In *Knowledge Solutions Tools, Methods, and Approaches to Drive Organizational Performance*, 1st ed.; Springer: Singapore, 2017; pp. 154–196.
31. Aguilar, L. The New Public Governance? Emerging perspectives on the theory and practice of public governance. *J. Public Gov. Policy Lat. Am. Rev.* **2015**, *1*, 126–134.
32. McBride, D.; Valencia, P.S.; Mejía, L.E. La Nueva Gobernanza Pública ¿Una Nueva Manera de Gestionar lo Público en el Siglo XXI? 2019. Available online: https://congreso.pucp.edu.pe/ciencias-gestion/wp-content/uploads/sites/54/2019/11/La-Nueva-Gobernanza-Pu%CC%81blica.-Congreso-de-Ciencias-de-la-Gestio%CC%81n.pdf (accessed on 10 August 2022).
33. Denhardt, R.B.; Denhardt, J.V. The new public service: Serving rather than steering. *Public Adm. Rev.* **2000**, *60*, 549–559. [CrossRef]
34. Lee, Y.; Schachter, H.L. Exploring the Relationship between Trust in Government and Citizen Participation. *Int. J. Public Adm.* **2019**, *42*, 405–416. [CrossRef]
35. Bekkers, V.; Edelenbos, J.; Nederhand, J.; Steijn, A.J.; Tummers, L.G.; Voorberg, J. The social innovation perspective in the public sector: Co-creation, self-organization and meta-governance. In *Innovation in the Public Sector: Linking Capacity and Leadership*, 2nd ed.; Palgrave McMillan: London, UK, 2014; pp. 223–244.
36. Wirtz, B.W.; Weyerer, J.C.; Rösch, M. Open government and citizen participation: An empirical analysis of citizen expectancy towards open government data. *Int. Rev. Adm. Sci.* **2017**, *85*, 566–586. [CrossRef]
37. Mees, H.; Uittenbroek, C.; Hegger, D.; Driessen, P. From citizen participation to government participation: An exploration of the roles of local governments in community initiatives for climate change adaptation in the Netherlands. *Environ. Policy Gov.* **2019**, *29*, 198–208. [CrossRef]
38. Agulló-Tomás, M.S.; Zorrilla-Muñoz, V.; Gómez, M.V. Género y evaluación de programas de apoyo para cuidadoras/es de mayores. *Rev. Prism. Soc.* **2018**, *21*, 391–415. Available online: https://revistaprismasocial.es/article/view/2469 (accessed on 10 August 2022).
39. Pérez, D.C.; de Miguel Molina, M.; Royo, R.C. La necesidad de la evaluación de políticas públicas como herramienta de impulso a la calidad democrática. In *La Regeneración del Sistema: Reflexiones en Torno a la Calidad Democrática, el Buen Gobierno y la Lucha Contra la Corrupción*, 1st ed.; AVAPOL: Valencia, Spain, 2015; pp. 215–235.
40. Gertler, P.J.; Martinez, S.; Premand, P.; Rawlings, L.B.; Vermeersch, C.M.J. *Impact Evaluation in Practice*, 2nd ed.; Inter-American Development Bank and World Bank: Washington, DC, USA, 2016. Available online: https://openknowledge.worldbank.org/handle/10986/25030 (accessed on 10 August 2022).

 *land*

Article

Real-Time Environmental Monitoring Platform for Wellness and Preventive Care in a Smart and Sustainable City with an Urban Landscape Perspective: The Case of Developing Countries

Victor Gonzalez *, Manuel Peralta, Juan Faxas-Guzmán and Yokasta García Frómeta

Pontificia Universidad Católica Madre y Maestra, Santo Domingo 10114, Dominican Republic
* Correspondence: vm.gonzalez@ce.pucmm.edu.do; Tel.: +1-809-654-0534

Abstract: Smart and sustainable communities seek to ensure comfortable and sustainable quality of life for community residents, the environment and the landscape. Pollution is a key factor affecting quality of life within a community. This research provides a detailed insight into a successfully developed and deployed framework for an environmental monitoring platform for an urban study to monitor, in real time, the air quality and noise level of two cities of the Dominican Republic—Santo Domingo and Santiago de Los Caballeros. This urban platform is based on a technology range, allowing for the integration of multiple environmental variables related to landscape and providing open data access to urban study and the community. Two case studies are presented: The first highlights how the platform can be used to understand the impact a natural event, for example, how dust landscapes (such as the Sahara) impact a community and the actions that can be taken for wellness and preventive care. The second case focuses on understanding how policies taken to prevent the spread of COVID-19 affect the air quality and noise level of the landscape and community. In the second case, the platform can be used to expand the view of decision makers in the urban landscape and communities that are affected.

Keywords: smart and sustainable city; landscape; urban studies; internet of things; air quality and noise monitoring; wellness and preventive care; COVID-19

Citation: Gonzalez, V.; Peralta, M.; Faxas-Guzmán, J.; Frómeta, Y.G. Real-Time Environmental Monitoring Platform for Wellness and Preventive Care in a Smart and Sustainable City with an Urban Landscape Perspective: The Case of Developing Countries. *Land* **2022**, *11*, 1635. https://doi.org/10.3390/land11101635

Academic Editors: Vanessa Zorrilla-Muñoz and Maria Silveria Agulló-Tomás

Received: 21 August 2022
Accepted: 20 September 2022
Published: 23 September 2022

Publisher's Note: MDPI stays neutral with regard to jurisdictional claims in published maps and institutional affiliations.

1. Introduction

Smart and sustainable cities are an expression of the multiple domains of urban life in which technology and policies can be applied to a community within their landscape [1]. The vision has progressed to focus on the alignment between policies related to human capital, education, economic development, territorial and governance, and how these could be improved with the use of ICT [2]. To become a smart and sustainable community, its development must combine ICT with territorial, human and social capital, along with broad economic policies [3]. Currently, the development of smart cities seeks to ensure a comfortable and sustainable quality of life for the residents of the community, the environment and their landscape [4]. This creates a bigger emphasis on the sustainability of the community and the surrounding territories [5].

One of the biggest threats to sustainability for both a territory and a community is the level of pollution that surrounds them. Urban communities tend to be affected by low air quality and high noise levels [6]. These pollutants have a direct effect on the wellbeing of a community—there is evidence that links them to an increase in deaths, admissions to hospitals, and general problems in the respiratory and immune systems of residents [7,8]. In addition, these contaminants have an impact on the territory as urbanization changes the land surface temperature, and this is critical for the decision-making process of sustainable and vibrant city development [9]. Scholars have recently focused on understanding the behavior of the pollutants as a preventive care assessment mechanism, after evidence of significant associations between the COVID-19 pandemic and these environmental indicators [10]. Other urban studies have also established links between how the pandemic has changed human behavior and how this has impacted the different environmental indicators in multiple cities around the world [11–15]. These environmental monitors are a critical component of urban and territorial management and development [16,17].

This study focuses on the development and deployment of an environmental monitoring platform framework for urban studies through low-cost sensors for smart and sustainable cities in developing countries. We propose an architectural design that connects different environmental sensors to an open data platform, proposing a framework on how the Internet of Things (IoT) can be used for pollutant monitoring for urban studies, targeting territorial, social and environmental sustainability. The study aims to initiate a discussion on how platforms can create awareness and motivate citizens to have a healthier lifestyle, taking into consideration the effect of pollution in their life and the surrounding territories. In addition, we seek to understand the behavior of the pollutants during a pandemic crisis, such as COVID-19, and their impact on a landscape. To achieve this, the platform was deployed in two cities of the Dominican Republic—Santo Domingo and Santiago de Los Caballeros—becoming the first real-time environmental monitor in the country.

This paper is organized as follows: In Section 2, a review of the literature is explored, building on the different dimensions that form our proposed framework. Section 3 presents the technology used to build the platform and its integration. In Section 4, we present our exploratory analysis on the behavior of the pollutants. We present two case studies and discuss the implications of the platforms. Finally, in Section 5, we summarize the article and present the limitations of our work. Furthermore, we highlight the most relevant findings and propose interesting questions for future research.

2. An Environmental Monitoring Platform for Urban Studies in a Smart and Sustainable City

The link between health complications and poor air quality or prolonged exposure to noise level has been extensively established [8,18–21]. Citizens' awareness of pollution levels and how to adapt their behavior as a response has become a new focus. In Oslo, Norway, a group of sensors were deployed in kindergartens to plan children's outdoor activities depending on the pollution levels [22]. Kumar et al. [23] measured the impact individual vehicles had during drop-off and pick-up time at a school compared to other times of the day by measuring the air quality at different points in the school with the purpose of promoting commuting in their student population. This shows the importance of environmental pollution awareness among citizens of a smart and sustainable community.

Some regions have been monitoring their air quality for multiple decades [24]. These platforms have been useful for developing prevention plans and ensuring the compliance of factories or industries with regard to their environmental regulations, to reduce the impact they have on a landscape [25,26]. In addition, monitoring air quality and noise level have been helpful in linking them with changes in human activity and their territorial expansion. Xu et al. [15] observed an improvement in air quality in different Chinese cities during confinement imposed because of the COVID-19 pandemic. Similar observations were established in multiple cities of India [14], Brazil [13] and the United States [18].

2.1. Internet of Things (IoT) in Environmental Monitoring for Urban Studies

To obtain the environmental data needed to observe changes in the contaminants, multiple monitoring stations are needed. Through IoT, these environmental sensors generate massive amounts of data that, if used properly, can enhance objective decision making. However, detailed observations of air quality on an urban scale are rare given the high cost of traditional monitoring stations [22], especially in developing countries.

Advances in sensor technology have allowed for the development of low-cost measurement equipment for the observation of small-scale spatial variability of pollutants [27].

These sensors, being smaller and cheaper, allow for the deployment of a higher-density network in urban spaces through IoT [28]. This advancement has allowed for the expansion of existing platforms for permanent [24] or transitory purposes on time-specific needs [29]. Besides the cost, these internet-enabled sensors allow for fast deployment and a customizable array of environmental variables depending on the requirements of each territory [30].

These IoT devices utilize electrochemical probes to detect, in real time, the concentration level of a gas. The material of the probe is designed to have a specific reaction depending on the pollutant and its concentration. For measuring particle concentration, different algorithms are used to determine the attenuation of a light signal due to the size of the particles [28]. However, the accuracy of these techniques has been questioned.

The state of development of these devices has improved significantly since their first introduction. A push from the industry and academia has improved both the accuracy and reliability of the sensors through studies that focus on understanding their performance under laboratory conditions and field studies [31]. Under laboratory conditions, these monitors tend to be more accurate than in the field [28]. If the sensors are used outside, they have better performance when exposed under higher concentrations of the contaminants [32]. With proper calibration, the IoT sensors can improve their performance and accuracy. However, these devices must be calibrated individually, since a unified calibration factor has not been identified [33]. New studies are focusing on understanding the long-term performance of these sensor networks not only to understand the impact the weather has on their performance, but also to assess how their performance changes through time [17].

Combining IoT with artificial neural networks (ANN) enables the possibility of predicting and forecasting environmental conditions at unmeasured points to create high-resolution pollution maps [34]. Using ANN for forecasting pollution reduces the complexity in understanding the interaction of each pollutant with other variables [35]. Since ANNs are problem-specific, a model cannot be easily replicated at other points or with other pollutants [34]. However, this is not a big inconvenience since sensor calibrations must be carried out individually [33].

2.2. Engaging with Environmental Data

Through an environmental monitoring platform, researchers have looked for the opportunity to establish a link between the landscape and the community so that they could monitor the contaminants that have a direct impact on their daily lives. This implies that engaging with different actors of society is important to obtain a holistic view and to better understand what problems are affecting the community and their surrounding territories. In turn, this will allow interventions to be established that have positive behavioral changes [36]. With this focus, different perspectives have been proposed and a citizens science approach has been widely accepted.

Using this approach and environmental monitoring through IoT can help new services to be developed for the benefit of a community by creating new partnerships [37]. This involvement helps us to understand the technology that forms the platform and develops trust [38], facilitating the adoption of the tools and the influence it has in the citizens' behavioral change. The communities that are more aware of their level of pollution exposure are more conscious of the health risks associated with that exposure [5] and their effect on the landscape.

Multiple studies have capitalized on the citizens science approach to engage with the community by promoting interactive quizzes and workshops [39], active participatory monitoring campaign [40,41], easy-to-understand maps [21], attractive visualization platforms [42], and mobile applications [43] to make sense of the environmental data.

2.3. Open Data

Open Data is defined as data that are available, accessible, and can be re-used and redistributed by anyone. This type of data must be available on a commonly used and machine-readable format. The license of the data must be allowed for re-use and redistribution so that the data can be repurposed for other services. Finally, there should be no restriction on who can use the data [44].

In a push for public sector transparency, governments have led the creation of Open Data Ecosystems with a government-created dataset as a shared resource. These ecosystems are valuable for government departments that can share the cost of data production and distribution. From a practical point of view, the success of open data arises when there is a collaboration between different actors and the use of the datasets [45]. There are multiple studies that describe the economic value of open data [45,46] as well as the social value that it creates [47].

From the context of smart and sustainable cities, an open data ecosystem is crucial for its development and engagement with citizens [48]. To develop the ecosystem, it is important to consider the movement of resources where demand encourages supply to create a sustainable relation between data suppliers, intermediaries and users [49]. Open Data go beyond accessibility and push towards capacity-building and training to assure the usage of the datasets. An Open Data Ecosystem incentivizes the creation of knowledge, value creation and self-development by providing data as a shared resource, pushing society to the next phase of super smart society, or Society 5.0 [45].

The purpose of deploying a network of sensors is to obtain real-time observations of the pollutants in a territory. Through an Open Data platform, citizens can engage with the environmental data to create value through new urban services.

2.4. Design and Implementation Framework

For an environmental monitoring platform for urban studies to successfully engage with the community, a holistic approach must be followed. Guided by the presented literature, this study proposes a framework with four dimensions: First, an internet-enabled network of sensors is needed. These sensors must be customizable so they can adapt to the requirements of the place where they are deployed and the capacity to measure the variables of interest [17,22,36]. The data collected must be placed in a reliable and secure infrastructure that is easily accessible by the community. Therefore, open data are considered as another dimension [50,51]. For the general public to make sense of the data, creating interactive, easy-to-understand visualizations of the territories impacted is important and used as a third dimension [52]. Finally, for citizens to go beyond the provided solutions, they need to develop the capacity to use the data and create innovative solutions for their own needs. Based on this perspective, the last dimension focuses on capacity building [3,16,43,53], as shown in Figure 1.

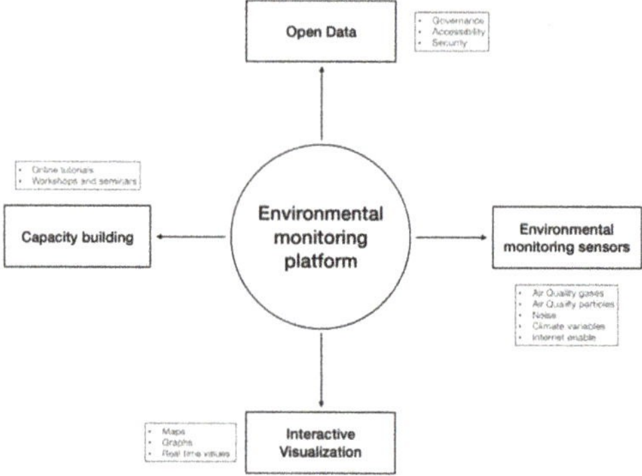

Figure 1. Multi-dimensional framework for an environmental monitoring platform.

3. Methodology

The platform was designed for the Dominican Republic, a country located in the Caribbean, forming part of the Greater Antilles. The country occupies the eastern part of the island of Hispaniola. The Dominican territory is divided into 32 provinces, of which the city of Santo Domingo is its capital and biggest city, and the city of Santiago de los Caballeros is the second largest in terms of population and economic contribution. The city of Santo Domingo has a population of more than 2 million inhabitants. The second largest is Santiago, a city in the center of the country, with great industrial and agricultural development and a population of approximately 1.5 million [54]. The platform was designed following the architecture described in this section.

3.1. Open Data platform

Our Open Data platform is based on a representational state transfer (REST) architectural pattern for application program interface (API) implementation. This is a popular approach in the development of web services, since it breaks one transaction into smaller parts, providing greater flexibility compared to other architectural patterns. This implementation utilizes multiple methodologies of the HTTP/HTTPS protocol, facilitating interactions with different programming languages [55]. The platform communicates with multiple services through the application programing interface (API) implemented. Some of these services come from the visualization platform, the citizens browser, our environmental monitoring sensors or other IoT devices. The development of the Open Data platform was released under a GNU GENERAL PUBLIC LICENSE and the code is available on GitHub [56].

For a GET request, the service accepts two parameters to delimit time and one parameter to specify a format. If no parameters are sent to the API, by default, the response will include all the data available in the JavaScript Object Notation (JSON) format. The platform can respond with a .CSV file if specified in the format parameter. If a specific timeframe is wanted, the platform supports start- and end-time parameters.

When making a POST request, the service must authenticate with the platform when sending the JSON with the information to be saved in the database. The device which authenticates the server responds with a "201 Created" if the data are saved correctly, or "204 No Content" if an error occurs. In case there is a failed authentication, the server responds with "400 Bad Request".

3.2. Environmental Monitoring Sensors

The environmental IoT sensor was divided into two layers: the hardware layer and the software layer. The hardware layer is based on the Libelium's Plug & Sense! line of products [57]. These are commercially available, tested, and calibrated environmental monitoring devices. The devices encapsulate a programmable Arduino-based microcontroller board in a waterproof and weatherproof plastic enclosure. All the sensor probes plug into the enclosure via a nine-pin waterproof connector, which facilitates initial deployments and future replacements. Figure 2 depicts the physical and virtual nodes in the system, along with their links and corresponding internal software components.

The network of sensors was deployed in a university campus in Santo Domingo and in Santiago de Los Caballeros. In both campuses, there was access to a reliable Wi-Fi network. To reduce the operational cost of the network of sensor infrastructure, we configured the devices to connect through the Wi-Fi___33 network of the university. For the monitoring sensor network, an exclusive service-set identifier (SSID) was configured. Having an exclusive SSID increases the security of the devices, as the network is only shared with the devices that connect to the platform.

The environmental monitoring platform, through the air quality IoT devices, measures four gaseous variables: carbon monoxide (CO), sulfur dioxide (SO_2), ozone (O_3), nitrogen dioxide (NO_2); three particle-matter variables: ultrafine particulate matter (PM1), fine particulate matter (PM2.5), and coarse particulate matter (PM10). These measurements are complemented with weather measurements: temperature, relative humidity, and barometric pressure. In the case of the air-quality-monitoring devices, the measurements of each variable were obtained every twenty minutes.

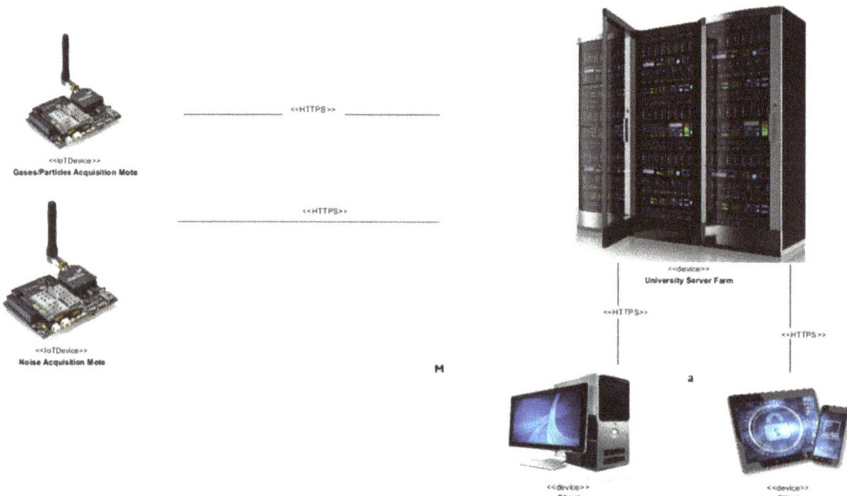

Figure 2. Deployment diagram of the physical and virtual nodes that compose the system.

To measure noise, the platform uses the noise IoT devices, which are equipped with high-caliber microphones. This variable is monitored every ten minutes following the IEC 61,672 standard, which obtains several noise data samples and averages these over the time-window in which the noise was produced. These devices also measure the complementary weather variables. All IoT devices track internal performance data. Table 1 shows the relation between variables, measuring unit and measuring period.

Table 1. Environmental variables with the unit and period in which they are measured.

Variable	Unit of Measurement	Measurement Period
Carbon monoxide (CO) Sulfur dioxide (SO_2) Ozone (O_3) Nitrogen dioxide (NO_2)	Parts per million (ppm)	20 min
Ultrafine particulate matter (PM1) Fine particulate matter (PM2.5) Coarse particulate matter (PM10)	Micrograms per cubic meter ($\mu g/m^3$)	20 min
Temperature Relative humidity Barometric pressure	Degrees Celsius (°C) Percentage (%) Pascal (Pa)	20 min
Noise	Additive decibels (dBA)	10 min
Battery capacity Battery voltage Battery charge current	Percentage (%) Volts (V) Milliampere (mA)	20 min

The software layer of the IoT devices was customized for the requirements of our platform. Each device measures their assigned variables following the period established in Table 1. After the data are acquired, the device sends the data to the Open Data platform via HTTP POST following the procedure mentioned in Section 3.1. After this event, the software agent goes to sleep and wakes up after the specified time period has elapsed. Figure 3 depicts the interaction between the software agents and the Open Data platform.

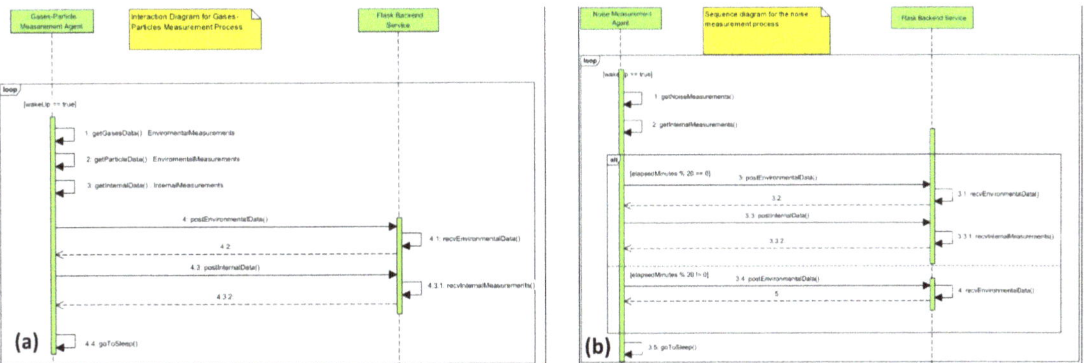

Figure 3. Sequence diagram between the OD platform and the IoT devices: (**a**) air quality IoT device; (**b**) noise IoT devices.

3.3. Determining Air Quality Index

The Air Quality Index can be determined for a short- or long-term duration [58]. The most popular algorithms follow a short-term duration, the Air Quality Index (AQI) algorithm developed by the United States Environmental Protection Agency [59] and the Common Air Quality Index (CAQI) developed in order to make the levels comparable between the cities of the European Union [60,61]. Both algorithms are similar in how they determine the index—they perform a linear interpolation between the edges of the classes. In the case of AQI, the classes are determined by the regulations of the United States, and in CAQI, by the regulations of the European Union. For each pollutant, a sub-index is determined and the final index is the highest sub-index. To determine each subscript, the following Equation (1) is used:

$$I = ((Ih - Il)/(Ch - Cl)) (C - Cl) + Il, \tag{1}$$

where

I is the index to determine;
C is the reading of the concentration of the contaminant;
Cl is the bottom edge of the class where C is found;
Ch is the top edge of the class where C is located;
Il is the value of the index that corresponds to Cl;
Ih is the value of the index that corresponds to Ch.

These indices have a short-term advantage when there is an increase in any of the pollutants, allowing the population to be alerted quickly [62]. The index does not need to consider all pollutants, so a platform can have sensors with different variables deployed within a city. The environmental monitoring platform implemented uses the AQI algorithm, since the Dominican Republic has similar regulations to the United States. The variables considered to determine the index are carbon monoxide (CO), nitrogen dioxide (NO_2), ozone (O_3), sulfur dioxide (SO_2) and particulate matter in two sizes (PM2.5 and PM10).

3.4. Interactive Visualization

This dimension should allow for the easy interaction of environmental data for exploration and analysis to a layperson user. For this purpose, three data dashboards were developed to simplify the interaction with multiple variables.

The Air Quality Dashboard was divided into three parts. The first part presents the last measured variables, using a gauge to represent the current air pollution level following the Air Quality Index (AQI) algorithm. This graph is followed by the current values of particulate matter at 1, 2.5 and 10 µm and the concentration of carbon monoxide, nitrogen dioxide, ozone and sulfur dioxide at ground level, as shown in Figure 4.

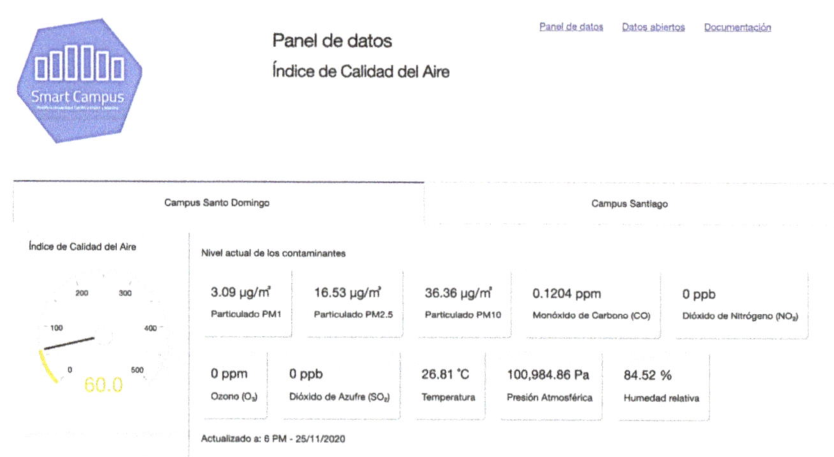

Figure 4. View of the Air Quality Dashboard, the last measurement section.

The second part of the dashboard is the exploration control panel. In this panel, the user can indicate the locations and the date range for the analysis. In addition, the user can manipulate how to aggregate the data for the analysis. Following the AQI algorithm, the data can be aggregated per month, week, day or hour.

The third part of the dashboard graphs the Air Quality Index along with the different contaminants according to the parameters selected by the user. All the graphs automatically update as a user changes any of the parameters. As an example, Figure 5a shows a time series of daily aggregate AQI from one point for December 2020, and Figure 5b shows a time series of hourly aggregate AQI from one point for the first seven days of January 2021.

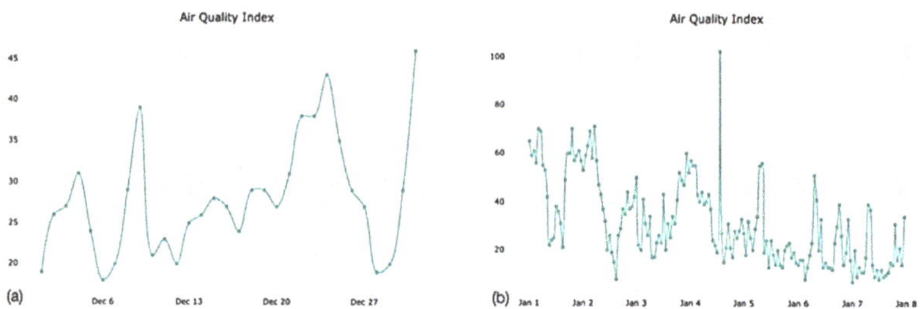

Figure 5. (**a**) Daily aggregate AQI time series for the Santo Domingo Campus Monitor for December 2020; (**b**) hourly aggregated AQI time series for the Santo Domingo Campus Monitor from 1 to 7 January 2021.

The second dashboard focuses on noise level and follows a similar design pattern to the Air Quality dashboard. As an example, Figure 6 shows an hourly aggregated noise level comparison of three monitors for the first work week of January 2021. The third dashboard shows the weather variables; the user can change the parameters following the same procedures of the two previous dashboards.

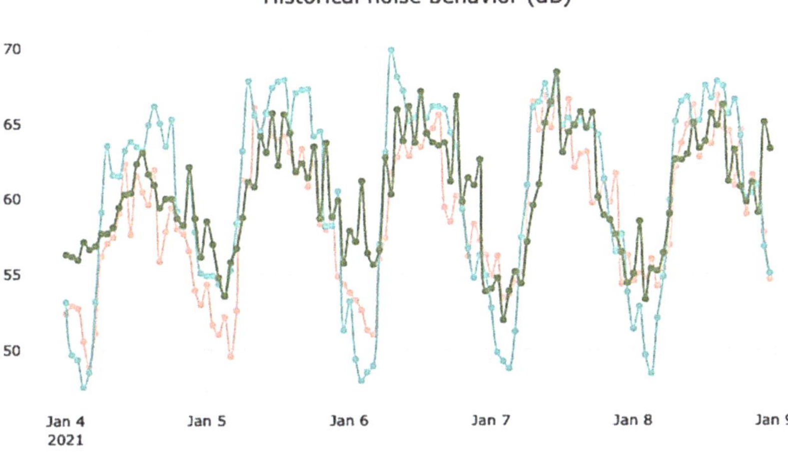

Figure 6. Hourly aggregated noise level time series between 4 and 8 January 2021.

3.5. Capacity Building

To motivate the usage of the platform, multiple strategies were developed. First, different tutorials were published on the website of the platform in the form of short videos and written explanations. These focused on how to use the different dashboards, the platform's API and the technology behind the IoT devices used to capture pollution levels [63]. In addition, we explain how this initiative aligns with the United Nations Sustainable Development Goals as a course of action for attracting new partners to expand the network of sensors. Furthermore, different workshops and seminars were developed targeting a wide range of audiences by customizing the content to their interests.

4. Results and Discussion

This section presents two case studies showing how the platform can be used to create awareness in a community with regard to the effect of pollution. For the first case study, we used data obtained from the platform limiting the dates between 22 and 24 June 2020. For the second case study, we extended the date range from 18 May 2020 to 26 January 2021.

4.1. Case 1: Sahara Dust

In this case study, we analyzed the impact of dust landscapes from the Sahara had on the air quality in Santo Domingo in 2020. This phenomenon is a recurrent event that happens every year, where the atmospheric difference in the Sahara Desert loads the dry air with fine particles that travel thousands of kilometers by air towards America. This phenomenon is known to be a health risk to the population as it can cause multiple respiratory issues, especially for sensitive groups [64].

The citizens of Santo Domingo experienced this event in 2020 between 22 and 24 June. On 22 June, the daily average Air Quality Index was 159 and, on 24 June, it was 153—both are considered to be unhealthy scores. The peak was reached on 23 June, where the daily average was 205, considered very unhealthy. This platform was able to register, for the first time, changes in the air quality of the Dominican Republic. Figure 7 shows the air quality fluctuation per hour while the Sahara dust cloud crossed the city. Figure 8 displays the fluctuation per hour of particles and gasses associated with this territorial dust. Both graphs provide a clear indication of the impact the Sahara dust phenomenon has on the air quality index, and how these fluctuations are dangerous for those who are exposed.

Even for short exposures, this level of contaminant can have a negative impact on the population's health and on the landscape.

Figure 7. Hourly aggregated air quality index time series between 22 and 24 June 2020.

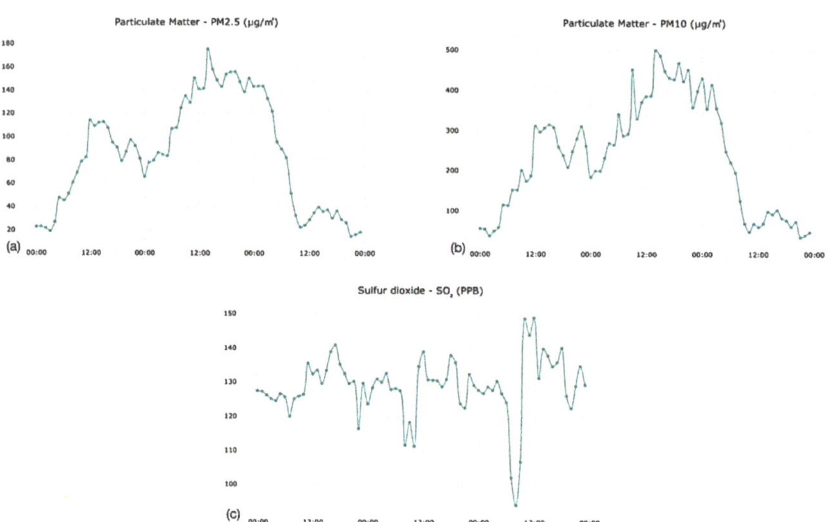

Figure 8. Hourly aggregated (**a**) fine particulate matter, (**b**) coarse particulate matter and (**c**) sulfur dioxide time series between 22 and 24 June 2020.

After the event, the platform was used to create awareness of the danger of the particulate matter to the respiratory system; multiple newspapers referenced the platform as a tool to monitor environmental conditions. The platform marks the first time that the changes caused by this event were registered. In future events, a comparison of magnitude will be possible. This allows for the community to gain an understanding of the changes in the territory and plan multiple urban studies.

4.2. Case 2: COVID-19 in Santo Domingo and Santiago de Los Caballeros

This case study aims to understand how the policies taken to prevent the spread of COVID-19 affect the air quality and noise level of the landscape and community of the two most important cities in the Dominican Republic. The first positive case of COVID-19 reported in the country was on 1 March 2020. With the first case, the Dominican government focused on preventing the spread of the virus in the country. Among the measures taken, the government declared a state of emergency and issued a nationwide curfew, affecting the two cities taken into consideration in this case study. From the first decree, establishing the curfew, the time range of the curfew was either reaffirmed or modified via decrees. Table 2 shows a summary of the changes in the time of the curfew by decree in the time frame of this study. All decrees found on the table were obtained from the Office of Legal Consult of the executive branch of the Dominican Republic [65].

Table 2. Curfew schedule established by the Dominican government.

Period	Decree	Start Date	End Date	Weekday Schedule	Weekend Schedule	Free Mobility
14	007-21	11 January 2021	26 January 2021	5:00 p.m.–5:00 a.m. (+1)	12:00 p.m.–5:00 a.m. (+1)	3 h
13	740-20	1 January 2021	10 January 2021	5:00 p.m.–5:00 a.m. (+1)	12:00 p.m.–5:00 a.m. (+1)	2 h (weekdays)
12	698-20	15 December 2020	31 December 2020	7:00 p.m.–5:00 a.m. (+1)	7:00 p.m.–5:00 a.m. (+1)	3 h
11	684-20	2 December 2020	14 December 2020	9:00 p.m.–5:00 a.m. (+1)	7:00 p.m.–5:00 a.m. (+1)	
10	619-20	12 November 2020	1 December 2020			
9	554-20	18 October 2020	11 November 2020			
8	504-20	28 September 2020	17 October 2020			
7	431-20	3 September 2020	27 September 2020	7:00 p.m.–5:00 a.m. (+1)	5:00 p.m.–5:00 a.m. (+1)	
6	298-20	9 August 2020	2 September 2020			
5	266-20	21 July 2020	9 August 2020			
4	No curfew	28 June 2020	20 July 2020			
3	214-20	14 June 2020	27 June 2020	8:00 p.m.–5:00 a.m. (+1)	8:00 p.m.–5:00 a.m. (+1)	
2	188-20	2 June 2020	13 June 2020	7:00 p.m.–5:00 a.m. (+1)	7:00 p.m.–5:00 a.m. (+1) Sat 5:00 p.m.–5:00 a.m. (+1) Sun	
1	161-20	18 May 2020	1 June 2020			

4.2.1. The Effect of the Curfew on Noise

The curfew has a clear effect on societal activities and the impact these activities have in surrounding territories. Several of these activities have shifted in time—businesses such as bars and restaurants were not able to have customers on site due to the state of emergency.

Using the Noise IoT devices, a variation in daily noise profile was detected. The data suggest the noise could be used to determine the levels of activity in an area. Most of the noise level is related to traffic, but a special situation is given in one facility located 60 m from a small commercial area consisting of several bars and restaurants. Figure 9a,b compare the average noise level by hour in two different periods at this location. During period 4, the state of emergency and any restriction was suspended. In period 5, the state of emergency and the curfew were re-established—the new restrictions allowed restaurants and bars to operate with customers on site; however, their operating hours were made to adapt to curfew schedules.

The graph shows an important difference with respect to night-time noise level, suggesting the activity of restaurants and bars. Additionally, we observed a peak during the curfew in the early afternoon during weekdays and around noon on weekends. This peak could be related to the concentration of citizens' activities as the curfew start time approached.

When the curfew time changed, a similar behavior was observed. Figure 9c,d show a comparison between periods 7 and 8. Period 8 has a more relaxed curfew by adding two hours for commercial and social activities. At the same location, the noise data clearly reflect how the new curfew prolonged the operating hours of the nearby businesses. This shows that using the platform behavioral changes of citizens can be monitored and analyzed, creating a valuable tool for a community when creating policies or studying the implication of a policy.

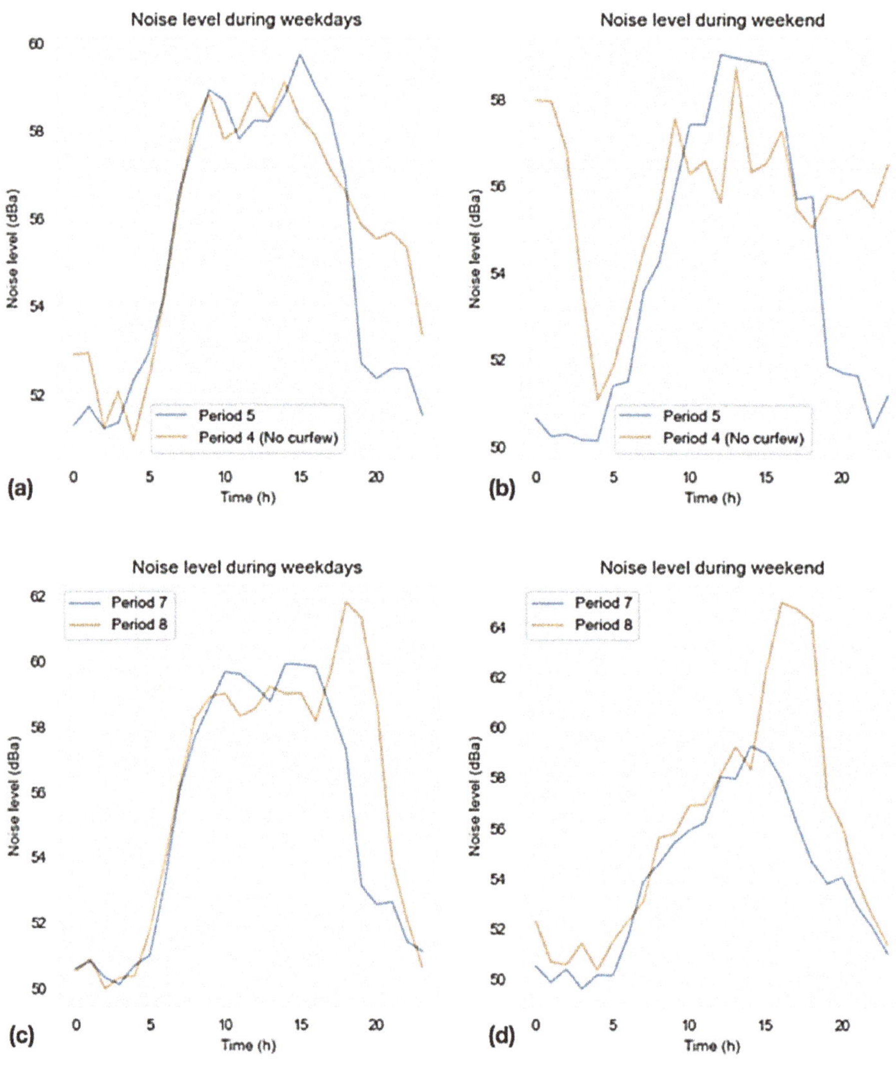

Figure 9. Comparison of the average noise level by hours (**a**) during the weekdays between a period with curfew and one without, (**b**) during the weekend between a period with curfew and one without, (**c**) during the weekdays between periods with different curfew schedules and (**d**) during the weekend between periods with different curfew schedules on a commercial area.

4.2.2. The Effect of the Curfew on Air Quality

In Santo Domingo, the level of air quality improved as the curfew was extended throughout the COVID-19 pandemic, as shown in Figure 10a. Comparing the different pollutants, we can see that the high levels of air quality index were due to dust particles (Figure 10b) in the first three months of the study and, in the latest month, were driven by carbon monoxide and nitrogen dioxide concentrations, as shown in Figure 10c,d.

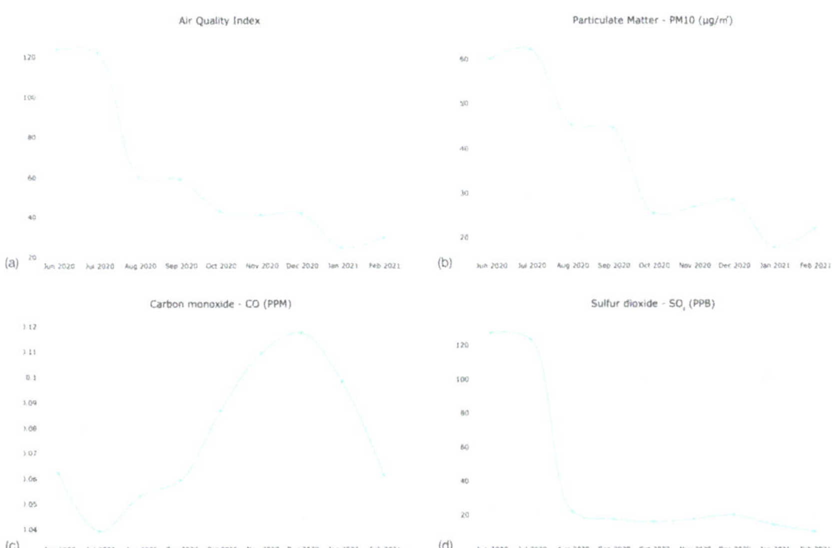

Figure 10. Monthly average in Santo Domingo of (**a**) air quality index, (**b**) PM 10, (**c**) carbon monoxide and (**d**) nitrogen dioxide as presented from the platform.

With the exception of naturally occurring events, a relationship can be seen between the curfew and pollutants. Figure 11a shows the hourly behavior of the air quality index from 28 to 29 September 2020. These days were chosen as they represent the citizens' behavior when the curfew was relaxed in period 8. The air quality index increased when the curfew ended and the index decreased as the curfew started. Comparing carbon monoxide levels during the strictest curfew, the weekends of period 13, Figure 11b shows how the concentrations of this gas increased when the curfew finished and decreased when it started between 2 and 3 January 2021.

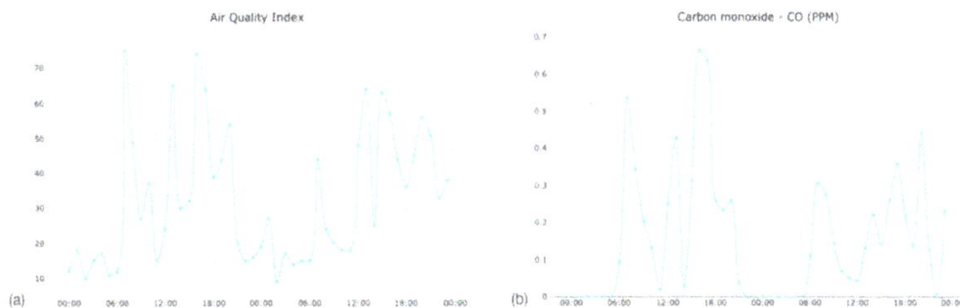

Figure 11. Hourly average in Santo Domingo of (**a**) air quality on 28 and 29 September 2020 and (**b**) carbon monoxide on 2 and 3 January 2021.

Similar to in the city of Santiago de Los Caballeros, the data suggest that a more relaxed curfew increased the level of pollution, reducing the air quality. Figure 12a shows the variability of certain gasses and fine particulates across the last seven periods. Sulfur dioxide average levels fell relatively slightly when compared to the rate at which carbon monoxide rose during the same periods. Similar to the levels of sulfur dioxide, the particulate matter data exhibit a downward tendency during the same period. Figure 12b shows how contamination associated with both types of particulate data becomes minimal specifically during the ninth and tenth periods and increases during the last two periods.

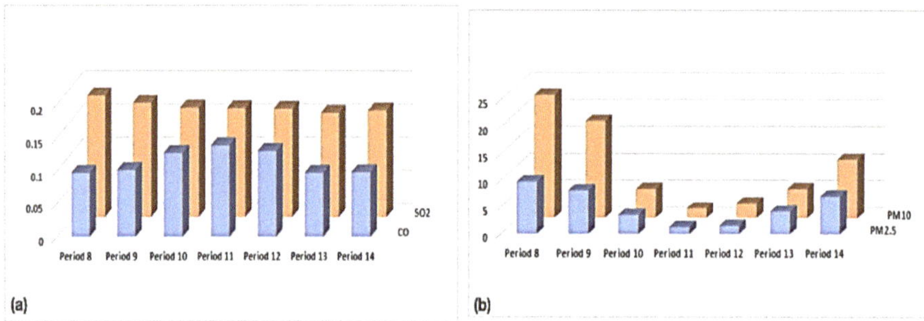

Figure 12. (**a**) Gas and (**b**) particulate pollution levels comparison in Santiago de Los Caballeros from period 8 to 14.

A further analysis of the behavior of carbon monoxide, a gas related to motor vehicle activities, during each day of the week shows an increase between periods 10 and 12. However, on average, the most contaminated weekday does change throughout all periods. In Figure 13a, the carbon monoxide averages reach the highest value during all measured Wednesdays. However, the highest variation is observed on Thursdays. The latter may signal that the average contamination on this weekday could have occurred in short bursts in which carbon monoxide levels rapidly rose and fell. This behavior was also observed, albeit in a less dramatic manner, on the measures taken over on Mondays.

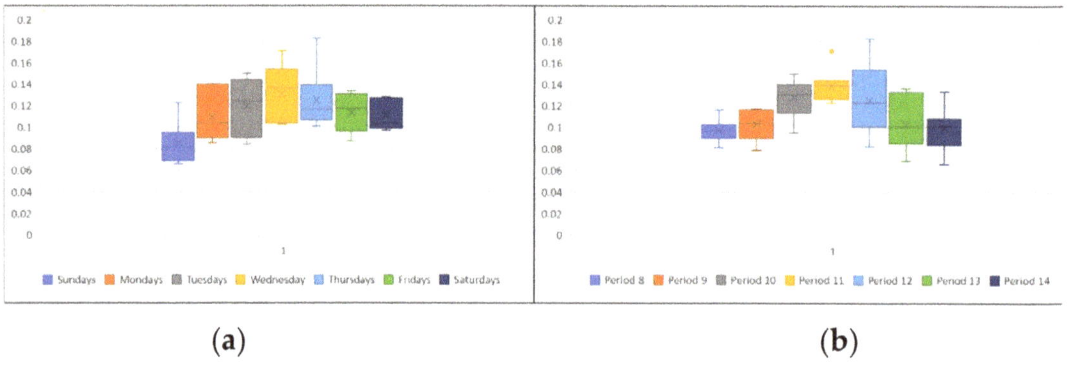

(**a**) (**b**)

Figure 13. Carbon monoxide minimum, maximum and average over (**a**) all weekdays and (**b**) periods in Santiago de Los Caballeros from period 8 to 14.

As shown in Figure 13b, the 11th period exhibits the highest average among all other periods. We can also see that the maximum in this period (0.1714 ppm) is a possible outlier as it falls outside the quartile in which the maximum should have fallen. This could be due to exceedingly high levels of pollution during relatively short periods.

Similar to noise, the air quality aspect of the platform can be used to assess the pollutant levels and their changes as policies are implemented and how they impact a landscape. With real-time monitoring of the pollutants, a community can better understand the impact human activity has on the environment and the territories they occupy.

5. Conclusions

There are many challenges when developing and deploying an environmental monitoring platform for air quality and noise level. On one hand, the platform must be flexible to interact with different environmental variables, accessible so that a community can easily use it, and secure to establish trust. Another aspect is how to interact directly with the community to create awareness of pollution and the impact they have in surrounding territories. Different strategies are required to reach the community. In this work, we present a framework for developing such a platform and how it can be used to understand both natural phenomena and human behavior impact on the pollutant and our territories.

Two case studies were presented; the first case study focused on how the platform can be used to understand the impact a natural event, for example, how dust landscapes impact a community and the actions that can be taken. This study analyzes the effect of the Sahara dust in Santo Domingo in 2020. The results clearly portray an increase in the Air Quality Index as the phenomenon crosses the city. An awareness campaign after the event allowed for citizens to understand the phenomenon and the danger of exposing themselves to the dust, incentivizing them to take action. With a monitoring platform, such as the one presented in this article, the effect of the phenomenon is recorded and allows for a comparison with future events for both policymaking and citizens' awareness. Multiple urban studies could be developed following these data.

The second case study portrays how changes in policies can have a direct effect on the pollutants that affect the community and landscape of a city. This enquiry explored how the policies taken to prevent the spread of COVID-19 in the Dominican Republic affected the air quality and noise level of Santo Domingo and Santiago de Los Caballeros, the two biggest cities in the country. The analysis showed an improvement in the air quality and a decrease in noise level as the curfew established by the government was stricter, limiting the freedom of movement in the population. As the restrictions were relaxed, an increase in the level of pollutants was recorded. This study reflects on the importance of environmental monitoring platforms for understanding the impact different policies have on the pollutants that surround a community and the landscape.

In the long term, environmental monitoring platforms allow for the assessment and impact of different recurring phenomena through time. The open access of the data encourages a culture of data-driven decision making, which improves the decision processes of a community. With the second case study, we were able to show how a curfew policy used for controlling the spread of COVID-19 affected the levels of air quality and noise in an urban space. If policymakers are data-driven, incorporating this dataset can help establish a holistic view that shows the effect the policy has on other aspects of community and the landscape.

An awareness campaign for the platform was launched where multiple actors showed interest in learning about the pollutant concentration, how to access the data and how to use the visualization apps. Even if the visualizations are thought to be intuitive, short sessions or tutorials are needed to reach a broader audience. A common concern was focused on the geographical area covered by the network of sensors and the expansion of the network in other localities. When deploying this type of platform, it is important to plan for geographical coverage and future expansion. The cost of sensors and their deployment is a major factor to consider in the expansion of the platform. The introduction of low-cost sensors allows for a tighter grid of measurement and a better understanding of how smaller communities in an urban space are affected by the pollutants.

Limitations and Future Work

This research is consistent with the development of smart and sustainable cities through an environmental monitoring platform for urban studies focused on air quality and noise level for cities in developing countries. However, this study has some limitations in the framework, the deployment of the platform and the case studies presented.

The framework provides different dimensions so that different components of the platform can operate. The framework does not expand on the mechanisms under which third-party applications can interact with the platform; this is an important limitation. The Open Data dimension must define the governance of the data in the platform as sensors from other partners are integrated to the platform and with the addition of new services from third parties.

Their deployment was focused on the two biggest cities of the Dominican Republic. The main limitation of the deployment is in regard to the number of sensors initially deployed, which did not cover the extension of each city. More sensors are needed to create a robust network that can measure the contaminants in each neighborhood of the cities. Furthermore, the platform is limited in terms of the functionalities related to IoT devices. To improve this issue, the platform could incorporate more lightweight application-level protocols such as constrained application protocol (CoAP). Adding this protocol to the platform would facilitate communication between the platform and low-cost sensors that are resource-constrained, which could facilitate the expansion of the platform.

Through an expansion of the platform in both sensors and functionalities, it would be of interest to investigate the value it creates on other territories and measure the trust of the community to the monitoring platform, especially those at risk. In addition, it is of interest to compare the behavior of the pollution in different parts of a city or country in order to develop more resilient communities.

From a case study perspective, multiple limitations exist. The short research period is one of those limitations. Further work is needed to understand the long-term performance of the platform, of pollutants and the usage given for other urban studies. For example, a multi-year study of the impact of the Sahara Dust in the air quality of Santo Domingo.

This study has focused on understanding the relation between urban conditions and a community and the impact they have on each other through the variables measured by the platform. This itself is a limitation, as the impact the platform has on the landscape is not considered; it is of interest in future studies to evaluate this through a life-cycle assessment of the platform.

In addition, the air quality sensors are limited to three types of particulate matter (PM1, PM2.5 and PM 10) and four gases (CO, NO_2, O_3 and SO_2). Even if these pollutants are the most critical for determining air quality, there are other gases that are not measured, and their impacts are not assessed on this study. Other studies should consider incorporating a broad variety of sensors to measure a broad variety of pollutants.

Author Contributions: Conceptualization, V.G. and J.F.-G.; methodology, V.G. and M.P.; software, V.G. and M.P.; validation, V.G., J.F.-G., M.P. and Y.G.F.; formal analysis, V.G., J.F.-G. and M.P.; investigation, V.G., M.P., J.F.-G. and Y.G.F.; resources, V.G., and M.P.; data curation, V.G.; writing—original draft preparation, V.G.; writing—review and editing, M.P. and Y.G.F.; visualization, V.G., J.F.-G., Y.G.F. and M.P.; supervision, V.G.; project administration, V.G.; funding acquisition, V.G. All authors have read and agreed to the published version of the manuscript.

Funding: This research received no external funding.

Data Availability Statement: The data that support the findings of this study are openly available in Zenodo at https://doi.org/10.5281/zenodo.7105153.

Acknowledgments: We would like to thank the support personnel that aided regarding the sensor-nodes' installation on site in both, Santo Domingo and Santiago de Los Caballeros.

Conflicts of Interest: The authors declare no conflict of interest.

References

1. Neirotti, P.; De Marco, A.; Cagliano, A.C.; Mangano, G.; Scorrano, F. Current trends in Smart City initiatives: Some stylised facts. *Cities* **2014**, *38*, 25–36. [CrossRef]
2. Kitchin, R. The real-time city? Big data and smart urbanism. *GeoJournal* **2014**, *79*, 1–14. [CrossRef]
3. Osman, A.M.S. A novel big data analytics framework for smart cities. *Future Gener. Comput. Syst.* **2019**, *91*, 620–633. [CrossRef]
4. Fialová, J.; Bamwesigye, D.; Łukaszkiewicz, J.; Fortuna-Antoszkiewicz, B. Smart Cities Landscape and Urban Planning for Sustainability in Brno City. *Land* **2021**, *10*, 870. [CrossRef]
5. Commodore, A.; Wilson, S.; Muhammad, O.; Svendsen, E.; Pearce, J. Community-based participatory research for the study of air pollution: A review of motivations, approaches, and outcomes. *Environ. Monit. Assess.* **2017**, *189*, 378. [CrossRef] [PubMed]
6. Baklanov, A.; Molina, L.T.; Gauss, M. Megacities, Air Quality and Climate. *Atmos. Environ.* **2016**, *126*, 235–249. [CrossRef]

7. Maesano, C.; Morel, G.; Matynia, A.; Ratsombath, N.; Bonnety, J.; Legros, G.; Da Costa, P.; Prud'Homme, J.; Annesi-Maesano, I. Impacts on human mortality due to reductions in PM10 concentrations through different traffic scenarios in Paris, France. *Sci. Total Environ.* **2020**, *698*, 134257. [CrossRef]

8. Araujo, L.N.; Belotti, J.T.; Alves, T.A.; Tadano, Y.D.S.; Siqueira, H. Ensemble method based on Artificial Neural Networks to estimate air pollution health risks. *Environ. Model. Softw.* **2020**, *123*, 104567. [CrossRef]

9. Ramaiah, M.; Avtar, R.; Rahman, M. Land Cover Influences on LST in Two Proposed Smart Cities of India: Comparative Analysis Using Spectral Indices. *Land* **2020**, *9*, 292. [CrossRef]

10. Yigitcanlar, T.; Kamruzzaman; Foth, M.; Sabatini-Marques, J.; da Costa, E.; Ioppolo, G. Can Cities Become Smart without Being Sustainable? A Systematic Review of the Literature. *Sustain. Cities Soc.* **2019**, *45*, 348–365. [CrossRef]

11. Zheng, S.; Huang, Y.; Sun, Y. Effects of Urban Form on Carbon Emissions in China: Implications for Low-Carbon Urban Planning. *Land* **2022**, *11*, 1343. [CrossRef]

12. Steigerwald, F.; Kossmann, M.; Schau-Noppel, H.; Buchholz, S.; Panferov, O. Delimitation of Urban Hot Spots and Rural Cold Air Formation Areas for Nocturnal Ventilation Studies Using Urban Climate Simulations. *Land* **2022**, *11*, 1330. [CrossRef]

13. Dantas, G.; Siciliano, B.; França, B.B.; da Silva, C.M.; Arbilla, G. The impact of COVID-19 partial lockdown on the air quality of the city of Rio de Janeiro, Brazil. *Sci. Total Environ.* **2020**, *729*, 139085. [CrossRef] [PubMed]

14. Sharma, S.; Zhang, M.; Anshika; Gao, J.; Zhang, H.; Kota, S.H. Effect of restricted emissions during COVID-19 on air quality in India. *Sci. Total Environ.* **2020**, *728*, 138878. [CrossRef]

15. Xu, K.; Cui, K.; Young, L.-H.; Wang, Y.-F.; Hsieh, Y.-K.; Wan, S.; Zhang, J. Air Quality Index, Indicatory Air Pollutants and Impact of COVID-19 Event on the Air Quality near Central China. *Aerosol Air Qual. Res.* **2020**, *20*, 1204–1221. [CrossRef]

16. Wu, M.; Yan, B.; Huang, Y.; Sarker, N.I. Big Data-Driven Urban Management: Potential for Urban Sustainability. *Land* **2022**, *11*, 680. [CrossRef]

17. Liu, X.; Jayaratne, R.; Thai, P.; Kuhn, T.; Zing, I.; Christensen, B.; Lamont, R.; Dunbabin, M.; Zhu, S.; Gao, J.; et al. Low-cost sensors as an alternative for long-term air quality monitoring. *Environ. Res.* **2020**, *185*, 109438. [CrossRef]

18. Bashir, M.F.; Ma, B.; Bilal; Komal, B.; Bashir, M.A.; Tan, D.; Bashir, M. Correlation between Climate Indicators and COVID-19 Pandemic in New York, USA. *Sci. Total Environ.* **2020**, *728*, 138835. [CrossRef]

19. Barceló, M.A.; Varga, D.; Tobias, A.; Diaz, J.; Linares, C.; Saez, M. Long term effects of traffic noise on mortality in the city of Barcelona, 2004–2007. *Environ. Res.* **2016**, *147*, 193–206. [CrossRef]

20. Lee, E.Y.; Jerrett, M.; Ross, Z.; Coogan, P.F.; Seto, E.Y. Assessment of traffic-related noise in three cities in the United States. *Environ. Res.* **2014**, *132*, 182–189. [CrossRef]

21. Silva, L.T.; Mendes, J.F. City Noise-Air: An environmental quality index for cities. *Sustain. Cities Soc.* **2012**, *4*, 1–11. [CrossRef]

22. Schneider, P.; Castell, N.; Vogt, M.; Dauge, F.R.; Lahoz, W.A.; Bartonova, A. Mapping urban air quality in near real-time using observations from low-cost sensors and model information. *Environ. Int.* **2017**, *106*, 234–247. [CrossRef] [PubMed]

23. Kumar, P.; Omidvarborna, H.; Pilla, F.; Lewin, N. A primary school driven initiative to influence commuting style for dropping-off and picking-up of pupils. *Sci. Total Environ.* **2020**, *727*, 138360. [CrossRef] [PubMed]

24. Riley, M.; Kirkwood, J.; Jiang, N.; Ross, G. Air Quality Monitoring in NSW: From Long Term Trend Monitoring to Integrated Urban Services. *Air Qual. Clim. Chang.* **2020**, *54*, 44–51.

25. Cujia, A.; Agudelo-Castañeda, D.; Pacheco-Bustos, C.; Teixeira, E.C. Forecast of PM10 time-series data: A study case in Caribbean cities. *Atmos. Pollut. Res.* **2019**, *10*, 2053–2062. [CrossRef]

26. Van Roode, S.; Ruiz-Aguilar, J.J.; González-Enrique, J.; Turias, I.J. An artificial neural network ensemble approach to generate air pollution maps. *Environ. Monit. Assess.* **2019**, *191*, 727. [CrossRef]

27. Borrego, C.; Costa, A.m.; Ginja, J.; Amorim, M.; Coutinho, M.; Karatzas, K.; Sioumis, T.; Katsifarakis, N.; Konstantinidis, K.; De Vito, S.; et al. Assessment of air quality microsensors versus reference methods: The EuNetAir joint exercise. *Atmos. Environ.* **2016**, *147*, 246–263. [CrossRef]

28. Castell, N.; Dauge, F.R.; Schneider, P.; Vogt, M.; Lerner, U.; Fishbain, B.; Broday, D.; Bartonova, A. Can Commercial Low-Cost Sensor Platforms Contribute to Air Quality Monitoring and Exposure Estimates? *Environ. Int.* **2017**, *99*, 293–302. [CrossRef]

29. Borge, R.; Narros, A.; Artíñano, B.; Yagüe, C.; Gómez-Moreno, F.J.; de la Paz, D.; Román-Cascón, C.; Díaz, E.; Maqueda, G.; Sastre, M.; et al. Assessment of microscale spatio-temporal variation of air pollution at an urban hotspot in Madrid (Spain) through an extensive field campaign. *Atmos. Environ.* **2016**, *140*, 432–445. [CrossRef]

30. Piedrahita, R.; Xiang, Y.; Masson, N.; Ortega, J.; Collier, A.; Jiang, Y.; Li, K.; Dick, R.P.; Lv, Q.; Hannigan, M.; et al. The next generation of low-cost personal air quality sensors for quantitative exposure monitoring. *Atmos. Meas. Tech.* **2014**, *7*, 3325–3336. [CrossRef]

31. Morawska, L.; Thai, P.K.; Liu, X.; Asumadu-Sakyi, A.; Ayoko, G.; Bartonova, A.; Bedini, A.; Chai, F.; Christensen, B.; Dunbabin, M.; et al. Applications of Low-Cost Sensing Technologies for Air Quality Monitoring and Exposure Assessment: How Far Have They Gone? *Environ. Int.* **2018**, *116*, 286–299. [CrossRef] [PubMed]

32. Sayahi, T.; Butterfield, A.; Kelly, K.E. Long-term field evaluation of the Plantower PMS low-cost particulate matter sensors. *Environ. Pollut.* **2019**, *245*, 932–940. [CrossRef] [PubMed]

33. De Vito, S.; Esposito, E.; Castell, N.; Schneider, P.; Bartonova, A. On the robustness of field calibration for smart air quality monitors. *Sensors Actuators B Chem.* **2020**, *310*, 127869. [CrossRef]

34. Cabaneros, S.M.; Calautit, J.K.; Hughes, B.R. A review of artificial neural network models for ambient air pollution prediction. *Environ. Model. Softw.* **2019**, *119*, 285–304. [CrossRef]
35. Mueller, S.F.; Mallard, J.W. Contributions of Natural Emissions to Ozone and PM2.5 as Simulated by the Community Multiscale Air Quality (CMAQ) Model. *Environ. Sci. Technol.* **2011**, *45*, 4817–4823. [CrossRef] [PubMed]
36. Snyder, E.G.; Watkins, T.H.; Solomon, P.A.; Thoma, E.D.; Williams, R.W.; Hagler, G.S.W.; Shelow, D.; Hindin, D.A.; Kilaru, V.J.; Preuss, P.W. The Changing Paradigm of Air Pollution Monitoring. *Environ. Sci. Technol.* **2013**, *47*, 11369–11377. [CrossRef]
37. Pritchard, H.; Gabrys, J. From Citizen Sensing to Collective Monitoring: Working through the Perceptive and Affective Problematics of Environmental Pollution. *GeoHumanities* **2016**, *2*, 354–371. [CrossRef]
38. Hubbell, B.J.; Kaufman, A.; Rivers, L.; Schulte, K.; Hagler, G.; Clougherty, J.; Cascio, W.; Costa, D. Understanding social and behavioral drivers and impacts of air quality sensor use. *Sci. Total Environ.* **2018**, *621*, 886–894. [CrossRef]
39. Mahajan, S.; Kumar, P.; Pinto, J.A.; Riccetti, A.; Schaaf, K.; Camprodon, G.; Smári, V.; Passani, A.; Forino, G. A citizen science approach for enhancing public understanding of air pollution. *Sustain. Cities Soc.* **2020**, *52*, 101800. [CrossRef]
40. Rickenbacker, H.; Brown, F.; Bilec, M. Creating environmental consciousness in underserved communities: Implementation and outcomes of community-based environmental justice and air pollution research. *Sustain. Cities Soc.* **2019**, *47*, 101473. [CrossRef]
41. Chen, L.-J.; Ho, Y.-H.; Lee, H.-C.; Wu, H.-C.; Liu, H.-M.; Hsieh, H.-H.; Huang, Y.-T.; Lung, S.-C.C. An Open Framework for Participatory PM2.5 Monitoring in Smart Cities. *IEEE Access* **2017**, *5*, 14441–14454. [CrossRef]
42. Mahajan, S.; Luo, C.-H.; Wu, D.-Y.; Chen, L.-J. From Do-It-Yourself (DIY) to Do-It-Together (DIT): Reflections on designing a citizen-driven air quality monitoring framework in Taiwan. *Sustain. Cities Soc.* **2020**, *66*, 102628. [CrossRef]
43. Che, W.; Frey, H.C.; Fung, J.C.; Ning, Z.; Qu, H.; Lo, H.K.; Chen, L.; Wong, T.-W.; Wong, M.K.; Lee, O.C.; et al. PRAISE-HK: A personalized real-time air quality informatics system for citizen participation in exposure and health risk management. *Sustain. Cities Soc.* **2020**, *54*, 101986. [CrossRef]
44. Open Data Handbook. Available online: https://opendatahandbook.org/ (accessed on 13 September 2022).
45. Nikiforova, A. Smarter Open Government Data for Society 5.0: Are Your Open Data Smart Enough? *Sensors* **2021**, *21*, 5204. [CrossRef] [PubMed]
46. Worthy, B. The impact of open data in the uk: Complex, unpredictable, and political. *Public Adm.* **2015**, *93*, 788–805. [CrossRef]
47. Sołtysik-Piorunkiewicz, A.; Zdonek, I. How Society 5.0 and Industry 4.0 Ideas Shape the Open Data Performance Expectancy. *Sustainability* **2021**, *13*, 917. [CrossRef]
48. Neves, F.T.; Neto, M.D.C.; Aparicio, M. The impacts of open data initiatives on smart cities: A framework for evaluation and monitoring. *Cities* **2020**, *106*, 102860. [CrossRef]
49. Welle Donker, F.; van Loenen, B. How to Assess the Success of the Open Data Ecosystem? *Int. J. Digit. Earth* **2017**, *10*, 284–306. [CrossRef]
50. Máchová, R.; Lněnička, M. Evaluating the Quality of Open Data Portals on the National Level. *J. Theor. Appl. Electron. Commer. Res.* **2017**, *12*, 21–41. [CrossRef]
51. Lnenicka, M.; Nikiforova, A. Transparency-by-Design: What Is the Role of Open Data Portals? *Telemat. Inform.* **2021**, *61*, 101605. [CrossRef]
52. Lněnička, M.; Machova, R.; Volejníková, J.; Linhartová, V.; Knezackova, R.; Hub, M. Enhancing transparency through open government data: The case of data portals and their features and capabilities. *Online Inf. Rev.* **2021**, *45*, 1021–1038. [CrossRef]
53. Roman, D.; Reeves, N.; Gonzalez, E.; Celino, I.; El Kader, S.A.; Turk, P.; Soylu, A.; Corcho, O.; Cedazo, R.; Calegari, G.R.; et al. An analysis of pollution Citizen Science projects from the perspective of Data Science and Open Science. *Data Technol. Appl.* **2021**, *55*, 622–642. [CrossRef]
54. Oficina Nacional de Estadística (ONE) Datos y Estadísticas. Available online: https://www.one.gob.do/datos-y-estadisticas/ (accessed on 9 September 2022).
55. Clemente, P.J.; Lozano-Tello, A. Model Driven Development Applied to Complex Event Processing for Near Real-Time Open Data. *Sensors* **2018**, *18*, 4125. [CrossRef] [PubMed]
56. Gonzalez, V. Smartcampus.pucmm: Plataforma de Monitoreo de Variables Medioambientales. Available online: https://github.com/victormlgh/smartcampus.pucmm (accessed on 13 September 2022).
57. Plug & Sense Waspmote Encapsulated Version Easyly Deploy IoT Networks. Available online: https://www.libelium.com/iot-products/plug-sense/ (accessed on 9 September 2022).
58. Mayer, H.; Kalberlah, F. Two impact related air quality indices as tools to assess the daily and long-term air pollution. *Int. J. Environ. Pollut.* **2009**, *36*, 19. [CrossRef]
59. United States Environmental Protection Agency. *Office of Air Quality Planning and Standards*; United States Environmental Protection Agency: Research Triangle Park, NC, USA, 1998. Available online: https://www3.epa.gov/airquality/index.html (accessed on 9 September 2022).
60. Van den Elshout, S.; Léger, K.; Nussio, F. Comparing Urban Air Quality in Europe in Real Time a Review of Existing Air Quality Indices and the Proposal of a Common Alternative. *Environ. Int.* **2008**, *34*, 720–726. [CrossRef]
61. van den Elshout, S.; Léger, K.; Heich, H. CAQI Common Air Quality Index—Update with PM2.5 and Sensitivity Analysis. *Sci. Total Environ.* **2014**, *488–489*, 461–468. [CrossRef]
62. Ruggieri, M.; Plaia, A. An aggregate AQI: Comparing different standardizations and introducing a variability index. *Sci. Total Environ.* **2012**, *420*, 263–272. [CrossRef]

63. Smart Campus—Pontificia Universidad Católica Madre y Maestra. Available online: https://smartcampus.pucmm.edu.do/ (accessed on 9 September 2022).
64. Wang, Q.; Gu, J.; Wang, X. The impact of Sahara dust on air quality and public health in European countries. *Atmos. Environ.* **2020**, *241*, 117771. [CrossRef]
65. Consulta Externa. Available online: http://www.consultoria.gov.do/consulta/ (accessed on 9 September 2022).

95

Article

Urban Ageing, Gender and the Value of the Local Environment: The Experience of Older Women in a Central Neighbourhood of Madrid, Spain

M. Victoria Gómez [1,*] and Irene Lebrusán [2]

[1] Facultad de Ciencias Sociales y Jurídicas, Universidad Carlos III de Madrid, 28903 Madrid, Spain
[2] Centro Internacional Sobre el Envejecimiento (CENIE), 37002 Salamanca, Spain
* Correspondence: mgomez@polsoc.uc3m.es

Abstract: Urban ageing is an emerging domain that mixes two challenges of current societies: the ageing of the population and the increasing urbanisation. While ageing in place has demonstrated numerous benefits, some social sectors question whether the city is the right environment for ageing, since cities are home to many of the social problems that characterise contemporary societies. Urban environments are widely described as rootless in most academic articles, with a focus on the impersonality, transience, and segmentation of links between city dwellers. However, this portrayal coexists with contrasting views of urban life that instead emphasise the importance of the local setting and other experiences of attachment to the place of residence. From the age and gender perspectives, in some urban areas, the neighbourhood plays a fundamental role in the lives of many older women, as a natural setting for interaction and an area conducive to collaborative relationships and practical and emotional support in times of need. This article analyses the role that the local space plays in the lives of older women, the value they attach to it and the meaning they attribute to neighbourhood relations in the local urban environment. In order to analyse this reality, the Universidad (Malasaña) neighbourhood, has been selected as a case study, a central area in Madrid (Spain) exposed to numerous processes of transformation, which shows the highest rate of residential mobility in the area. Despite this reality, in a context marked by new difficulties, the conclusion shows that elderly women have a strong attachment to their neighbourhood. This attachment is not necessarily characterised by deep friendships, but by the existence of a significant social network that responds in case of need.

Keywords: ageing in place; place attachment; older women; neighbourhood

Citation: Gómez, M.V.; Lebrusán, I. Urban Ageing, Gender and the Value of the Local Environment: The Experience of Older Women in a Central Neighbourhood of Madrid, Spain. *Land* **2022**, *11*, 1456. https://doi.org/10.3390/land11091456

Academic Editors: Vanessa Zorrilla-Muñoz, Maria Silveria Agulló-Tomás, Eduardo Fernandez, Blanca Criado Quesada, Sonia De Lucas Santos and Jesus Cuadrado Rojo

Received: 16 July 2022
Accepted: 28 August 2022
Published: 1 September 2022

Publisher's Note: MDPI stays neutral with regard to jurisdictional claims in published maps and institutional affiliations.

1. Introduction

Urban ageing is an emerging domain that mixes two challenges of current societies: the ageing of the population and the increasing urbanisation. While ageing in place has demonstrated numerous benefits, the media and some sections of civil society question whether the urban realm is the right environment for it, considering that cities are not friendly spaces for the elderly [1].

In fact, cities are often scenery of segregation, inequality, fragmentation, and atomisation [2], in addition to the generation of enormous problems related to the environment, climate change, and urban health [2–4]. The contemporary city is often described in terms of rootlessness and liquidity focusing on the impersonality, transience, and segmentation of ties between urban dwellers [5,6]. While these problems cannot be denied, this portrayal coexists with other views of urban life that emphasize the importance of the local environment and the positive effects of rootedness in the place of residence. This latter angle is advantageous to understanding ageing in place and its social benefits.

Although the demographic reality of ageing has been widely studied, the urban ageing dimension has not received the attention it deserves in this context [7,8]. Opening up avenues of research in this field not only reinforces the importance of urban space in the quality of ageing, but also underpins key knowledge for implementing improvements with regard to the issue. The current in ageing in place studies is to advocate the beneficial effects of older people remaining in a familiar setting for as long as possible, based on the notion that an accessible, familiar, and safe environment can reinforce independence and improve well-being at this stage [9–13]. As Herbert notes, connections formed between people and places are associated with quality and well-being in later life [14]. However, for this to be possible, it is important to foster an integrative spatial configuration since the physical and spatial context influences people throughout their life course [15–17], but in old age, it is even more important than ever and comes to determine how we age and how we respond to illness [12,18]. Several studies have shown how frail, elderly people are able to remain independent due to the benefits of living in a familiar environment, while depression and anxiety are more prevalent in the absence of settings that encourage socialisation, such as public spaces, parks, and gardens [19]. The sum of all these factors gives rise to new and very different ways of experiencing old age, linked to the desire for autonomy and permanence in a familiar social environment [11,20].

It is thus recognised that ageing is a heterogeneous process as the well-being of the elderly depends on multiple factors [21,22], among which not only their lifestyles, but also their spatial relationships are important. In fact, the public space and the environment outside the home play an important role in the independence of the elderly, allowing them to age in the way they choose [12,13]. Furthermore, from a policy perspective and at the international level, the importance of ensuring an enabling and supportive environment to achieve the highest possible level of health and well-being for the elderly has begun to receive much attention, such as in the Madrid International Plan of Action on Aging and in the Political Declaration adopted at the Second World Assembly on Aging in April 2002 [21]. Recently, age was a key aspect of the Sustainable Development Goals, and urban ageing received specific consideration, whereby target 11.7 states that, by 2030, cities should provide universal access to safe, inclusive, green, and public spaces, with a focus on vulnerable populations and specifically including older people. This international commitment is a major step forward because it shows that policymakers are finally being encouraged to address the importance of the local environment for the quality of ageing [23].

These approaches necessarily imply that cities must be prepared to respond to the needs of a growing number of older people in order to provide a good quality of life. These are requirements that, to date, Spanish cities have failed to meet [24,25].

When coming to gender, it has been proved that conceptions, experiences, and uses of space are different for men and women [26], with the consequent differentiated impact on knowledge, behaviour, and lifestyles [27]. Urban studies that look at gender differences have a long academic history revealing how the design and planning of cities have systematically ignored women's urban trajectories, needs, and itineraries [26,28–32] on the basis of an analytical perspective that not only contributed to the perception of females as exceptions or residuals in relation to males but was also an epistemological obstacle to understanding the global logic of women's behaviour [33].

The gender perspective is particularly useful for understanding the new dimensions of the importance of ageing in place and for looking at how interactions unfold in the local space where people carry out their daily activities, relate with others, and come to consolidate identities linked to their neighbourhoods. The neighbourhood can facilitate meetings, social relations, and the reception of support, physical or emotional [12]. In this sense, ageing in place is related to the sense of identity both through independence and autonomy and through the roles of relationship and solidarity in the places where people live [34]. Despite the certain consensus on the loss of community relations at the local level, some neighbourhoods still maintain them, and their residents attach significant value to these, particularly those who have lived there for a long time, as tends to be the case with older women who have a life expectancy longer than that of men. As Gómez and Álvarez [35] point out, the predominance of individualistic tendencies that favour disunity sometimes leads to a failure to appreciate the value of relational networks and the importance that community discourse and practices attain in certain contexts. Along these lines, Blokland and Nast [36] note that despite all the changes experienced in terms of the dissolution of ties, the significance of the local community has not been lost.

It is, therefore, necessary to pay attention to the relationships that people build in their neighbourhoods and to the material and symbolic links they forge with the environment in which they live. Certain circumstances can help to strengthen these relationships, to create a certain atmosphere of comfort [36,37], and to initiate or consolidate lines of exchange based on the establishment of weak, low-intensity ties which, despite their fragility, contribute to social cohesion due to their capacity to create 'bridging' social capital between distant groups [37,38]. Even so, the knowledge that is forged over time, when obligations generate more intense ties with the area of residence, is an especially key aspect in the case of older women [14,23].

On the other hand, when these spaces are subject to processes of change, as is the case with gentrification, there is a significant disruption of the everyday coordinates of the elderly. The problems and consequences associated with gentrification processes have been directly or indirectly addressed in academic studies for decades. However, despite their major relevance, the gentrification perspective tends to neglect the everyday interactions and small-scale changes that constitute and shape the experience of living in a neighbourhood [39].

This article reflects on the role that the local urban space plays in the lives of older women, taking into account the importance of ageing in the place where they spend their lives, not only from the point of view of physical health and the recognition of its relevance in terms of the Sustainable Development Goals, but also from people's relational needs and the links associated with the place of residence [34]. For this purpose, empirical material is used that was collected in the Universidad (Malasaña) neighbourhood of Madrid between 2015 and 2017, an urban area in transition in which the processes of gentrification and touristification have defined new vital realities for older women but which also allows us to apprehend the strength of the links generated through years of living in this local area.

The research is based on the following premises:

1. Most Spanish population prefer to grow old independently, in an environment that is known and, whenever possible, in their own homes [11].
2. As referred to before, the urban dimension of ageing has received less attention than other related aspects.
3. Even though women are living longer than men and their place of attachment has been considered higher [14,40]. This dimension in the context of ageing from a gender perspective has received little attention.

The article is therefore in line with the current stream of research on how the process of growing old is experienced in the local urban environment by women and the importance of attachment to space in old age.

The paper is organized as follows. After this introduction, Section 2 explains the materials and methods used in the analysis. Section 3 presents the results and findings, and Section 4 discusses them. Finally, Section 5 wraps up with the conclusions, limitations, and the main lines of future work.

2. Methodology: A Qualitative Analysis in Madrid, Spain

2.1. The Framework

This research is comprised of a wide-ranging project on place attachment based on two inner neighbourhoods belonging to two cities within Europe: Kumpula in Helsinki (Finland) and Universidad in Madrid (Spain) which was mainly developed between 2015 and 2017. Elements of major interest converge in the Spanish local urban space analysed. In contrast to the attachment to the area among older women who have spent most of their lives in the neighbourhood, there is a great deal of mobility on the part of newer, younger residents, together with the usual tensions that accompany local processes of urban change and the presence of women in a country with a very high level of life expectancy. While the referred project allowed us to investigate different aspects of the feeling of belonging to the local area, both the focus on the specific perspective of older women and the influence of neighbourhood changes on their experience of ageing in place remained unexplored.

Accordingly, the aim of this article is to analyse how attachment to place is constructed and reinforced in the process of ageing in an urban setting from a gender perspective. To better understand the interrelations between the ageing in place construction, place attachment, and formation and consolidation of ties in inner-city areas in the case of a big city like Madrid, a case-study approach was selected. It allows for a detailed analysis of the 'holistic and meaningful characteristics of real-life events', such as neighbourhood transformation [41] as well as considers how external factors and events influenced the construction of place attachment while ageing. Besides this approach enabled us to explore a complex set of experiences in detail and to recount the role of life events and neighbourhood change over time, specifically in urban areas in transition [42].

Madrid is the capital and biggest city of Spain (3,286,662 inhabitants in 2022), and therefore the one with the largest number of elderly people (666,599 people are over 65 years old). This city has 21 districts which are further subdivided into 131 neighbourhoods. The districts are territorial divisions of the municipality, equipped with decentralised management bodies to facilitate the governance of such a big city. These districts are very different from an urban perspective (as a result of different construction stages) but also in terms of wealth and quality of life [25].

As in other historic cities, central areas are exposed to numerous processes of transformation that generate tensions and conflicts. The Centro district [Id.1 in Figure 1] is immersed in a process of gentrification and touristification and has the highest net migration rate[1]. in Madrid City [see Table 1, column "net migration"] which means that it has a high population turnover, with an inflow of 109.06 and an outflow of a population of 40.36. Within Distrito Centro (Id.1 in Figure 1), the Universidad neighbourhood [Figure 2] is considered in this research as an illustrative cultural case [43,44] of the challenges faced by women in the process of ageing in place.

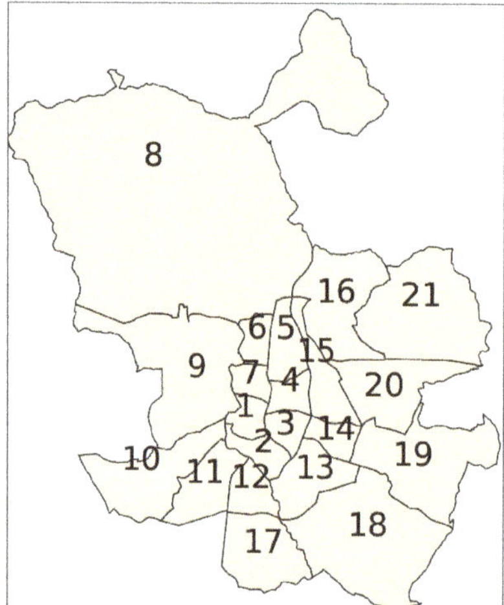

Figure 1. Map of the 21 districts division of the city of Madrid. The correspondence between the Id number and the name of the district can be found in Table 1.

Table 1. * Migration growth in 2020. Rates per 1000 inhabitants. Total population, elderly population, percentage of elderly population over total population, percentage of elderly women over elderly population, net migration, per district, Madrid. Source: Madrid City Council, 2020 and 2022.

	Total Population [1]	Elderly Population [2]	% of Elderly Over [1]	% of Elderly Women Over [2]	Net Migration *
01. Centro	139,682	21,914	15.7	60.3	30.0
02. Arganzuela	152,638	31,212	20.4	60.7	0.6
03. Retiro	117,672	31,268	26.6	61.5	0.8
04. Salamanca	145,457	34,895	24.0	62.8	13.4
05. Chamartín	144,371	34,286	23.7	62.1	2.2
06. Tetuán	157,433	30,366	19.3	63.2	8.2
07. Chamberí	137,287	33,522	24.4	64.0	7.5
08. Fuencarral-El Pardo	246,281	52,925	21.5	58.9	−5.6
09. Moncloa-Aravaca	120,360	26,685	22.2	60.5	2.3
10. Latina	237,048	57,637	24.3	60.6	1.5
11. Carabanchel	25,5514	47,869	18.7	61.4	3.5
12. Usera	140,808	23,503	16.7	61.3	4.2
13. Puente de Vallecas	235,638	41,860	17.8	60.6	2.9
14. Moratalaz	92,390	24,370	26.4	61.5	−7.5
15. Ciudad Lineal	213,905	49,225	23.0	62.1	1.4
16. Hortaleza	195,017	36,491	18.7	58.9	−2.1
17. Villaverde	153,829	25,662	16.7	60.1	2.4
18. Villa de Vallecas	114,817	14,990	13.1	58.7	−5.7
19. Vicálvaro	79,328	10,805	13.6	58.3	5.8
20. San Blas-Canillejas	158,783	28,027	17.7	60.5	−1.1
21. Barajas	48,404	9087	18.8	57.2	−5.8
Ciudad de Madrid	*3,286,662*	*666,599*	*20.3*	*61.0*	*2.8*

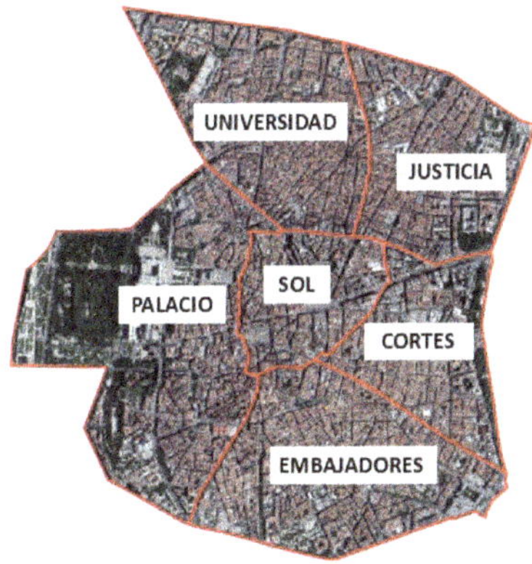

Figure 2. Division of the Distrito Centro into neighbourhoods and location of the University neighbourhood. Source: Author's adaptation of the map of the city council of madrid [45].

2.2. Methods and Data Collection

Qualitative case studies originate from the particular way of looking at the case under investigation as a whole: its context and its boundaries, with intensive analysis of the case or collective cases, and always under the conception of their idiosyncrasy and limited generalising ambition. This approach allows researchers to effectively understand how a social setting operates or functions given the extremely rich, detailed, and in-depth data which characterises the type of information gathered [46]. Within this framework and in order to capture elements of an experiential nature, the empirical approach was to use qualitative methods, mainly in-depth interviews [47] which are viewed as sub-units of the case study.

In the context of the overall project of which this research is part, 39 in-depth interviews (N = 39) were conducted with different profiles of residents in the neighbourhood, until the principle of discursive saturation was attained. An important part of these interviews (N = 18) was oriented towards gathering information about older women in the area, which opened the possibility of observing and tracing the concrete experiences of this segment of the population. As in the general project, the initial contact was made through snowball sampling.

In addition to the actors themselves (the elderly women), we considered of great relevance the inclusion of the testimony of three women below our age-scope, considering them as key informants (E11, E15, and E17). As pointed out by Valles [48] key informants are people who provide information directly relevant to the objectives of the study and who are selected because they occupy a unique position in the community, group, or institution [15: 213]. These key informants were residents in the area, women, and considered a trusted reference by the neighbours. They become an important source of information, helping to understand some of the dynamics in the area and, besides, serving as a trustworthy reference when starting more difficult interviews or treating some topics. So, for example, some elderly women interviewed were more prone to open up and share some opinions after knowing these three people were included and agreed to be part of the analysis. Regarding the profile of these key informants, one held the position of president of one of the most prestigious neighbourhood associations in the area. In a second case, the interviewee held a position of great responsibility in a music school that is deeply rooted in the area and highly valued by residents. In the third case, the key informant, in addition to having been born and lived all her life in the neighbourhood, was the owner of a well-known commercial establishment considered "traditional" that was highly appreciated by the elderly women in the area.

In the analysis of the experience of the elderly women, we selected the interviews based on two principles: that the interviewees were over 65 years of age or were approaching that age and had been living in the neighbourhood for more than 15 years. The anonymous information of the selected participants can be found in Table 2.

In all cases, the interview format was adapted to the comprehension and narrative competencies of each interviewee, with a combination of structured and semi-structured questions. These narrative competencies did not depend on the level of education, but on more individual, subjective characteristics. When necessary, we tried to follow the stimulus-response approach, reformulating the questions, pointing out recent events in the area, or connecting with the specific characteristics of the neighbourhood. This has been pointed out as the best way to approach the elderly [10], as allowed us to extract the maximum richness by adapting to the communicative style of each interviewee. To this end, the specific analysis of the universe of older women was complemented with the help of two neighbourhood centres: the Espacio Pozas (Red Cross), a social centre highly valued by the residents of the neighbourhood, and the Municipal Senior Citizens' Centre which were also the places where part of the interviews was carried out.

In the case of the key informants, it was always attempted to combine the collection of information relating to the more particular aspects of life and activities in the neighbourhood in each case, with the broader and more generalist view derived from the specific position of each of the three interviewees.

The interviews lasted between 1 and 3 h (average 1.5 h) and all of them were digitally recorded with the participants' consent.

Table 2. Qualitative interviews analysed.

Code	Age	Years Living in the Neighbourhood	Country of Origin (if Not Born in Spain)
E1	54	15	
E2	62	7	Dominican Republic
E3	68	68	
E4	62	33	
E5	65	22	
E6	77	54	
E7	70	5	Dominican Republic
E8	66	39	Colombia
E9	86	86	
E10	87	60	
E11	58	34 (Key informant)	
E12	63	16	
E13	58	33	
E14	85	16	
E15	53	Key informant	Venezuela
E16	59	30	
E17	53	53 (Key informant)	
E18	77	56	

2.3. Data Analysis

The qualitative analysis process consisted of different steps: first, transcriptions of the interviews were made, which allowed them to be put into context. During this process, the transcripts were checked against the fieldwork notes. Second, all interviews were coded and analysed using the analysis tool ATLAS.ti.

The data consist of a detailed amount of qualitative information to which ethnographic content analysis has been applied. This differs from classical content analysis in that it involves a redefinition of qualitative positions and emphasises the reflexive analysis of the documents produced after transcription, going beyond quantitative description in an attempt to delve deeper into the understanding of meanings [49]. The inclination towards content analysis, as opposed to discourse analysis, was a considered decision justified by the nuances of interviewing older people and the interests of the research, emphasising that the interpretative analysis carried out does not presuppose the need to unravel hidden structures, but rather aims to identify and categorise elements (themes, patterns, contents) and explore their connections, their regularity or rarity, and their genesis [27: 387] in line with the research context. The definition of the analysis categories can be seen in Table 3.

When considered necessary, some quantitative data was provided in order to illustrate particular findings. Sources were pointed in each case.

Table 3. Summary of the categories and subcategories obtained from the ethnographic content analysis of the responses of the older women interviewed.

	Subcategory 1.1. The real name of the neighbourhood
Category 1: Knowledge of the environment.	Subcategory 1.2. Links with the environment that transcend individual comfort
	Subcategory 1.3. The neighbourhood as a village
	Subcategory 1.4. Relationships between residents
Category 2: Experiences of the past in the neighbourhood	Subcategory 2.1. Problems and difficulties of the recent past
	Subcategory 2.2. The strength of neighbourhood relationships from decades past
Category 3: Obstacles arising from the transformation of the neighbourhood at present	Subcategory 3.1. Differential characteristics of the neighbourhood
	Subcategory 3.2. Physical impact of the transformation of the area
	Subcategory 3.3. Relational impact of the area's transformation
Category 4: Promotion of the social and associative network	Subcategory 4.1. Public spaces aimed to promote social interactions and associative networks

2.4. Ethical Aspects

Regarding the ethical aspects of the research, the subjects were assured that the information they shared would be anonymous, that their recordings would not be made available to others, and that the data would be published in such a way that they could not be identified.

3. Results

This section presents the qualitative analysis of the interviews. The first piece includes key information on the neighbourhood under study that serves as a contextual reference for the results. The analysis is articulated on the basis of the categories and subcategories previously defined as key significant elements (Table 3).

3.1. The Neighbourhood

The Universidad neighbourhood in the Centro district is an example of social change and diversity. It is a dense area, with 353 inhabitants per hectare, as opposed to 270 in the Centro district, and 55 people per hectare in the city of Madrid as a whole [45]. Around a quarter of the residents of Universidad (24.5%) are foreigners, a very similar percentage to that of the Centro district (25.8%) (Madrid City Council, 2022). With an abundant social fabric, the area is noted for its associative networks, which organise their own local festivals, run campaigns related to the public interest, and foster improvements to the area and for the residents, even articulating "neighbour-helping-neighbour" informal organisations (for example, time banks, a reciprocity-based work trading system in which hours are the currency).

Of artisan and working-class origins, the neighbourhood has experienced major social problems in the recent past, such as inadequate and unhealthy housing, deteriorated residential buildings, and, in the 1980s and 1990s, a boom in drug consumption and trafficking. More recently, in the first decade of this century, it began to experience a process of gentrification, and came to be viewed as an underground and bohemian neighbourhood, attracting artists and other creative groups [39,50]. Thus, it is now home to a mix of former residents—of modest socio-economic status—and more affluent, more highly educated newcomers (artists, students, and professionals). The incorporation of new residents and the conversion of the neighbourhood into a hub for nightlife and alternative leisure has led to a rapid increase in rent and property values[2], while the housing stock has undergone significant improvements, largely thanks to public investment in refurbishment. From a morphological point of view, the growth of the neighbourhood has been shaping a compact structure, with an abundance of small commercial premises on the ground floors of buildings and streets that lead to squares, which are the most significant spaces in terms of socialising, leisure, strolling, and play [32].

As explained in the Methodology section, the analysis is articulated around three explanatory axes each of which is broken down into subcategories that exemplify the different aspects that the women highlighted in the interviews: (i) The importance of knowledge of the environment and the desire to remain in the known space; (ii) the construction of the neighbourhood as a confluence of several pasts; and (iii) in contrast to the desire to grow old in the place, the difficulties associated with change, neighbourhood rotation and appropriation of public space.

3.2. Knowledge of the Environment

The importance of familiarity with the space in which older women live is a recurring theme in the interviews. They want to continue living in the space to which they feel attached, a feeling that tends to increase the longer they live in the neighbourhood. In a way, they feel it is theirs, which manifests itself in different ways.

3.2.1. The Real Name of the Neighbourhood

The first manifestation of the importance of the neighbourhood's history is in how the interviewees refer to it by its traditional name. Even though since 1845 its official administrative name has been Universidad, it is popularly known as Malasaña, a name that especially spread from the 1980s onwards due to the so-called *movida madrileña* [3], which was very much focused around this area. However, its historical name, which is the one used by some of the interviewees, was "Barrio Maravillas", named after Las Maravillas Monastery, which housed a Virgin of the same name.

> *"Malasaña no, Maravillas... I don't know why, I don't like the name Malasaña" (E6, 77 years).*

> *"I don't like the fact that they call it Malasaña. For me... Maravillas" (E8, 66 years).*

3.2.2. Links with the Environment That Transcend Individual Comfort

The interviews show that the sense of neighbourhood is far removed from administrative delimitations and official nomenclature, and is instead constructed throughout the lifecycle, in relation to experiences. Most of the older women emphasise the comfort that the neighbourhood offers them, despite the difficulties accessing their own homes and other urban planning problems (absence of lifts and narrow pavements, as will be seen below). It is a symbolic comfort that is closely related to the space they are familiar with and the ties they have forged over time.

> *"I wouldn't change it for anything, this neighbourhood. Because I like it. It's a neighbourhood that I've always liked (E10, 87 years old).*

> *"Look, I have a flat in another neighbourhood. It's nicer than this flat because this one is on the fourth floor without a lift. You get the idea, 90 steps. Well, I don't know. I go out, I think... the stones know me, I go out and... It's not that I'm bad in the head, no. It's just that I like the neighbourhood more than the other place. It's just that I like the neighbourhood more than anything else" (E9, 86 years old).*

3.2.3. The Neighbourhood as a Village

The women translate the feeling of attachment to the neighbourhood in a very expressive way by referring to it as a village. In their minds, there is a strong parallelism between the relationships within this neighbourhood and the idealisms that are more typically associated with villages or smaller communities and never imagined in the central area of a large city:

> *"To me my neighbourhood, because it is my neighbourhood, seems like a village. I go down to the street, you greet people, you drink your coffee or have a beer, you greet someone*

*on one side or someone on the other side, someone says: "Let's see if you can help me
with this", or "Hey, what do you know about this?" In other words, it's very family-like,
very family-like. (...) if you go somewhere twice to have a coffee, the third time they ask:
"How are you?" I mean, it seems to me... a luxury. A luxury, because if you go three days
without being seen somewhere, the fifth time they ask you: "Did something happen to
you?" And I think that's very important because that's human relations (...) (E11, 58
years old).*

*"I'm very happy with the neighbourhood, I love it (...) It doesn't feel like a neighbourhood,
it feels like a village" (E13, 58 years old).*

3.2.4. Relationships between Residents

In the same sense, the interviewees constantly refer to the interweaving of local links
and ties between residents, to the feeling of being recognised and getting to know others
when one goes out into the street through interactions that may not be very profound, but
do have strong symbolism and an impact on everyday life:

*"You go to the bakery and one day you don't, but on the tenth, eh, well already, whether
you like it or not, well you establish a relationship with the person who's selling, or with
the owner or... I don't know. Well, for example, then in Espíritu Santo, I've been buying
bread there for a hundred years, so... with the owner, the people who work there, well, you
create a tie. Not a friendship like that, but "how are you?", in other words, normal things,
I think" (E1, 54 years old).*

Beyond the apparent laxity that seems to characterise these social relations, they can
become powerful support networks in case of need. Neighbourhood solidarity manifests
itself in very different ways:

*"In fact, now that I've had a knee problem, three people who didn't know me at all offered
me crutches on the way to the three shops: three people who didn't know me at all. Being
from the neighbourhood (...) It's a very... special place. Yes, I went to the churreria[4] (
...) and there was an old lady who said "Oh, well, I've got some [crutches]" (...) And
then I went to the market and again [a stranger offered her crutches], and at the drugstore
as well. So, people are very supportive in that sense" (E1, 54 years old).*

*"I was living with my brother. My brother had mental issues. My brother burnt down my
flat... my brother. And everybody helped me.... Everybody helped me. They brought me
food; they went to collect my clothes to wash them. Everybody helped me (she emphasises).
Up to here. The depression I had, they found out about it here [senior citizens' centre]
and they brought me here and my depression went away. Because everyone helped me...
So, I tell you that there are very kind-hearted people here. Everyone helped me, everyone,
my neighbours... everyone helped me, I'm serious. And you can see that it was a tough
problem, but for me, I didn't have any problem being alone or anything, not at all. And
I'm telling you, even the "queers" living above gave us clothes and everything to dress
up in (she emphasises), that's all I'm saying. Everyone turned out in force! Because I was
left with nothing, of course" (E13, 58 years old).*

This solidarity and support in cases of need do not necessarily mean that the neighbourhood is the scene of intense friendships among older women, but rather that informal,
everyday processes seem to predominate, sometimes leading to friendship but generally of
a laxer and elastic nature [51]. However, practices clearly associated with trust can be detected, with older women frequently exchanging flat keys, which implies the psychological
security that someone will be able to get in should there be an emergency:

*"My neighbour Paquita has them [home keys], yes, and now I have hers" (E9, 86
years old).*

These behaviours would corroborate the importance of the general climate of trust which,
from Jacobs' (1961) perspective, characterises coexistence in the city and which is made up

of "many and very light contacts established on its pavements", most of them apparently trivial, but whose result is "a sense of public identity between people, a network and fabric of mutual (public) respect and trust, and also a guarantee of mutual assistance in case the neighbourhood needs it, the neighbourhood in general or one particular local resident".

3.3. Experiences of the Past in the Neighbourhood

The strength of past experiences helps to understand the feelings that the interviewees express about their involvement in the neighbourhood. The relationships and attitudes of the older women are the result of previous experiences and derive from trajectories that are linked to the neighbourhood's own history. As we will see below, they are experiences of opposite nature that nevertheless form an inseparable part of women's memories.

3.3.1. Problems and Difficulties of the Recent Past

The presence of important social problems such as drug trafficking and consumption emerged in the neighbourhood in the late 80s and early 90s. It was a time when drug use spread in the cities, manifesting itself in a large number of neighbourhoods and causing multiple problems of all kinds, including the death of many young people who were unable to overcome their addiction. The central location of the neighbourhood under study also favoured the presence of drug dealers and consumers. This problem is repeatedly mentioned in older women's accounts, as well as the strategies that sought to minimise its impact on the daily life of the local community, especially children:

"Well, look, when I came here, things were a bit bad because there were a lot of drugs. It was a bit... to go out with the kids and stuff... the square was a bit difficult, because there was a lot of... a lot of drug dealing. At night you went out feeling very scared because they were on all the corners and all the poor people were lying around in the corners. Now that doesn't happen anymore. You used to be sleeping at night and you'd hear: "help!" because they'd mugged one guy, they'd mugged another. And that doesn't happen anymore... the neighbourhood has got much better now. When I came to live here it was a bit terrible, it was a bit... scary" (E13, 58 years old).

"In the past it was a problem. It bothered us because they [drug addicts] used to climb up the stairs, into our building. They came in a lot, you know? And they [police and emergency health workers] took more than one out... they took one away half dead. Because they used to climb up to the top floor where I live, and of course, you could see that they were pricking themselves and... that's a time when I was quite... yes" (E3, 68 years old).

"You see that you can't take them [the children] to the parks in the neighbourhood because they are full of syringes—that's the first thing my children learnt: you mustn't touch a syringe; if you see something red, don't touch it, because it's blood. In other words, very heavy situations... The junkie problem ended, it ended, but they were very dark times for this neighbourhood... A guy with a chainsaw chasing another guy (...) Well, there were quite a few stories... Drug dealing, drug trafficking. And a lot of people from this neighbourhood died [because of drugs], kids you knew..." (E11, 58 years old).

3.3.2. The Strength of Neighbourhood Relationships from Decades Past

The harsh images of the past are intertwined in the older women's memories with those of an earlier time when neighbourhood ties were considered much stronger and more entrenched than they are today, and when convivial relations within the neighbourhood communities were much more intense. The older women recall the custom, still present today in small Spanish villages, of taking the chairs out "al fresco" (in the cool) in summer at night to avoid the heat inside their flats and to interact with their neighbours.

"As for neighbourliness, I can tell you, for example, in my building we were all... well, families before. And also, very close families, that's the truth, which is not the case now" (E3, 68 years old).

"I liked the way it used to be more than now. It was more familiar. I liked it better before (...). Because we were all, as they say, as if we were family. People stayed in the neighbourhoods longer" (E9, 86 years old).

"Even me, when I came to live there, my house was never locked at any time. You pulled a cord and opened the door and whoever needed something came in (...) there was a very good relationship. Even when you start to... when you're working, when you don't have anyone to leave [your kids] with in the afternoon and so on, they [neighbours] took care of them. And... There were a lot of relationships, not now because of the rotation of people" (E11, 58 years old).

This led to a strengthening of neighbourhood relations and a greater sense of community and trust, which seems to have suffered in recent times due to the continuous turnover of residents (tenants, who do not set foot in the neighbourhood and are less interested in sharing their daily lives) but also due to the reconceptualisation of public space by the municipal administration itself.

3.4. Obstacles Arising from the Transformation of the Neighbourhood at Present

As we have seen above, interviewees like the neighbourhood. They like it to the extent that they compare it to other areas of Madrid that may offer alternative amenities and advantages but lack the vibrancy and attractiveness of Universidad. However, the impact of the transformations experienced by the area ends up becoming obstacles of a different nature (physical and relational) to the permanence of the older residents.

3.4.1. Differential Characteristics of the Neighbourhood

Relatively frequently, the women interviewed draw contrasts between the vitality of their neighbourhood and other areas of Madrid that they feel are far removed from their daily customs and practices. From this perspective, they often compare the possibilities for interaction in their area with the restricted and limited nature that characterises other neighbourhoods of the city.

"Here you always saw people in the street. You always saw people. And I like meeting people. And not in a lonely neighbourhood where you can get scared" (E18, 77 years old).

"It's... a neighbourhood like my neighbourhood was when I was little. Well, the people, um, they all know each other, they talk to each other from the windows. I talk to my neighbour from the window, she's an old lady.... It's a different life to the neighbourhoods now. I, for example, have gone to visit my old neighbourhood and (...) it's sad, you don't see anyone, everyone is stuck in their seat... in their own house. It's very sad, and here everyone is in the street, everyone is socialising, everyone is... do you understand? That... we all know each other" (E13, 58 years old).

3.4.2. Physical Impact of the Area's Transformation

It seems that in the face of the social vitality, municipal policies have encouraged the privatisation and invasion of public space, which tends to eliminate interactions that are not subject to economic exchange:

"Well, I don't know if it's because the Mayor has found it more convenient to permit outdoor bar terraces, because there are taxes that have to be paid. And then, the children have little space to play in" (E18, 77 years old).

In fact, the ratio between benches in the public space and outdoor terraces (private) of the Centro district is one of the lowest in the city in Madrid (Figure 3), which perverts the use of public space, occupying the available space for commercial and economic purposes. At this point, it is worth recalling the importance of the benches in relation to the local daily life trajectories of the older inhabitants of the neighbourhood.

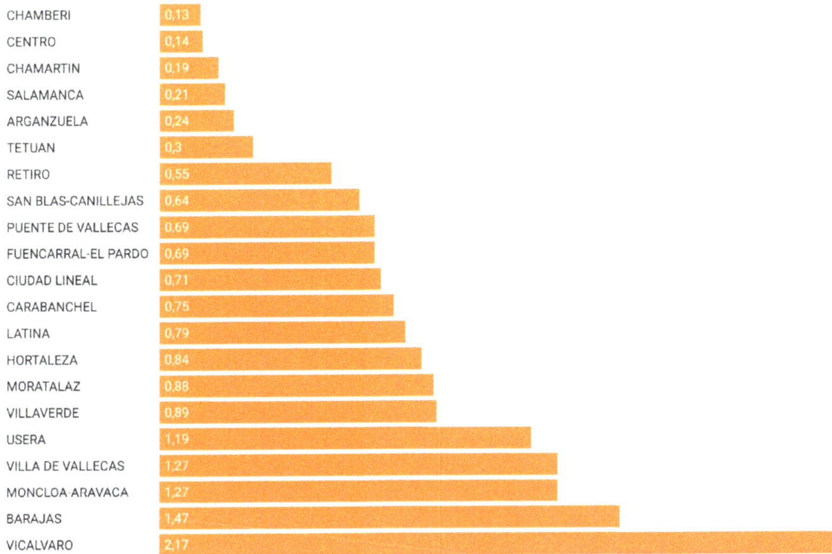

Figure 3. Ratio of benches per terrace seat per district in Madrid, 2022. Source: Madrid City Council, 2022.

The perception of neglect by the municipal administration among local residents is particularly noticeable in terms of cleanliness and upkeep of the district. Being a central area with narrow streets, graffiti (and, in its less artistic manifestation, tagging) floods its walls, disfiguring the facades of traditional buildings and proving very costly to remove:

"Graffiti is horrible. They have ruined the streets. I'm telling you the truth. You see a street that has just been painted... There's the ice factory, which is now a residential building. What they paint it with...!" (E9, 86 years old).

"They really should have punished those who do it [graffiti]. It's a shame, because look, first of all, not respecting, because if you pay for your façade, it hurts your soul the next day when you find it... That's not respecting" (E13, 58 years old).

The neighbourhood grew throughout the 19th century and was consolidated in the 20th, and its density and physical structure play an important role in the lives of the interviewed women. Its squares are a fundamental space for socialising and coexistence, and are recognised as such [32], despite the fact that the urban furniture is often unsatisfactory, as we have seen concerning the benches, and sometimes seems to be designed to be inhospitable, thus dissuading people from using it:

"Squares are fundamental. And the squares and whatever is put in those squares, because it is not the same to provide a bench without a backrest, which isn't inviting to sit on, as it is to have a bench with a backrest, so you can get sun in winter and shade in summer" (E11, 58 years old).

On the other hand, in addition to the relatively frequent absence of lifts in buildings, there is also a lack of universal accessibility, with small pavements that are hard for more than one person to walk on at a time and prevent residents with mobility problems from getting around.

"At the same time, it is a difficult neighbourhood, because there are a lot of buildings without lifts, with sloping streets" (E15, 53 years old).

"The design, the layout of very narrow sidewalks. There are people who use wheelchairs, that's obvious and we can't deny it. And this neighbourhood, for disabled people... disabled

people and even mums with pushchairs, it has many barriers, it has many barriers" (E11, 58 years old).

"I don't like the streets being so narrow. I don't like them so narrow. Especially the pavement. In order for a person with a shopping trolley to pass, you have to stick very close to the wall, that's why there are a lot of arguments between people, a lot of them. Because there are people who have no patience (...) I don't like that, for them to be so narrow" (E16, 59 years old).

3.4.3. Relational Impact of the Area's Transformation

Beyond the problems mentioned above, the process of gentrification and touristification that the central area of Madrid in general and Universidad (Malasaña) in particular are experiencing represents a particularly alarming threat to the social lives of older women in the neighbourhood. In addition to the mobility resulting from the permanent rotation of tenants who live in the area for shorter and shorter periods of time, there is also the impact on local shops, a resource that older women identify with, but which is gradually disappearing. Small neighbourhood shops have not only covered the residents' needs but have also been a fundamental axis of social relations.

"We are 15 people in our building, and only 2 or 3 of us are permanent... They are students who are just passing through... during the year, the people in flats might change twice. So...". (E5, 65 years old).

"It's changed because now the flats are rented and that's it, they're different. Do you understand? Well, maybe they are foreigners who come and stay for a year or half a year. They leave. Others come. And so..." (E10, 87 years old).

"It makes me sad that the shops that used to be there are no longer there (...) It was like a family relationship" (E18, 77 years old).

"Shops have changed a lot. A few years ago, shops were... I had a shop next to my building, and well, it was like my mother and father. I would take my eldest to school and leave the little one with them when it was cold. I would say: "Go on, stay with him". There was also a churreria in Calle Escorial that also did the same for you. I'd say: "Go on, stay with the boy, I'm going to take the other one" and it might be raining, I'd call the shop assistant: "Hey, come upstairs for a moment, (...) I'm going to take the boy to school" and the girl would go upstairs. It was a grocer's store, but they closed down. I miss the shops a lot, because it was... trust... there's no more, there's no more (...). The shops are not like they used to be. And I do miss that a lot" (E13, 58 years old).

The accelerated process of change imposed by these recent dynamics has obvious consequences, from the difficulty obtaining everyday products (with the effects that longer distances and the need to travel have on older people and those with mobility difficulties) to the potential loss of neighbourhood identity. The occupation of these local shops by bars, restaurants, and specialised shops has altered the type of activity in the neighbourhood and does not meet the sociability needs of these older women residents.

3.5. Promotion of the Social and Associative Network

At the same time as they talk about the problems resulting from the transformation of the neighbourhood, the older women convey very positive feelings when they refer to some of the community centres that offer a wide range of activities and become real meeting, exchange, and socialisation spaces.

Public Spaces Aimed to Promote Social Interactions and Associative Network

Centres such as Espacio Pozas (a public centre managed by Red Cross) and the Municipal Centre for the Elderly are cited by the interviewees as opportunities to maintain and reinforce the community spirit by offering spaces for the elderly to forge relationships, in the same sense that Bosch-Farré et al. [34] point out as places that create community.

"This year I found this centre, which is wonderful (Espacio Pozas)" (E5, 65 years old).

"Yes, for example, Espacio Pozas seemed to me to be an extraordinary place to be able to relate to different cultures, and to share, that seems to me to be very, very, very important. And to admit that there are people who are different. I mean, I think that's very, very important" (E6, 77 years old).

The neighbourhood is changing and so are the forms of relationships, although certain venues are resisting where the importance of social relations for community well-being is emphasized.

4. Discussion

Based on the conception of ageing as a complex and heterogeneous process, the entire body of research currently encompassed under the concept of ageing in place, as we have been pointing out, coincides in defending the need for older people to continue in the area in which they reside, are familiar with, and feel recognised in, for as long as possible. If the physical environment affects people throughout their lives [15,16], its influence is even greater in old age, conditioning the ageing process and even the response to illness [18].

Most of the older women in the Universidad neighbourhood coincide in highlighting the harmony and affinity they feel towards the social environment in which they have spent a significant part of their lives. They generally value the characteristics of an urban fabric in which squares and public spaces are fundamental elements of everyday relationships. This, however, conflicts with the lack of an age perspective in the configuration of space, as can be seen, for example, in the aforementioned small size of sidewalks which, in addition to prioritising the use of private cars, hinders pedestrian use. This creates barriers not only for older people with mobility problems, but also for care in public spaces and social interactions. On the other hand, the lack of accessibility in residential buildings themselves and in the immediate surroundings, as already demonstrated in previous research [11,20,52,53], hinders not only social relations in old age, but also the development of everyday life itself.

Without idealising a past in which negative memories are interwoven with very positive ones, the older women seem to enjoy the loose configuration of neighbourly relations that evokes Granovetter's conceptualisation of weak ties [38] and connects with the atmosphere of local comfort described by Blokland and Nast [36]. However, these low-intensity relationships prove strong enough to provide support in times of need, when the circumstances demand it, in the sense that Torres [51] attributes to "elastic ties", which can be activated in certain situations.

The research shows, however, that the process of change to which the neighbourhood is currently being subjected—in terms of gentrification and touristification—has a decisive influence on the daily lives of older women, who are helpless witnesses to the transformation of their environment and their potential relationships.

Gentrification and touristification, as well as a greater inclination on the part of the municipal public administration to make the urban space economically profitable (by granting excessive space to bars and outdoor cafés, for example) reinforce the use of the area as a platform for night-time leisure, privatise the use of public space (thus displacing uses and users) and have repercussions on the cleanliness of the neighbourhood. However, the main threat to the local social fabric, according to the older women, comes from the high turnover of residents (who do not take root in the neighbourhood and who will be succeeded by new arrivals in a short period of time), as well as the gradual disappearance of local commerce, a key element in generating a sense of belonging and attachment to the area. The limited permanence in the neighbourhood of the new residents generates distance between them and the older women, although this segregation is in no way equivalent to intergenerational confrontation, according to the interviewees themselves.

Along the same lines, the change in the population has had an impact on the neighbourhood's small shops, which have shifted towards catering to the demands of gentrifiers and tourists, ignoring the daily needs of older residents (making it even harder for them to remain in the neighbourhood) and causing the relationships between residents and shopkeepers, built up over time, to disappear. This affects not only the intensity of coexistence, which is transcendental when it comes to forging ties and building a neighbourhood, but also removes a potential resource for detecting loneliness or vulnerability. This shows that the transformation is not only physical and economic, but also cultural, relational, and symbolic [52].

In the light of international standards and commitments, these outcomes are relevant in the context of the role of age and especially urban ageing in achieving the SDGs, which require appropriate environments to ensure the health and well-being of older people. From this perspective, it is a priority for older people to feel a sense of ownership of the space in which they live to prevent them from becoming prisoners in their own homes [15] and experiencing old age as something negative. The potential lack of integration of older people in the urban space is therefore not the result of individual choice (theory of disengagement), or even the loss of physical capacity and motor skills but is largely a consequence of the inadequate configuration of that space to accommodate the different needs not only of use, but of users [20].

Appropriate measures, therefore, need to be implemented to minimise the impact of gentrification and resident turnover. This includes all the side effects that they entail in local environments such as the one investigated, in which attachment to the place and social relationships with fellow residents play a transcendental role [14] in the daily lives of the elderly, as opposed to the likelihood of experiencing exclusion that some authors link to the ageing process [53,54].

5. Conclusions

The increase in longevity and the growth of cities in Spain is leading us towards a very specific future: most of the elderly of the future will be urban. For older people, staying in a familiar place has many positive effects and can be the key to better health and greater social integration. It also satisfies a desire to remain part of a community that knows them and of which they feel part, and in which neighbourhood ties, a source of solidarity and sociability, are an essential aspect. However, for all these benefits to be possible, there need to be spaces that favour these neighbourly relations and foster intergenerational relations and the opportunity to generate ties with the new neighbours who are arriving in the neighbourhood.

The case study carried out in Universidad (Malasaña), a neighbourhood in central Madrid, reveals the importance of the local space and the relationships and experiences of the older women who live there, the strength of past collective experience in shaping the present and the trends that have been threatening to become established in the area for some time, with the consequent weakening of older women's social relationships, of their familiarity with the space and therefore, of their well-being and health.

Furthermore, and contrary to the idea put forward by certain sectors that the city is not a suitable place to grow old, the women interviewed show a strong attachment to their local space. They have been defending the construction of a neighbourhood and community over many years, which is not necessarily characterised by deep friendships, but by the existence of a social network that responds in case of need.

The research, however, includes a number of limitations. As explained above, case studies allow for an in-depth exploration of the object of research and provide detailed and nuanced information. On the downside, however, there is a limited generalisation of the results obtained. Even so, and with due caution, the findings could be extrapolated to other central areas of historic cities with a certain community tradition but currently under stress from gentrification processes in which there is a high rate of population turnover accompanied by the progressive disappearance of local commerce. In future research, the authors plan to carry out the empirical work necessary to update the research, focusing exclusively on older women. This will make it possible not only to check whether the trends detected in the period covered by this article have continued subsequently, but also the effect of the COVID-19 pandemic on the daily life and intensity of women's relationships in the neighbourhood.

In the future, it would be useful to extend this research by comparing Universidad (Malasaña) with other areas of Madrid or with neighbourhoods in other cities that are undergoing the same process. On the other hand, it should be noted that urban ageing from a women's perspective is a highly relevant area of research that still needs to be deepened and clarified in some aspects.

Author Contributions: Conceptualization, M.V.G.; methodology, M.V.G. and I.L.; formal analysis, M.V.G. and I.L.; investigation, M.V.G. and I.L.; resources, M.V.G. and I.L.; data curation, M.V.G. and I.L.; writing—original draft preparation, M.V.G. and I.L.; writing—review and editing, M.V.G. and I.L.; visualization, M.V.G. and I.L.; supervision, M.V.G. and I.L. project administration, M.V.G. and I.L. All authors have read and agreed to the published version of the manuscript.

Funding: This research received no external funding.

Institutional Review Board Statement: The study was conducted in accordance with the Declaration of Helsinki.

Informed Consent Statement: Informed consent was obtained from all subjects involved in the study.

Data Availability Statement: Not applicable.

Acknowledgments: The authors would like to thank the residents of the Universidad (Malasaña) neighbourhood for their extraordinary generosity and the time they have devoted to the research. The author would like to thank also to José Ariza de la Cruz, author of Figure 3.

Conflicts of Interest: The authors declare no conflict of interest.

Notes

[1] This is the number of immigrants (people moving into the district) minus the number of emigrants (people moving out of the district) in the previous year, divided by the person-years lived by the population of the receiving district over that period.

[2] According to Madrid City Council, the price per square metre of second-hand housing was 5282 euros, with the city average being 3732 euros per square metre (Source: data bank, Madrid City Council. Viewed in May 2021).

[3] La Movida was a social, artistic, cultural, and transgressive movement, led by young people, in which music played a fundamental role. It paved the way towards ideological and sexual freedom, and drug consumption featured prominently. It happened during the Spanish transition to democracy after the death of the dictator Franco in 1975.

[4] Shop or stall selling churros: long fritters made with flour and water.

References

1. Smedley, T. Are Urban Environments Best for an Ageing Population? The Guard. 2012. Available online: https://www.theguardian.com/sustainable-business/blog/urban-environments-ageing-population-design. (accessed on 10 June 2022).
2. Gómez, M.V.; Kuronen, M. Interpreting Vulnerabilities Facing Women in Urban Life: A Case Study in Madrid, Spain. In *Women, Vulnerabilities and Welfare Service Systems*; Kuronen, M., Virokannas, E., Salovaara, U., Eds.; Routledge: Oxfordshire, UK, 2020; pp. 39–52, ISBN 9780429276910.
3. Lebrusán, I.; Toutouh, J. Car Restriction Policies for Better Urban Health: A Low Emission Zone in Madrid, Spain. *Air Qual. Atmos. Health* **2021**, *14*, 333–342. [CrossRef]
4. Lebrusán, I.; Toutouh, J. ¿Son Efectivas Las Zonas de Bajas Emisiones? El Caso Del Centro de Madrid. *Papeles Econ. Española* **2022**, *171*, 159–173.
5. Ascher, F. *Los Nuevos Principios Del Urbanismo. El Fin de Las Ciudades No Está a La Orden Del Día*; Alianza Editorial: Madrid, Spain, 2007; ISBN 978-84-206-4198-0.
6. Bauman, Z. *Vida Líquida*; Paidós: Barcelona, Spain, 2006; ISBN 9788449324543.
7. Garay Villegas, S.; Montes de Oca Zavala, V. La Vejez En México: Una Mirada General Sobre La Situación Socioeconómica y Familiar de Los Hombres y Mujeres Adultos Mayores. *Univ. Autónoma Nuevo León* **2011**, *13*, 143–165.
8. Pelaez, E.; Molinatti, F. Evolución de La Segregación Residencial y Condiciones de Habitabilidad de Las Viviendas de Los Adultos Mayores En La Ciudad de Córdoba, Argentina. *Márgenes. Espac. Arte Y Soc.* **2016**, *9*, 7. [CrossRef]
9. Nordin, N.; Nakamura, H. The Influence of the Objective and Subjective Physical Neighbourhood Environment on the Physical Activity of Older Adults: A Case Study in the Malaysian Neighbourhoods of Johor Bahru. *Sustainability* **2020**, *12*, 1760. [CrossRef]
10. Lebrusán Murillo, I. La Vivienda En La Vejez: Problemas y Estrategias Para Envejecer En Sociedad. Doctoral Dissertation, Universidad Complutense de Madrid, Madrid, Spain, 2017.

11. Lebrusán Murillo, I. *La Vivienda En La Vejez*; Colección Politeya; Editorial CSIC Consejo Superior de Investigaciones Científicas: Madrid, Spain, 2019; ISBN 9788400105471.

12. Rojo-Pérez, F.; Fernández-Mayoralas, G.; Rodríguez, V.; Lardiés-Bosque, R.; Prieto-Flores, M.E.; Gallardo-Peralta, L.P.; Molina-Martçinez, M.A.; Rodriguez-Blazquez, C.; Forjaz, M.J.; Schettini, R. Contextos Residenciales, Envejecimiento Activo y Calidad de Vida. Un Análisis a Microescala En España. In *Población y territorio. España tras la crisis de 2008*; Sempere Souvannavong, J.D., Cortés Samper, C., Ernesto Cutillas, O., Valero Escandell, J.R., Eds.; Editorial Comares: Granada, Spain, 2020; pp. 191–208.

13. Zorrilla Muñoz, V.; Agulló Tomás, M.S.; García Sedano, T. La Vivienda y Su Entorno Social: Análisis Cuantitativo Desde Las Personas Mayores de 50 Años. *Reis Rev. Española Investig. Sociol.* **2020**, *170*, 137–153.

14. Herbert, A. What Role Does Rural Place Play in the Lives of Mid-Life Women in Sweden and Ireland? *Societies* **2020**, *10*, 84. [CrossRef]

15. Phillipson, C. Developing Age-Friendly Communities: New Approaches to Growing Old in Urban Environments. In *Handbook of the Sociology of Aging*; Angel, J., Settersten, R., Eds.; Springer Nature: New York, NY, USA, 2011; pp. 279–296.

16. Wahl, H.W.; Weisman, G.D. Environmental Gerontology at the Beginning of the New Millennium: Reflections on Its Historical, Empirical, and Theoretical Development. *Gerontologist* **2003**, *43*, 616–627. [CrossRef]

17. Ortega-Gil, M.; Mata García, A.; ElHichou-Ahmed, C. The Effect of Ageing, Gender and Environmental Problems in Subjective Well-Being. *Land* **2021**, *10*, 1314. [CrossRef]

18. World Health Organization Active Ageing: A Policy Framework. Available online: https://apps.who.int/iris/handle/10665/67215 (accessed on 7 June 2022).

19. Pozo, E.; Gómez, M.V.; Higueras, E. Las Zonas Verdes y La Población Mayor En Madrid: Bienestar, Salud Mental e Inclusión. In *Resiliencia: Espacios de adaptación de nuestras ciudades a los nuevos retos urbanos*; Hernández-Aja, A., Viedma, A., Díez, A., Eds.; Universidad Politécnica de Madrid: Madrid, Spain, 2021; pp. 19–36. ISBN 978-84-9728-596-4.

20. Lebrusán, I. Personas Mayores En Situación de Riesgo Residencial Extremo En España: Un Diagnóstico Regional. *Rev. EURE - Rev. Estud. Urbano Reg.* **2022**, *48*, 145. [CrossRef]

21. Solanes Mullor, J. The Right to Housing and the Protection of Family Life and Vulnerable Groups: European Judicial Activism. In *The Right to Family Life in the European Union*; Rouletdge: Oxfordshire, UK, 2016; pp. 228–245.

22. Agulló-Tomás, M.S.; Zorrilla- Muñoz, V.; Gómez-García, M.V. Aproximación Socio-Espacial Al Envejecimiento y a Los Programas Para Cuidadoras/Es de Mayores. *Rev. INFAD Psicol. Int. J. Dev. Educ. Psychol.* **2019**, *2*, 211–228. [CrossRef]

23. Zhang, Q.; Yung, E.H.K.; Chan, E.H.W. Meshing Sustainability with Satisfaction: An Investigation of Residents' Perceptions in Three Different Neighbourhoods in Chengdu, China. *Land* **2021**, *10*, 1280. [CrossRef]

24. Lebrusán, I. No Es Ciudad Para Viejos?: Los ODS y La Experiencia de Envejecer En Las Ciudades Españolas. In *Políticas Urbanas y localización de los Objetivos de Desarrollo Sostenible: Teoría y Práctica*; González Medina, M., De Gregorio Hurtado, S., Eds.; Tirant lo Blanch: Valencia, Spain, 2021; pp. 359–383. ISBN 978-84-1397-235-0.

25. Lebrusán, I.; Toutouh, J. Smart City Tools to Evaluate Age-Healthy Environments. *Commun. Comput. Inf. Sci.* **2021**, *1359*, 285–301. [CrossRef]

26. McDowell, L. Towards an Understanding of the Gender Division of Urban Space: *Environ. Plan. D Soc. Sp.* **1983**, *1*, 59–72. [CrossRef]

27. Durán, M.-A. La Ciudad Compartida: Urbanismo y Movimientos Sociales. In *Género y política urbana. Arquitectura y urbanismo desde la perspectiva de género*; Serrano, B., Mateo, C., Rubio, A., Eds.; Instituto Valenciano De La Edificación: Valencia, Spain, 2017; pp. 33–58.

28. Sánchez de Madariaga, I. From Women in Transport to Gender in Transport: Challenging Conceptual Frameworks for Improved Policymaking. *J. Int. Aff.* **2013**, *67*, 43–65.

29. Álvarez, N. El Espacio Urbano Como Condición Social. La Experiencia de La Mujer En La Ciudad Contemporánea. In *Género y política urbana. Arquitectura y urbanismo desde la perspectiva de género*; Serrano, B., Mateo, C., Rubio, A., Eds.; Instituto Valenciano De La Edificación: Valencia, Spain, 2017; pp. 159–176.

30. Gómez, M.V. Una Mirada Sociológica a La Movilidad Sostenible: Género y Exclusión Social. In *Movilidad urbana sostenible y acción administrativa: Perspectiva social, estrategias jurídicas y políticas públicas de movilidad en el medio urbano*; Antonio Fortes Martín (dir.), Ed.; Aranzadi Thomson Reuters: Cizur Menor, Spain, 2019; pp. 497–523, ISBN 978-84-1309-839-5.

31. Red2Red. *La Evaluación de Impacto En Función Del Género En El Transporte y La Movilidad*; Emakunde: Instituto Vasco de la Mujer: Vitoria-Gasteiz, Spain, 2013.

32. Kuurne (Ketokivi), K.; Gómez, M.V. Feeling at Home in the Neighborhood: Belonging, the House and the Plaza in Helsinki and Madrid. *City Community* **2019**, *18*, 213–237. [CrossRef]

33. Tobío, C. Estructura Urbana, Movilidad y Género En La Ciudad Moderna. In Proceedings of the Conferencia en la Escuela de Verano Jaime Vera, Galapagar, Madrid, Spain, June 1995.

34. Bosch-Farré, C.; Malagón-Aguilera, M.C.; Ballester-Ferrando, D.; Bertran-Noguer, C.; Bonmatí-Tomàs, A.; Gelabert-Vilella, S.; Juvinyà-Canal, D. Healthy Ageing in Place: Enablers and Barriers from the Perspective of the Elderly. A Qualitative Study. *Int. J. Environ. Res. Public Health* **2020**, *17*, 1–23. [CrossRef]

35. Gómez García, M.V.; Alvarez Dorronsoro, J. *El Cambio Social En La Era de La Incertidumbre: Una Reflexión Sobre Teoría Social*; Talasa: Madrid, Spain, 2013; ISBN 978-84-96266-44-5.

36. Blokland, T.; Nast, J. From Public Familiarity to Comfort Zone: The Relevance of Absent Ties for Belonging in Berlin's Mixed Neighbourhoods. *Int. J. Urban Reg. Res.* **2014**, *38*, 1142–1159. [CrossRef]
37. Brey, E.; Gómez, M.V.; Domínguez, M. El Rol de Las Redes de Apoyo En Mujeres de Barrios Vulnerables de La Comunidad de Madrid. *Rev. Española Sociol..* forthcoming..
38. Granovetter, M.S. The Strength of Weak Ties. *Am. J. Sociol.* **1973**, *25*, 1360–1380. [CrossRef]
39. Zukin, S. *Naked City: The Death and Life of Authentic Urban Places*; Oxford University Press: New York, NY, USA, 2010.
40. Kamalipour, H.; Yeganeh, A.J.; Alalhesabi, M. Predictors of Place Attachment in Urban Residential Environments: A Residential Complex Case Study. *Procedia - Soc. Behav. Sci.* **2012**, *35*, 459–467. [CrossRef]
41. Yin, R.K. Research Design Issues in Using the Case Study Method to Study Management Information Systems. *he Inf. Syst. Res. Chall. Qual. Res. Methods* **1989**, *1*, 1–6.
42. Feagin, J.R.; Orum, A.M.; Sjoberg, G. *A Case for the Case Study*; UNC Press Books: Chapel Hill, NC, USA, 1991.
43. Hernández Sampieri, R.; Fernández Collado, C.; Baptista Lucio, P. *Metodología de La Investigación*; McGraw-Hill Interamericana: Mexico, Mexico, 2014.
44. Yin, R.K. *Case Study Methodology. Case Study Research Design and Methods*; Sage Publications: Thousand Oaks, CA, USA, 2003; ISBN 076192552X.
45. Ayuntamiento de Madrid Subdirección General de Estadística. Available online: https://www.madrid.es/portales/munimadrid/es/Inicio/El-Ayuntamiento/Estadistica/Areas-de-informacion-estadistica/Demografia-y-poblacion/Cifras-de-poblacion-y-censos-demograficos-/Padron-Municipal-de-Habitantes-%28Explotacion-Estadistica%29?vgnextfmt=detNa (accessed on 17 July 2022).
46. Berg, B.L. *Qualitative Research Methods for the Social Sciences*, 3rd ed.; Allyn and Bacon: Needham Heighs, Massachussetts, 1997.
47. Valles, M.S. *Entrevistas Cualitativas*; Centro de Investigaciones Sociológicas: Madrid, Spain, 2002; ISBN 9788474763423.
48. Valles, M.S. *Técnicas Cualitativas de Investigación Social. Reflexión Metodológica y Práctica Profesional*; Editorial Síntesis: Madrid, Spain, 1999; ISBN 84-773-449-5.
49. Cea D´Ancona, M.A. *Metodología Cuantitativa. Estrategias y Técnicas de Investigación Social*; Izquierdo Escribano, A., Leal Maldonado, J., Ramos Torre, R., Eds.; Editorial Síntesis: Madrid, Spain, 2001; ISBN 8477384207.
50. Florida, R. *The Rise of the Creative Class: And How It's Transforming Work, Leisure, Community and Everyday Life*; Basic Books: New York, NY, USA, 2002; ISBN 9780465024766.
51. Torres, S. On Elastic Ties: Distance and Intimacy in Social Relationships. *Sociol. Sci.* **2019**, *6*, 235–263. [CrossRef]
52. Sequera, J. A 50 Años Del Nacimiento Del Concepto 'Gentrificación'. La Mirada Anglosajona. *Rev. Bibliográfica Geogr. y Ciencias Soc.* **2015**, *21*, 742–798. [CrossRef]
53. Callander, E.J.; Schofield, D.J.; Shrestha, R.N. Multiple Disadvantages Among Older Citizens: What a Multidimensional Measure of Poverty Can Show. *J. Aging Soc. Policy* **2012**, *24*, 368–383. [CrossRef] [PubMed]
54. Sacker, A.; Ross, A.; MacLeod, C.A.; Netuveli, G.; Windle, G. Health and Social Exclusion in Older Age: Evidence from Understanding Society, the UK Household Longitudinal Study. *J. Epidemiol. Community Health* **2017**, *71*, 681–690. [CrossRef]

Article

Reconstructing the Social Image of Older Women and Ageing: The Transformative Power of the Narrative Set in the Local Context

Paquita Sanvicen-Torné [1,*], Ieva Stončikaitė [2], Anna Soldevila-Benet [3] and Fidel Molina-Luque [1]

[1] Department of Sociology, GESEC-INDEST, University of Lleida, 25003 Lleida, Spain; fidel.molinaluque@udl.cat
[2] Department of English and Linguistics, DEDAL-LIT, University of Lleida, 25003 Lleida, Spain; ieva.stoncikaite@udl.cat
[3] Department of Pedagogy, GESEC-INDEST, University of Lleida, 25003 Lleida, Spain; anna.soldevila@udl.cat
* Correspondence: paquita.sanvicen@udl.cat

Citation: Sanvicen-Torné, P.;
Stončikaitė, I.; Soldevila-Benet, A.;
Molina-Luque, F. Reconstructing the
Social Image of Older Women and
Ageing: The Transformative Power of
the Narrative Set in the Local Context.
Land **2022**, *11*, 1057. https://doi.org/
10.3390/land11071057

Academic Editors:
Vanessa Zorrilla-Muñoz, Maria
Silveria Agulló-Tomás,
Eduardo Fernandez, Blanca
Criado Quesada, Sonia De
Lucas Santos and Jesus
Cuadrado Rojo

Received: 30 April 2022
Accepted: 2 July 2022
Published: 12 July 2022

Publisher's Note: MDPI stays neutral
with regard to jurisdictional claims in
published maps and institutional affiliations.

Abstract: This case study reveals that age-related areas are the least desirable professional future options for many university students in social work degree programmes. One of the possible causes is the negative social labelling of older age, especially pronounced in respect of older women. Additionally, there is a poor and limited educational approach towards later life and growing older inside and outside the educational settings. This article focuses on the social construction of older age from gender and double theoretical perspectives. In particular, it centers on the pillars of education and profiguration. For educational and analytical purposes, these aspects are approached in the classroom setting from a critical perspective by using the in-depth reading of a book that is set in the local context, in particular, the city of Lleida (Spain). It presents the results of the content analysis and reflections written by 170 first-year university students taking a degree course in social work, and the outcomes of the subsequent classroom discussions with the author of the book. The study results show that better knowledge about the complexities of ageing and later life can lead to the reconstruction of the students' viewpoints about older age, help foster critical thinking, and defy age-related stereotypes, beliefs, and prejudices.

Keywords: older women; social imaginary; active ageing; profiguration; gender; narrative; local territorial context; social sustainability; prejudices

1. Introduction

The narrative has a transformative power both in the person who writes it and in the person who reads it. The latter aspect is the main focus of this article. The transforming power of the narrative is closely related to the educational power, especially taking into account that education occurs between permanence and change. On the one hand, education as a socialisation process implies the transmission of values, norms, and cultural contents that are transmitted from one generation to another/older adults to younger people. Hence, seen from the socio-anthropological perspective, education can be considered as permanence and the common guiding line of society and culture. However, during the teaching-learning processes, new discoveries, inventions, and innovations occur that are linked (or not) with what has traditionally been transmitted. Both young people and adults live in a society marked by continuous and accelerated changes that affect our everyday experiences and life stages. From a gerontological perspective, the changes that are already taking place in the ageing process will become even more evident in the coming years, described as the "new old age" [1]. Consequently, different alterations also take place in educational and socialisation processes. Relatedly, the narrative also identifies and represents arguments, thoughts, experiences and imaginations that relate to both permanence and change. When narratives are expressed, either orally or in written way, or even through other artistic and creative manifestations, they acquire a transforming component of attention, reflection, teaching-learning, contemplation, affection, and humanity [2]. This case study focuses on these aspects and incorporates reflective work based on narratives and critical approaches towards later life.

From a sociological lens, narrative gerontology and literary analysis show that we can learn about the complexity of the ageing process and its social nature from the stories that people tell or write [3–9]. This article shows the importance of fostering intergenerational relationships as well as reading and reflecting on narratives as ways to build a more sustainable society, in which the interdependence between different generations implies enrichment and solidarity and helps challenge age-based discrimination or ageism. Breaking negative stereotypes about old age and gender and building a more positive image of later life, both from the point of view of young people and older people, favours self-esteem, quality of life, and people's physical and psychological wellbeing [10].

The guiding thread of this case study is profiguration, a new concept in sociology and the social sciences that refers to the social agreement between different generations. Profiguration implies interdependence and socialisation, and respect and support with the aim of enriching intergenerational relationships and social connections. It also helps improve social participation, creative actions, social innovation and active ageing as it is characterized by participatory, collaborative, and comprehensive education that promotes positive change and challenges social marginalisation and ageism [7,11,12]. Working on intergenerationality as based on reading and reflection tasks in university settings is an opportunity that provides students with an important base from which to critically reflect on their beliefs and attitudes in relation to the different ages of life and ageing. These exercises also help share and create new knowledge among students.

As will be explained in more detail in the following section, we observed conscious distancing attitudes of the first year of social work degree students towards older adults. After the appropriate reflection and discussions, we developed the activity that has become the object of this study and the basis of this article. At the same time, we wondered what beliefs and stereotypes about being older have taken root in the education of young people. How have these prejudices been built? What notions and archetypes of older people, especially older women, have been assimilated and constructed? Does knowing alternative models of ageing, such healthy and active ageing, change younger people's perspective towards older people and later life? Does reading and reflection on narratives about ageing, which are different from the stereotypical images of older age, favour positive images of growing older in younger people? Can guided and reflective work in university classrooms act as an instrument to overcome ageism and build a more age-friendly and inclusive society? These questions have been answered in this case study.

2. Materials and Methods

To facilitate the understanding of the methods, a diagram has been prepared that graphically illustrates the procedure and its parts (see Supplementary Materials—Annex S1). The following section explains the whole process in more detail.

2.1. Initial Data That Motivated the Development of the Case Study Task: Intergenerational Work Based on the Reading and Reflection of a Narrative

For several years, the first-year students of the social work degree enrolled in sociology have been invited to voluntarily answer a questionnaire at the beginning of the course. The aim was to get to know them better, their previous studies, interests and knowledge, and their experiences related to the topics of the course. They were also asked to express their opinion about the profession they have chosen, their future plans as well as their personal and professional interests. One of the questions in the questionnaire asked them to explain in detail the fields in which they would like to work in the near future. Relatedly, there was another question asking them to indicate which areas they would not like to work in. During the consecutive years, the area related to older people was among the most undesirable future career options among the first-year students of the social work degree. We systematically analysed the results of these questionnaires to detect the learning/training needs, improve pedagogical aspects and the image of later life, and overcome ageism. The answers to the questions (In which field/s would you like to work when you finish your degree? Why? In which field/s would you not like to work when you finish your degree? Why?) are shown in Tables 1–3. Table 1 collects the quantitative computation of the answers, in which the reference is made to advanced age and older adults, also referred to as senior citizens.

Table 1. List of students' responses on whether they would like to work in areas related to old age and older adults.

Academic Year	Enrolled Students in the Subject	Collected Questionnaires (Received Answers/%)	Interested in Working with Older People	Not Interested in Working with Older People
2020–2021	99	68 (68.68%)	6 (8.82%)	13 (19.12%)
2021–2022	100	64 (64%)	4 (6.25%)	14 (21.87%)

Table 2. Students' positive motivations.

2020–2021	2021–2022
I want to help older people in care homes	I have always wanted to help older people. I have chosen this degree specifically for this reason
Older people catch my attention, I'm curious, I find them interesting	Since I was little, I have been interested in knowing the stories of older adults and wanted to help them improve their situation/circumstances
Because being with them makes me happy	
It is the group of people with which I would feel better	
Older people are underestimated and deserve more respect	

Table 3. Students' main arguments not to work with older people.

2020–2021	2021–2022
It is the area where I would least see myself working in. I wouldn't be able handle it in a right way	I would not know how to manage it since I would relate it to my everyday life
It doesn't appeal to me and I don't think I would ever feel comfortable working with older people	In my life I have not had contact with older people. I hardly got to know my own grandparents.
It is an area in which I would surely have a very bad time	I don't feel empowered because I haven't had any close contact with older people. I would not know how to think or how to express myself or speak to older adults
From my experience, I think it requires a very hard emotional sacrifice	Grandparents are the best company, but dealing with older people and their problems would affect me too much
I wouldn't have a good time	I would take their problems to my personal sphere/life

Table 3. *Cont.*

2020–2021	2021–2022
I think it would be very hard to manage it	I don't have the 'right' personality traits to deal with them
I think I am not mentally prepared to face the situations that occur in this area	I don't think I could contribute anything to this area
Working with older people is already a very 'common' area with which we live day by day, I would like to work in areas that are more 'taboo' and which are not that talked about	It does not appeal to me, I find other areas more interesting
I have been working in this sector all my life and I would like to open up new horizons	I am not motivated
I think it is a very dependent 'guild'	I am not interested
I don't want to work in a nursing home	I am not interested

It must be taken into account that answering the questionnaire at the beginning of the course has been optional/voluntary (until now). This explains why only 68.68% and 64% of the total number of students enrolled answered it. Yet, it is significant to observe that only 19 (6 + 13) and 18 (4 + 14) students referred to older people in their responses. The rest mentioned other areas of interest, such as mental health, childhood, justice, social services, among others. This reveals that students are quite distant from the real needs of older people and their challenges. The percentage of those students who showed no interest in working with older people and in care homes did not reach 10% in any academic year. On the other hand, those who explicitly stated that they would not like to work with older people reached 20% and even exceeds this percentage. The distance between those who explicitly denied working with older people increased in the academic year 2021–2022 compared to the previous course 2020–2021, reaching 21.87%. Table 2 shows the main motivations and arguments of those students who stated that they did not want to work with older adults or in related areas:

The results reveal that the student main motivation is to help, which appears in three of the seven responses. The second motivational category is related to emotions, such as 'it makes me happy, 'it is where I would feel better', 'it is underestimated, it deserves respect'. There is a repeated temporal reference that indicates a vital implication linked to the socialisation received throughout life: 'I have always wanted to help', 'since I was little'. Other answers could be categorized as discovering the 'unknown' or the 'little known' sphere, knowing older people's stories, 'they call my attention, I'm curious … ' In other words, the sense of altruism, empathy, positive experiences, as well as the desire to help and better understand older adults are among the main motivations of the social work students who answered the questionnaires at the beginning of the course.

Table 3 collects the written explanations and the main arguments of those students who said that they did not want to work with older people in their future.

The results reveal how the student personal life experiences and previous knowledge influence their future decisions and are also the possible causes of the refusal to work with older adults. The emotional impact linked to one's personal experience (or the lack of it) and the socially constructed image about later life appears in 16 out of 22 responses (72.72%), and affects students' perceptions about old age and their future work preferences. These and similar notions are also echoed in the students' written reflections that are evidenced in the development of the reading guide (see Supplementary Materials—Annex S2), which demonstrates the growing need to promote activities that stimulate intergenerational relationships and mutual knowledge in order to overcome ageism. As research shows, breaking stereotypes and negative notions about later life entails a change in the social paradigm and fosters active and healthy ageing, intergenerational dialogue, and a more inclusive society [12,13].

2.2. Reflections Prior to the Study Based on the Analysis of the Previous Answers

Students' refusal to work with older people raised a question whether there was a (co)relation of different causes of a social nature that related to their formal, non-formal, and informal education. We also took into account the influence of social stereotypes created by the media on ageing and the type of social value attributed to old age; the lack of education and reflection about the process of getting older; the social limitation of daily relationships between young and old people, which are often reduced to dependency relationships; and a lack of more positive images of ageing and models of active older men and women beyond the traditional roles of grandfather/grandmother or caregivers. Since the collected data reveal low student motivation to work with older adults, which is one of our educational challenges, during academic year 2020–2021, we decided to approach the subject from more innovative, humanistic, and proximity methodologies. The choice of a specific book about ageing and older women was chosen to provide students with new notions about later life and improve educational opportunities.

2.3. The Case Study: Tasks on a Specific Reading to Overcome Age-Related Stereotypes

The specific use of narratives as a working method and as a training object is already widely documented in various research projects and studies that demonstrate its potential, interest and didactic and educational validity [2]. The contributions of the narrative to the study of older age also reveal the construction of positive images of later life, specifically from a gender perspective [13–15]. To address the causal elements that determine the images that young people have about ageing and older women, we focused on narrative analysis with a focus on older women. We incorporated a reading guide (see Supplementary Materials—Annex S2) that helps students critically read the book. As educators and professors, we also defend the need and importance of developing narrative inquiry, reading, and reflective and comprehensive methods as teaching and training tools in higher education settings.

The current digital world, characterised by the speed of information, the simplification of messages, the obvious displacement of the written form by visual content and the predominance of emotion over reason, greatly affects the construction of our social reality, opinions, and attitudes. Impatient immediacy and imminence bring added difficulties and accelerate the transformation of habits, making it more difficult to dedicate time, concentration and attention to observe, read and understand reality. The notion of 'learning to be, to make known, to live together' [16,17] implies the habit and ability to think critically, which is not innate but learned and developed throughout one's lifespan [18]. For this reason, in our teaching practices and academic evaluation we incorporate reading practices—essays, biographical, and literary narratives—that help promote knowledge of the social world and the local contexts. We used the guide that has been prepared specifically for this reading from reflective, experiential and critical approaches. Students were also asked to write a letter to the author of the book expressing their opinion, questions or concerns, which helped foster individual creativity and critical thinking (see Supplementary Materials—Annex S1).

For this case study, we selected a recently published fictional book *Casa Yé-ye* [19], which addresses active ageing experiences as inspired by real life situations of older women, who live in city of Lleida, which is familiar to the students enrolled in the University of Lleida. The book portrays women over 70 that do not conform to stereotypical notions of old age and the narrative of decline, according to which older people are seen as frail, dependent, and inactive [14]. The four older women defy stereotypes related to old age and the traditional family models, and choose to live together. This choice of life gives them more freedom to make decisions, which is often denied to older people. The male characters, who belong to the same age group as the heroines, also actively participate in life and challenge the negative notions of later years. The text is written in a slightly humorous and witty tone, and contains some dramatic points, as well as certain but uncomfortable truths about ageing that involve topics such as death, suicide, sexuality, family relationships, physical or mental health, loneliness and the loss of love. The book brings readers closer to the dynamics of old age in a positive, constructive and innovative way.

The daily experiences of the older characters, as described in the book, were read and discussed with the first-year students. After the in-depth reading and analysis of the text, students showed a significant change in their attitudes and perspectives on later life. Working with literary narratives with intergenerational groups facilitates profigurative socialisation and interaction, and helps overcome age discrimination. The educational and transformative power of the narrative based in a local context and, in particular, the reconstruction of the social image of older women, also helps break the prejudices and alters students' future professional preferences. Many of the university students, as will be shown, claimed that, initially, they did not think about later life, ageing, or intergenerational relationships. However, after the critical reading of the book, they started to think more about ageing and some of them even considered working with older people. The results of this study show that working on intergenerational relationships through narrative inquiry can become a basis from which to reflect on beliefs and attitudes related to different stages of life, create and share new knowledge, and improve intergenerational relationships.

The literary and cultural analysis of the fictional narrative, together with the debates, provide us with relevant information that reveals the current social conceptions and beliefs about old age among first-year university students. At the same time, it allows us to see a more specific social image of older women, the perception of different ages and the ideas of what is considered as socially appropriate or inappropriate depending on one's age [20,21]. The in-depth analysis of the text, along with subsequent discussions by one of the authors of this article (educational psychologist), professors (sociologists), and first-year social work students, represents a rich source of data that offers not only a valuable exchange of information, but also invites us to question old age, the current images of ageing women and gender dynamics.

The gender aspect deserves a special attention in this study as it reveals how older women are perceived in the local context, which is still predominantly sexist and heteropatriarchal. In addition, the analysis of this work reveals other relevant issues in relation to age and ageing, such as the loneliness of old age, the concern to maintain one's autonomy until death, the difficulty in establishing new social relationships, pain, disability, and sexuality, among other significant topics linked to later life and its challenges. The following section explains in more detail the work methods of this study.

Casa Yé-ye and the Practical Application of the Reading Guide

This case study was developed in three phases that involved the interaction between educators, students and the author of the book. In each of them an ad hoc didactic and pedagogical methodology was used to achieve the study objectives.

(a)   First phase: The selection of books and the elaboration of the reading guide

The first phase was focused on the decision to prepare an academic task that involved reflective and creative learning based on a reading that portrayed ageing of women in a more positive light. The choice of the book *Casa Yé-ye*, which was published before the start of the academic year 2020–2021, was based on the aspects discussed in the previous section: the focus on older women; the local and familiar settings; daily events and local scenarios; the fusion of both fiction and reality and a deep observation and lived experiences of ageing. As the author of the book, who is also a professor and researcher in areas related to gender and ageing, explains:

> *After more than two decades of teaching subjects about active ageing and later life, and given a structural trend towards increased life expectancy, especially pronounced in older women, I have observed an urgent need to transfer my knowledge by using alternative ways. I wanted to bring my own experiences and observations closer to people in order to break the beliefs, stereotypes, prejudices, and negative attitudes about older age. The experiential literary narrative was chosen as an effective tool to share my previous expert*

knowledge and personal reflections to foster social change, intergenerational dialogue, and voice out the complex dynamics of growing older. The book Casa Yé-ye emerged precisely from these motivations and personal and professional experiences.

The decision to develop a reading guide and the distribution of roles was shared with the author and the educators. Although the author of the book is also a professor at the University of Lleida, she teaches the social education degree. Thus, she had no connection with the social work students and could not influence their opinions and the content of their reflective writing. During the 2020–2021 and 2021–2022 academic courses, she used the same reading guide (see Supplementary Materials—Annex S2) and methodology. The objective of the reading guide was to foster critical thinking skills and analysis of social aspects and notions about old age, especially older women. Specifically, it was intended to trigger students' thoughts and reflections on ageing that, in this case study, are addressed transversally as part of the program that involves life long learning, social inequalities, and gender dynamics. It was also intended to make students more aware of the heterogeneity of ageing and bring them closer to active and healthy ageing models. The reading guide consisted of eight sections on which the students had to give their opinion: (1) Reflection on the title of the book and its meaning; (2) Identification of the relevant ideas of each chapter of the book; (3) Reflection on the themes and problems presented in the book; (4) Reflection on the main characters; (5) The reflection on whether what is represented in the book is reality or fiction; (6) Reflection on students' prior knowledge of the topics explored in the book and how reading may have contributed to the construction of knowledge; (7) Reflection on possible contributions to the area of social work; (8) Personal assessment and final reflection in the form of a written letter addressed to the author of the book. The letter was intended to serve as a development of critical thinking skills for the students, as well as a guiding tool for the author to prepare feedback (see the content of the complete reading guide in Supplementary Materials—Annex S2).

(b)　Second phase: Reading, reflecting and writing

The second phase involved the process of reading the book, the reflection, the analysis of the content and the subsequent written work that included critical approaches to the book and its main characters. The reading and orientation tasks were performed individually by each of the students.

(c)　Third phase: Tasks, evaluation, feedback and meeting with the author

An assessment rubric was used to analyse students' written assignments for each section. Assignments and letters were also read by the author of the book to prepare comments and feedback for the students. The author did not participate in the evaluation process and did not have information about the students' grades. The final grade was shared only between the teacher and the student. The most significant aspects that were taken into account were the reflections of the students, the questions raised and the subsequent dialogue with the author and the students, which was scheduled once the semester was finished. This final dialogue was not part of the evaluation process and was voluntary. The meeting was supposed to be held in person, however, due to the restrictions imposed by the COVID-19 outbreak, it was held out online.

2.4. Method of Analysis of Written Content of the Reading Guides

In the previous section, we explained the methodology used in each phase, which involved the reading guide and feedback. In this section we highlight the method we used to perform the content analysis of the 170 reflective tasks, which are the basis of this article. Van-Dijk [22] and Critical Discourse Analysis (CDA) were used as a methodological approach to analyse task content and develop critical thinking skills. Before starting the general reading, a preliminary reading has been carried out to erase any personal data or identification of authorship. Likewise, all tasks have been anonymised by adapting a specific coding for data protection. For the content analysis, the parts of the tasks that focused on reflection and argumentation have been identified.

2.4.1. Parts of the Tasks Selected for Analysis

Out of the 8 sections of the reading guide, we selected 6 for content analysis because they were the ones in which the students had to express their opinions: (1) Reflection and argumentation on the interest in the topics and problems presented in the book; (2) Reflection on the main characters; (3) Reflection on whether what is represented in the book is reality or fiction; (4) Reflection on the previous knowledge of the students on the topics explored and how reading the book has contributed to the construction of knowledge; (5) Reflection on possible contributions for social work professionals; (6) Personal assessment and letter to the author.

2.4.2. Elements of the Analysed Content

The reading and analysis of the content of these 6 sections was based on the identification of the content of the general categories as shown in Table 4:

Table 4. General categories.

Categories:
Reflections on the effect of the known environment
Construction of the social image of older age and the ways it is perceived
Characteristics attributed to older age in general and older women in particular
Descriptors of older age and older women before and after the reading of the book
Opinion about older women and the ageing process before and after the reading
Existence of explicit gender or feminist perspectives in the book analysis
Elements of surprise found in the book
Elements of self-reflection on grandmothers/grandfathers before and after the reading
Elements of reflection from a professional (social worker) point of view
Reflection on how the students imagine their own later life

3. Results

3.1. Participants

The tasks explained in the previous sections correspond to those elaborated by 170 students: 80 students of the academic year 2020–2021 and 90 students of the academic year 2021–2022.

3.2. Contributions of Reading and Reflection

In this section we present the results of the content analysis that has been organized thematically, as shown in the following sections.

3.2.1. Relevance of the Local Context

As described previously, one of the reasons for choosing the book *Casa Yé-ye* was the fact that it is set in the local area that is familiar to the students. The environment and the elements described in the book portray the bar, the cafeteria, the gym, the nightclubs, the streets and the neighbourhoods familiar to everyone who knows the city of Lleida. There is also a strong emphasis on the social aspect of Lleida, which is a peripheral, medium-sized, inland city, and evokes an image of 'Spain where nothing ever happens', as sociologist Sergio Andrés describes it [23]. Very often, literature suggested to university students is written by national or international authors and there is a lack of visibility of local authors that address the local issues and minority groups. Lleida is the capital city of a large territorial area with rural characteristics. The city of Lleida is a member of the Network Educator City and it is also an age-friendly city. It is the capital of a diverse territory of 11 interrelated regions in two large geographical areas: the plain area and the Pyrenees. According to data from 2021, its population varies from 3.945 inhabitants of the less populated region with a population density of 9.2 inhabitants per square meter, to 211,609 inhabitants of the most populous area, which includes Lleida with 140,080 inhabitants (151.5 inhabitants per square km). Regarding the age-related factors, more than 20% of people are over 65. The ageing rates are between 113 and 207.9, and the age-dependency rates are between 21.3 and 40.3.

We prioritised the local context in this study, especially taking into account that students are trained for the social work profession that involves close human interactions in local contexts. In addition, many of them were born in the city of Lleida, study there or know the city quite well. We also thought that their own personal experiences in the environment familiar to them influenced their opinion about their future professional choices. We aimed to investigate whether the students noticed that the book was set in the local context and whether or not this generated any interest in the reading. The results affirm that the familiarity and proximity with the environment provides an important value for student. For example, one of the students SO8701 mentioned that "the book is set in Lleida and I find it fantastic since I had never read anything that was set in my city". Andrés [23] explains the reasons for invisibility of medium-sized cities such as Lleida:

> *Medium and small cities do not seem to have been as narrated, told about and sung about as other spaces and places. The medium-sized city and its intermediate territories have been assigned a symbolic image of a 'no man's land.' They are not considered attractive enough to be the protagonists of stories, and it seems that nothing special ever happens in those places* [3] (pp. 205–206).

Yet, significant and special events do occur daily in medium-sized cities, as described in the book *Casa Yé-ye*, which shows active lifestyles of older people. As one of the SO9801 students observed, "everything that happens, happens in the city of Lleida, and that makes everything familiar to you, and you can relate and imagine the story as if it were more real". Another student SO6901 pointed out that "seeing typical places in Lleida has allowed me to become more familiar with the book and experience each situation in much more detail." Student SO9201 also noted that the local setting gave the story more credibility, familiarity, and value: "It's very exciting to be able to read about familiar areas as it helps you imagine the action in a more realistic way." The atmosphere that portrays the common and familiar places of the book also generates more empathy towards the main characters, whom some students imagine as real people that they could meet one day in the city of Lleida. As student SO6412 said, "we have been able to get closer to the characters, as if someone we knew was explaining their own story to us". In the same way, the student SO9501 observed that the story "has made me connect faster with the protagonists, I have imagined them in the cafeteria and in the Ekke gym, I have imagined the Coquettes at night walking down the street looking for a pub".

3.2.2. The Social Images of Older People

This case study also intended to identify the preconceived social image that students had about old age in general and older women in particular. Age is not only a matter of adding biological years to one's lifespan, but it is also a socio-cultural construction [24]. Social age [25] is a series of descriptions of what one can and cannot do at each age, and it changes according to different historical periods and social milieu, thus being a product of social interaction and culture. Just as culture defines the differences between men and women that occur throughout life, so does the process of ageing [26]. We began the analysis by examining the concept of ageing in a broad sense to identify the general stereotypes attributed to older people, especially older women. It should be noted that students have never been asked to incorporate a gender or feminist perspective in the reading and preparation of their tasks. However, it was significant to see if they referred to gender dynamics and differences in relation to older age.

It is also worth a note that the majority of social work students are female. In 2020–2021 there were 13.40% male students and in 2021–2022 the number increased to 20.98%. Taking into account the necessary anonymisation and data protection, we have specifically studied and identified significant gender differences in content analysis. Although it is important to investigate the social construction that students have about older women, we decided to focus on how young people internalised these notions. The analysis shows that all the answers incorporate reflections on the images of later life as acquired throughout social interaction and education. We also tried to identify the stereotypes that young people had about older age and ageing. The content analysis reveals four thematic fields. Three of them are recurrent in a large number of tasks: (1) silence and inhibition; (2) the negative view of old age; (3) the antagonistic construction of old age versus youth; (4) the positive image of old age and ageing. The fact that the latter appears in very few responses is highly significant.

(a)   A positive view of old age

There are few student responses that show a positive view of old age and ageing. In all cases, the constructive evaluation of later life derives from a positive experience in the family, at work or during an internship, or from the image of an older woman as a caregiver. The positive elements related to old age are linked to values such as respect, gratitude, kindness, experience, wisdom, care, and education. As student SO9401 argued

> *I am surprised by the whole context surrounding old age. This group of people are my greatest weakness, my weak point, I admire all older people for their long history, for the backpack they carry on their backs, I would like to be able to work with this group. I think that working with them would bring me happiness and personal satisfaction.*

Another SO8512 student also noted that: "I really appreciate older people, both in my family and with whom I live in my daily work. They are generous, kind and loving; likewise teachers, psychologists." Similarly, student SO8112 highlighted that: "I have a very positive perception of older people, it is a group that I have always liked and admired (. . .) Most of them are people who have had a thousand experiences and it is interesting to take their opinions into account".

(b)   Silence and inhibition

Although today's society is characterized by the so-called longevity revolution [27], the writings of many young people demonstrate a sense of detachment, social distance and "not wanting to look" at older people. In their writings, many of the students argued that ageing begins with retirement, around 65 or 67, and they think that nothing interesting happens after that phase of life or turning point. The idea of vital emptiness—of 'stopping' and 'disappearance'—after retirement is quite recurrent in the students' reflections. These notions reveal that students have been educated with the notion that people, upon reaching the retirement stage, cease to have their own will, interests, and social identity. For example, student SO9101 said that "many young people and adults (...) are a little afraid of getting old because they think they will lose part of their freedom and their way of being and doing things." Many students also thought that old age is a taboo subject, which is evidenced in the following examples. For instance, student SO8701 observed that "society transmits to us a type of old age and ideas about it that make us think that later life is a topic that should avoided to talk about". Similarly, student SO9501 pointed out that "old age is a topic that does not interest young people. Many times we tend not to talk about it, we label it as a taboo subject, we do not give it importance or we do not understand that stage." Repetitions such as 'not wanting to talk about it', 'being a taboo subject', 'it is not given importance', 'it is not understood' are also common in students' reflections. The students' responses also show that old age is a stage they do not think about. For example, student SO4301 pointed out that "another negative connotation that we attribute to this stage is the perception of retirement as a void that cannot be filled and for which no one feels prepared." Another student SO401 observed that

> *In our society, getting older is a taboo subject. It is not common to talk much about older age, but the reality is that it will come to all of us anyways, no one can escape this final stage of life. Why don't we talk about older age and getting older? I think it is because*

society is afraid—just as death is a taboo subject, older age is also associated with death. Since this stage of life is the closest to death, we tend to avoid conversations that involve these two concepts: older age and death.

Additionally, the invisibility and conscious social uprooting of old age is one of the causes that generate fear of ageing among young people and adults. The following examples exemplify this idea. Student SO9801, for instance, questioned:

Why are we so afraid of reaching older age? Because it is a subject that nobody likes to talk about, a taboo subject. We associate the fact that a person can have a very busy life with a teenager, and perceive them as young; however, we associate a retired person with a person who does nothing all day long, which leaves older adults out of any consideration and interest.

Similarly, another student SO901 reflected that:

People see older age in a negative way or are afraid of reaching it. It may be due to the influence of the media and the cosmetic and aesthetic industries, such as the miracles of anti-wrinkle creams, aesthetic operations, etc. It is also because of the fear of physical changes—when you are older you have a higher risk of suffering from diseases, you are closer to death, loneliness, a fear of not feeling useful, not accepting your age and social pressure.

Student SO9501 also made a similar comment by stating that: "We tend to think that when a person is older, he/she has fulfilled his/her 'obligations' and now it is his/her turn to rest and disconnect from everything around him/her. It's like nullifying him/her as a person."

(c)   Old age versus youth: an antagonistic social construction

It is increasingly necessary to promote intergenerational relationships to overcome stereotypes about old age and "learn to be and learn to know" the complex, heterogeneous and diverse process of ageing [11,12]. Such notions are also highlighted in the students' responses. For example, student SO7912 expressed that "old age is the last stage of human life, it is the moment in which you begin to say goodbye to everything you have lived through." Without exception, all students were surprised by certain behaviour patterns of the main characters of the book, who were labelled as 'juvenile' because they acted in a socially constructed way as 'young people'. They were amazed that the heroines openly talked about sex, that one of the female characters was transsexual, and that other characters enjoyed smoking marijuana. According to student SO8501,

Society has a perspective of sex deprivation of older people, that is to say, that they do not make love, that when they reach a certain age, they stop having sexual desires. It is clear that no matter how much sexual the activity decreases, sexual desire and the wish to feel loved does not disappear. It is a basic need that we as human beings have.

Similarly, student SO8601 pointed out that "sexuality is usually one of the main stigmas about older people. This book shows us that this is not the case: the protagonists enjoy sex in a free and impartial way. They even talk about sex without having any doubts and openly explain possible sexual problems." By the same token, student SO401 commented that: "It seems that in older age sex no longer exists. The coquettes demonstrate that sex is not only for young people." Additionally, student SO9501 argued that: "Margot [one of the older female characters] is shown smoking a joint to ease her pain. I had not internalised this aspect either, I must admit that I was not aware that older people would take this type of substance." These comments show that youth is conceived as a stage full of vitality and activity, while later life is linked to the end of opportunities, sexual desire, and active lifestyles, as revealed in the narrative of decline [14]. These aspects, according to students, are more visible in older women due to the double standard of ageing [28], which explains how the disadvantages linked to physical appearance and sexual attractiveness can lead to a double categorization, segregation and marginalization of older women. Student SO4301 commented that:

Older age is a stage of anguish due to the losses or difficulties and, consequently, a decrease in self-esteem, like in the case with Lily [one of the older female protagonists]. She finds herself in a continual confrontation between buying clothing accessories to enhance a slender youthful figure and disguising the ravages of time that affect her body and make her curves fade away.

Student SO401 also observed that "it seemed that the author was describing young people, but in reality the characters were not young." According to student SO801, "I loved the characters, they are like a group of friends from now, but much older and, logically, more experienced." Following the same line of thought, student SO4301 contended that:

In short, the main characters have a lifestyle that is similar to mine, and I am twenty years old. I also I share a flat with other students, I go to the gym, I go to pubs, and I drink alcohol, just like them. But the difference is that they could be my grandmothers. It would seem surreal to me to think that I could meet my grandmother in the gym or at a party, but it is possible because, in the gym, I have seen older people who are doing exercises, but with less physical effort.

Some students expressed a great interest in older age and offered interesting and though-provoking reflections and conclusions. For instance, student SO4301 commented that:

Old age has traditionally been related to negative notions, and linked to dependency and progressive physical incapacity and weakness, while youth has always been glorified. People and families need to plan and, above all, accept this stage of life (...) it is not just a stage, but a progressive process that depends on the entire life cycle.

Student SO4901 also noted that:

The story deals with the problem of older age from a positive and dynamic perspective. It employs adjectives that are usually related to youth, but in this book, it shows us how you can still live among friends, go out for drinks, show off, flirt, smoke joints, etc. What strikes me most about the subject of older age is the hiding of the effects of ageing. It's like the youth has totally taken over the stage. (...) The obsession with beauty and the hypersexualisation (through social networks), together with the digital divide, leave older people out of the game. Young people have the responsibility to give older adults their deserved place in family and society.

(d)   A negative view of old age

As explained above, the majority of students showed a negative view of old age. We grouped the elements that appear in their explanations into several characteristics and roles.

d.1.   Negative characteristics

The negative notions about old age, as shown in students' writings, include: dependency, loneliness, sadness, annoyance, inactivity, illness, and the end of life. The students' reflections reveal the idea that as people age, they are perceived as less human and are seen as 'others' whose lives no longer matter much [14,29]. Viewing older citizens as problems for health care, the economy, and future generations further strengthens age discrimination and negative notions about old age. Furthermore, people tend to judge the value of other people based on how much they add to the betterment of society: those who no longer contribute to society "are not considered full members of society" [29]. These and similar ideas are expressed in students' writings. For example, student SO6412 stated that:

Many people think that older people are a burden and, thus, do not want to spend time with ageing people. (...) People do not want to be aware of the existence of older adults and there is also a great controversy about their retirement pensions. But younger people do not think about the cultural and historical enrichment that they provide us with—older people are our history with legs, names, and surnames.

Similarly, student SO8601 pointed out that "the majority of the population believes that older age and dependency are closely linked". Relatedly, student SO1612 commented that:

For me, growing up means starting to be alone or lonely, since I relate it to death. When people get older, they begin to experience the deaths of their friends, family members or other important people to them. Once they get here [older age], they see themselves as a burden to their family members and that's when the feeling of loneliness begins.

The student SO8612 also mentioned that:

A person who is over 70, for the rest of the people, especially for young people, is perceived as a passive person whose life is practically over. There is a lack of experiences to live and, in general, we associate older age with the end of life.

d.2.   Roles of old age viewed negatively

The students also pointed out to the obligations and social pressure in later life, such as taking care of grandchildren, not being able to decide for themselves and doing "what is expected of them" instead of doing what older adults want. Student SO4512 wrote that:

In the book, I see a call, which I would name a call for older people's freedom.' Let me explain it better: when you get older, it seems that you are deprived of everything because you are old, which is synonymous with typical obligations that correspond to a certain age. I want to highlight the fact that the canons of beauty oppress us, women, until the last stage of life. In the book, it is shown how the three female protagonists feel vulnerable because of their physical appearance and how it changes over time. However, they are women who have high self-esteem and there is nothing that could stop them.

Student SO9001 also explained that:

An older woman today has lived through the times of sexual repression in a severe way. Hence, as shown in the book, the decision to live with her female friends and stop being a super-grandmother makes her feel bad. It becomes even more difficult to the heroine when she wants to tell her family that she wants to fall in love again. This shows that there is still a lot of struggle and the need to eliminate the consequences of sexual repression.

The influence of heteropatriarchy conditions the way we judge older people and generates prejudices and stereotypes attributed to different female social roles. In the past, if older women did not embody socially acceptable roles as caregivers or grandmothers, they were not given visibility and value in society [30]. These roles confined them to domestic obligations, thus limiting their freedom and autonomy in later life [31,32]. Although today many women and older women in particular oppose to these socially constructed roles and expectations, many of them continue with caretaking duties, making older women important figures in multigenerational family relationships. The students realise the lack of role models for older women and understand that their own prejudices also limit them. Mary, one of the protagonists of the book, reflects on her lived experiences, which are based on her relationship with her grandchildren. One of the SO1012 students noted that:

She [Mary] gives voice to many older women who are in the same situation as her. The heroine is a woman who was married for many years to the same man and was completely taken away her freedom. She represents the figure of a grandmother who helps with everything that is needed. It is very positive that her opinion is shown in the book, which reveals that she is tired and she does not want to do it because of obligation.

Similarly, the student SO8401 observed that:

Today society expects grandmothers to take care of their grandchildren, as if they had nothing else to do, as if it was their obligation. Today many families live at the expense of older people, their help, and their pensions. And let's not talk about the loneliness older adults feel.

Student SO5212 also noticed that:

The most endearing character in the book is Mary, who embodies the vast majority of today's grandmothers. She represents the role of a woman as a caregiver who takes care of her grandchildren and becomes a grandmother-slave. She actually has no obligation to take care of her grandchildren.

Thus, student SO4012 highlighted that she felt identified with the female protagonist when comparing her with her own grandmother and her lived experiences:

I have seen my grandmother reflected in Mary's character because she also had to take care of my cousins when they were little because my uncles didn't have time to do it. My grandmother, just like the protagonist in the book, many times wanted to flee from that reality and start a new stage in life. However, the fact that she come from a family of farmers and has been taught the traditional roles since she was a child, stopped her from taking decisions on more than one occasion.

Additionally, student SO5201 highlighted the existing loneliness and boredom that many older people face today, especially older women:

We associate growing older with a stage of loneliness, dependency, illness, and boredom. If I analyse these aspects in more detail, I can see my grandmother reflected. Luckily, she is living with us and I think that she does feel lonely. Of course, her life mainly involves being at home, watching TV, sewing, walking, calling on the phone, receiving visits from family or friends, etc. But I think that other older people, who are in care homes, hospitals, or living alone in a house, must feel even lonelier, more sick, and discouraged.

It is also worth a note that the absence of proximity and direct contact between young people and older people has a negative effect on the construction of the image of ageing. Two of the students, a male and a female, expressed the need to focus on a didactic, narrative and profigurative approach to education to better understand the process of growing older. Student SO3901 argued that:

Personally, I openly admit that I have always said that the last group I would like to work with in my future is older people. I know, and I understand, that it is a comment that can generate controversy and that I can turn people against me, but I firmly believe so. It's not that grandmothers and grandfathers bore me, quite the opposite. Luckily, I have three of the four grandparents still alive and I try, whenever I can, to visit them during the week. But that is not an obstacle for me to reaffirm what I have said previously. I have always considered myself a practical person, which does not mean I am cold and insensitive, and that is why I prefer to focus my efforts on future generations. Obviously, everyday life should be made easier for older adults and there are many tasks to be done in order to take care of them, but I think that—right now—my character traits and personality fit more with young people. I also know, based on my first hand experiences, that this point of view is also shared by many of my classmates and university friends. And it is not the argument itself, but rather the idea. That is why I am concerned about the fact that older individuals are the forgotten group of society. They are the ones who have made our starting point. We now walk on the foundations they had built and we owe them everything. Without them we would be nothing today.

Similarly, student SO6412 pointed out that:

Older age evokes tenderness, but also sadness and loneliness. The maximum contact I have had with her has been through observing older people from my town, many of whom do not leave their houses and do not interact with others. They enclose themselves at home and do not give life a chance, they think that their lives have already ended. My only grandfather that is still alive is not like that, he started traveling when he was seventy years old, he walks, he studies, he clings to life with all his might. I always tell myself that when I reach his age I want to be like him, leave the idea that 'I'm too old' aside and live as I have always lived. (...) It is true that I would not like to work with older people, perhaps because I have never been able to experience the moment of visiting my grandmother for a meal or listening to her stories, or playing cards with her. I have never been able to have the first hand grandparent-granddaughter relationship, and I have only experienced it by listening to the anecdotes and stories told by my friends. I feel that because of this lack of contact I don't have the tools or I don't know how to relate to older people. I also don't know how they are. I guess I feel the same way as adults who say they that don't know how to deal with teenagers. Maybe because I'm afraid to deal with older individuals and start creating a bond that I, probably, will never have.

3.2.3. Overcoming Prior Beliefs about Older Adults

The quantitative analysis of the tasks shows that more than 90% of the students, after reading and doing the exercises, have changed their previous notions about old age and incorporated new aspects about ageing. The results affirm that careful and guided reading and, above all, the possibility of reflecting on it, helped students to question the previous ideas about old age and gender. Student SO9801, who has worked in a nursing home, commented that:

This book has allowed me to see that there are many different ways to live later life. The author speaks of an active, free, happy, and crazy older age … She depicts an uncommon image of older age that nobody is used to (…) getting older can be boring or you can feel like a coquette, you can live your life to the fullest and be happy.

Another student, SO9401, also stated that "the lifestyle of the characters in the book is breaking stereotypes. It is a style that does not match with the typical notions of older people that we are used to seeing around us." Similarly, student SO1201 pointed out that:

I had a very different perspective about older age, based on what I have seen and what my grandparents have told me. I thought that when we reach that stage of life, the only things that happen to us are a physical damage, a loss of memory, there is no desire to live, to do things, to go to parties, to leave home, to show off, etc.

Students' responses show that they have dismantled previous myths about old age and created a new view of ageing from a different perspective. However, the fact that students' grandmothers regularly appear in their reflections on later life reveals the gendered nature of old age. The reports also demonstrate that later life can be an active stage in life that adds to quality of life [21,33]. For example, student SO9201 pointed out that "the book has made me understand that old age can be lived in a more fun way. Knowing this, I can accompany my grandmother (online) during this stage of her life and be able to better understand and help her." Relatedly, student SO8501 wrote that:

My grandmother has spent her whole life taking care of her grandchildren, and nobody ever asked if she wanted to do it. The book has made me realise that we have to take action and help her see that she should think more about herself and her wellbeing, because she has already taken enough care of everyone and has never dedicated time to herself. I think that many grandchildren and parents experience this type of situation but we do nothing about it because we are afraid to create discussions within family.

We have classified the students' comments that referred not only to older people, but, specifically, to older women in different sections of the reading guide. In the section where students had to choose the main characters that had impacted the most, everyone (100%) mentioned the female protagonists Lily and Margot—strong, independent and confident older women. The elements that surprised them the most were linked to the deconstruction of stereotypes about older women, as has been mentioned in the previous sections. Student SO9801 mentioned that:

The book gives us a very different perspective of grandmothers. It shows us how older women can have a youthful spirit and a lot of courage. The moments that stand out in the book are the parties, the sharing of a flat, going to pubs, putting on make-up, sex, drugs... The book brings you closer to an atypical reality of growing older, it breaks with the typical notion of later life that is mainly constructed around the idea of taking care of grandchildren.

Student S09201 also highlighted sexual diversity in later life and stated that being a transsexual older woman often implied invisibility and forgetfulness:

I would like to give transsexual older women a voice, it seemed like a very interesting topic (. . .) I came to the conclusion that transsexual women should be much more normalised in their older age, I think that they only make themselves known during their youth and people forget about this group of people as time goes by and as they get older.

By the same token, student SO8701 mentioned that:

It is interesting to see that the protagonists still have sexual desires, since many older women tend to lose sexual attraction, for various reasons, such as menopause and because there is a decrease in oestrogen levels.

The adjectives and related synonyms used by students to describe older women were grouped into the antonyms of the social imaginary of later life: strength, freedom, decision, joy, courage, pursuing dreams, creating projects, standing up and challenging the traditional social norms. For example, student SO401 noted that the book shows older women who have dreams and want to have fun later in life: "being older does not mean one cannot have fun in life." Similarly, student SO1201 highlighted that age does not determine the way we feel about ourselves: "reading shows that being older does not depend on age: you can be over 60 or 70, or even 80, and not consider yourself old. In this regard, student SO4301 argued that "the coquettes [older women in the book] show us that the negative social connotation of the retirement stage is not real. They show that it is time to enjoy free time while bolstering their collective self-esteem to combat a negative view of later life."

It is also worth mentioning that some of the students, without having been previously consulted or guided, pointed out the positive effects of working with narratives. According to them, they had not thought about old age, but the reading guide and class discussions allowed them to discover new approaches to narrative inquiry and imagine their (future) later years. For example, student SO8601 commented that reading "has given me space to break many of the stigmas I had about old age. I would like my old age to be similar to what is described in the book, and I would like to break the stereotypes related to later life." Another SO9401 student stated that:

For me, older people are not less important that other age groups, and they should not be an invisible group within society. On the contrary, I consider older adults as the most important individuals. An older person is an individual who is in the last stage of life. Older people bring with them all the knowledge, experiences, skills, and learning; they are wise and we can learn a lot from them. Older people also have fewer obligations and have the opportunity to enhance their both physical and emotional health, recover or start new hobbies, and participate in society with the aim to improve their quality life and wellbeing. This is how I want my old age to be.

The book's in-depth reading and critical reflections have also helped other students discover the unknown aspects about ageing and, as they stated, overcome their own fears about later life and ageing. As student SO8701 argued:

The reading of the book helped me to overcome the fear of growing older and showed that age as a category does not matter. If you did not fall in love when you were young, you can still fall in love when you are older, you should not be afraid of not being able to find the love of your life because many things will happen to us during our life span. Similarly, many of us are afraid of death or growing older, and I am one of them. However, the book helped me not to feel so overwhelmed, because if you enjoy your every day as if it was the last day, you will not have that worry, because you will have done everything you wanted and even more than that.

In the initial section, we showed the quantitative and argumentative predominance of the students' reluctance towards the idea of working with older people in care homes and related areas and services. However, the written tasks and the students' comments on the ways in which older women and older men experience active ageing show a significant change in their responses and reflections about older age. Many of the students have changed their initial thoughts. For example, student SO9201 commented that "although we do not want to work in areas related to old age, at some point in our professional career we will meet older people. Therefore, we must treat them in a certain way and be more patient and open-minded." Similar ideas are also expressed by student SO8201, who pointed out that "I believe that every social worker should know different things related to old age, for example, the potentialities of older people, the ways in which they physically age and their state of health. The book helped me see old age in a different way." Another SO9101 student stated that:

> *Thanks to the reading of this book, my perspective towards older age has changed in a positive way. Now I see that being older does not mean not being able to do anything. That is why this reading has helped me change my mind about social work and older adults. From the very beginning I was very sure that I did not want to work with older people because I saw older age as the last stage before death. I thought that working with them would be very monotonous. Now I no longer look at later life with the same eyes. I think that many projects can be carried out with older people that do not focus on their decline or only on their needs.*

In a similar vein, the student SO201 wrote that:

> *Thanks to this reading, now I want to learn more about older people. Until now it was the group I was the least interested in, but after reading the book I have realised that older people can teach us many things about life and I would like to delve deeper into it.*

4. Discussion

We began this article by considering the possible causes of the negative considerations about older age and students' reluctance to work with older people as a sum of social factors in formal, non-formal and informal education. We also looked into what beliefs and stereotypes about later life have been internalised by young people. We questioned how and why some of them had prejudices and negative notions of later years of life. We also wanted to know if alternative models of ageing, such as healthy and active ageing, could generate positive changes in the images of older age, and if reading and reflective work could favour the reconstruction of later life. The content analysis and the examples provided in this article show that there is a sum of causes of a social nature related to the social construction of old age. The results of the questionnaires indicate that we are facing an educational challenge that will impact the future generations. The longevity revolution [27], the feminisation of old age (since older women outlive older men), and the prolonged years of retirement create new paradigms and, therefore, a need for a conscious intergenerational education that could foster more positive aspects of ageing and old age [1,12].

It is also worth noting the students' image of old age as the last stage of life. Many students think that reaching retirement implies approaching the end of life or an empty life without projects or activities. Their answers show the stereotyping of older people and a clear differentiation between one's chronological age and social age. As in Van Gennep's rites of passage [34], the 'cut' that defines an administrative and sociological situation, such as retirement, is identified with passivity, the end of the vital process and the taboo of speaking about death. The idea of old age as the "end " must be demystified and integrated into the processes of profigurative reflection and intergenerational relationship, which would enable a broader and more holistic view of later life. Fostering different and more positive perspectives about ageing is especially relevant for older women because they live longer than men and are subject to the double standard of female ageing [28], higher levels of loneliness and poverty, lack of civic and social participation, and more financial burdens [24,32,35–37]. According to Nerenberg, "discrimination and disadvantage associated with both gender and age combine to compromise older women's ability to achieve or maintain self-sufficiency and make them more likely to be poor and/or dependent" [38] (p. 4).

The study also reveals how narrative inquiry can give an opportunity to tell one's life story. For example, students' surprise that older age is not an asexual phase in life is an example of the importance of what Bodil Hansen Blix calls untold and unexplained narratives [4]. Neither grandmothers nor grandfathers speak at home, out of fear or out of shame, about their sexual options or needs. The narrative voice facilitates this disinhibiting and gives visibility to older age and sexuality that is being presented from a more positive point of view [39]. The reflective use of narratives as an educational resource also fosters intergenerational education and profigurative socialisation. In fact, one of the main motivations in our initial university education is to overcome ageism by working on literary narratives with intergenerational groups at the local level (the characters in the book, the author of the book and the university students).

The educational and transforming power of the narrative set in a local context and, particularly, the reconstruction of the social image of being an older woman, also helps to break social prejudices and enables changes in students' attitudes and their future professional orientation. Many of them affirmed that they did think about active ageing or intergenerational relations, but after the reading exercises they started to consider working with older people as one of the possibilities or even as the main choice. Working on intergenerationality through narrative also helped reflect on beliefs and attitudes in relation to the different ages of life and gave an opportunity to share and create new knowledge. The literary and cultural analysis of the book, along with students' comments and feedback, provided us with relevant information regarding the notions and social beliefs about old age and ageing and, in particular, the social image of older women [15,40]. The class discussions and group debates served not only as a rich database and an exchange of information, but also allowed students to question the beliefs about old age and gender, especially in relation to a local context that is still predominantly sexist and heteropatriarchal. In addition, the study results reveal the often hidden aspects about ageing that would have remained invisible if they had not been collected through the narrative analysis and class discussions. The students reflected on loneliness in old age, the concern to maintain one's autonomy until death, the difficulty of establishing new social relationships after certain ages, pain and disability, among other important topics. The class exercises also helped students to think more about the relationships between young people and older adults and the need to foster intergeneration dialogue, social integration and profiguration that occurs by promoting education (formal, non-formal and informal) [6,7,12].

This case study also reveals that narratives can have a more powerful impact than scientific empirical data and demonstrates the importance of incorporating flexible reading guides to strengthen critical thinking skills and interdisciplinary education that leads to lifelong learning [41]. The results also show the need to incorporate the gender perspective in higher education and the study of the ageing process. The use of narratives as a didactic, formative, and socialising tool also implies dialogue, empathy, understanding, and solidarity, which involve the creation and sharing of new knowledge. Ageing, within the framework of intergenerational relations, has an alternative, enriching and socialising approach. In sum, the narrative connects the reader and the writer, fosters critical thinking, and contributes to raising awareness about ageism and the ways to overcome it.

5. Conclusions

One of the limitations of this case study is that is it only focused on the responses and work of the students enrolled in a specific course. In the future, we would like to extend the study and repeat the activity based on a different narrative of similar characteristics at the end of the course degree. Another limitation is that the questionnaires were carried out during the COVID-19 pandemic, which might have affected the students' responses due to negative images of ageing shown on the media that involved suffering, death, chaos, and the fear of ageing. Another limitation is that the questionnaire, introduced at the beginning of the course, was voluntary and, therefore, we could not take into account all the responses of enrolled students. These aspects will be improved in the near future and upcoming research. Taking these limitations into account, the results of the study provide useful and rich data to foster profiguration and age-friendly society and education.

The case study has also revealed that is important to take into account how older people are seen by young people. The study outcomes show that, on the one hand, there is awareness of the stigmatisation of old age created during upbringing and education, which makes younger people see older people as belonging to the retirement stage or the end of life. On the other hand, there is a permanent and constant idealisation of youth. Students are aware of the existing stereotypes about old age and they continue to construct later life as the opposite of the active and autonomous life that belongs to young people. Therefore, there is a need to provide students with deeper knowledge of the heterogeneity of all stages of life via a humanistic educational approach, intergenerational exchanges, and examples from familiar environments, gender relations, and local narratives (written in their mother tongues). Intergenerational reading would also allow breaking with stereotypes about later life and offer valuable data for future research. The incorporation of didactic and intergenerational work of proximity through the reading and analysis of narratives can help promote attitudes for action.

The scarcity of the gender perspective that is found in student reflections is also striking. It seems that the students start university studies having been educated in aesthetic feminism: with the politically correct usage of masculine/feminine/neutral or generic articles, but without openly clarifying whether they refer to men or women. In fact, 80% of the social work students are female and, theoretically, they are part of the generation that is well educated in gender-related issues and is aware of feminism. Taking into account that ageing affects not only older people but every one of us, the construction of an active and healthy ageing from the narrative and the gender perspectives must be part of the curricular content not only in sociology-and gerontology-related disciplines, but also in other degrees and courses. Moreover, aspects related to ageing are only part of optional subjects unlike the topics related to childhood, adolescence and youth, which are better integrated in core subjects.

This case study also demonstrates that the significance of giving more importance to the power of narratives in university education, since they transform viewpoints and promote attitudes that trigger action and change misconceptions. In fact, narratives can have a more powerful dimension in terms of personal impact than scientific research because they are presented as literary models that make us think critically. However, in an educational and socialisation context, 'disoriented' and unguided reading is not valid and empowering. The use of reading guides, which are open and flexible, help strengthen critical thinking and foster interdisciplinarity and intergeneration dialogue. Additionally, initial students' training should also trigger an interest in lifelong learning and how to age actively. On this matter, the data obtained from this study offer better clues about how to teach ageing and challenge ageism among university students. We have identified three main modalities of action: (a) to reflect on the students' training as professionals, which makes them more aware about the need to have better knowledge about later life beyond the stereotyped image of older age, (b) to overcome ageism, since it determines the ways the students perceive and treat older people, and (c) to make later life and ageing more visible by helping the students understand that there are many different ways of growing older.

The study results also reveal that the local context is not perceived as explanatory enough. Giving more emphasis and importance to local issues would help identify and think critically about global problems, such as globalisation and internationalisation. This study shows that the local context, seen from a micro perspective, allows students to identify the guidelines that are later developed and consolidated at a global level, which García-Canclini [42] refers to as a 'glocalised' context. Additionally, the references to local context and the gender dimension help provide space for intersubjectivity and reflection from a local perspective. This article also aimed to demonstrate that the use of narratives as a didactic, formative, and socialising tool implies proper individual characteristics and identifications, such as dialogue, empathy, understanding, solidarity, and sharing between those who are vulnerable and interdependent. Ageing, seen from the lens of intergenerational relations and profiguration, offers an alternative, enriching, and socialising approach to later life that can help improve emotional health and wellbeing of older adults. The use of narrative allows students to extend their reflections to other groups and alter their perceptions, which is needed in order to achieve social change and betterment. The in-depth reading helps connect with the readers, invite students to think critically, and overcome ageism.

In order to make significant changes in the social and cultural perception of ageing and to undo the narrative of decline [14], more action is needed beyond lectures or classroom sessions with university students. Scientific articles do not provide the necessary awareness, attitudes, values and norms about old age; thus, it is necessary to explore other ways of teaching and learning. Profiguration, narrative inquiry, and bringing together different generations to critically engage, discuss, and better understand the complexities of ageing is crucial to challenge ageism, give older adults a voice, and promote different perspectives on later life. Finally, although we started this section by listing the limitations, we would like to end on a positive note and hope that there will be more studies that use critical narrative approaches to challenge ageism, build a more sustainable and age-friendly society, and foster intergenerational dialogue. We plan to continue our research by using the same narrative and reading guide, while adding other readings related to ageing and content analysis in the near future.

Supplementary Materials: The following supporting information can be downloaded at: https://www.mdpi.com/article/10.3390/land11071057/s1, Annex S1: A Reflective Reading Guide; Annex S2: Analysis Scheme.

Author Contributions: Conceptualization, F.M.-L. and A.S.-B.; methodology, P.S.-T. and A.S.-B.; validation, F.M.-L., A.S.-B., P.S.-T. and I.S.; formal analysis, P.S.-T., A.S.-B. and F.M.-L.; investigation F.M.-L., A.S.-B. and P.S.-T.; data curation, I.S.; writing—original draft preparation, P.S.-T., A.S.-B. and F.M.-L.; writing—review and editing, P.S.-T. and I.S.; visualization, P.S.-T. and I.S.; supervision, F.M.-L., A.S.-B., I.S. and P.S.-T.; project administration, P.S.-T. and I.S. All authors have read and agreed to the published version of the manuscript.

Funding: This study is part of the ECAVINAR research project "Ageing, Quality of Life and Creativity through Narrative" (FFI2016-79666-TR) funded by the Spanish Ministry of Economy and Competitiveness MINECO. It has also received funding from the ASISA Foundation Chair of "Health, Education and Quality of Life".

Institutional Review Board Statement: The study was conducted in accordance with the protocol for the Processing of Personal Data in Research and Transfer Activities (code PO 019) of the University of Lleida (Data Protection Office, General Secretariat of University of Lleida).

Informed Consent Statement: Informed consent was obtained from all subjects involved in the study.

Data Availability Statement: Both the book *Casa Yé-ye* and the tasks were written in Catalan (the mother tongue of the author and the students). The book's online version is currently being prepared for open-access. The tasks of the students are stored in the repository of the activities of the subjects of the two courses.

Acknowledgments: We thank all the students for the work they have carried out developing the tasks with reflection, sincerity and critical thinking. All individuals included in this section have consented to the acknowledgement.

Conflicts of Interest: The authors declare no conflict of interest.

References

1. Yanguas, J. *Pasos Hacia una Nueva Vejez*; Planeta: Barcelona, Spain, 2021.
2. González, P.C.; Alba, B.G. *El Uso de las Narrativas en la Enseñanza Universitaria. Experiencias Docentes y Perspectivas Metodológicas*; Octaedro: Barcelona, Spain, 2020.
3. Barri, E.; Vibe Skagen, M. *Literature and Ageing*; D.S. Brewer: London, UK, 2020.
4. Blix, B. The importance of untold and unherad stories in narrative gerontology: Reflections on a field still in the making from a narrative gerontologist in the making. *Narrat. Work. Issues Investig. Interv.* **2016**, *2*, 28–49. Available online: https://journals.lib.unb.ca/index.php/NW/article/view/25799 (accessed on 12 February 2022).
5. Casado-Gual, N.; Dominguez-Rué, E.; Oró-Piqueras., M. *Re-Discovering Age(ing): Narratives of Mentorship (Aging Studis)*; Transcrip Verlag: Bielefeld, Germany, 2021.
6. Molina-Luque, F.; Casado-Gual, N.; Stoncikaité, I. University stakeholders, intergenerational relationships and lifelong learning: A European case study. *Educ. Gerontol.* **2018**, *44*, 744–752. [CrossRef]
7. Molina-Luque, F.; Stončikaité, I.; Torres-González, T.; Sanvicén-Torné, P. Profiguration, Active Ageing, and Creativity: Keys for Quality of Life and Overcoming Ageism. *Int. J. Environ. Res. Public Health* **2022**, *19*, 1564. [CrossRef] [PubMed]
8. Stončikaité, I. Age, Gender & Feminism: Filling the Gap From Literary and Cultural Perspectives. *Gend. A Výzkum/Gend. Res.* **2021**, *22*, 59–77.
9. Stončikaité, I. Ageing, creativity and memory: The evolution of Erica Jong´s literary career. *Life Writ.* **2019**, *16*, 25–36. [CrossRef]
10. Zorrilla-Muñoz, V.; Agulló-Tomás, M.S.; Rodríguez-Blázquez, C.; Ayala, A.; Fernandez-Mayoralas, G.; Forjaz, M.J. Ageing Perception as a Key Predictor of Self-Rated Health by Rural Older People. A Study with Gender and Inclusive Perspectives. *Land* **2022**, *11*, 323. [CrossRef]
11. Molina-Luque, F. The Art of Living as a Community: Profiguration, Sustainability, and Social Development in Rapa Nui. *Sustainability* **2020**, *12*, 6798. [CrossRef]
12. Molina-Luque, F. *El Nuevo Contrato Social Entre Generaciones. Elogio de la Profiguración*, 1st ed.; La Catarata: Madrid, Spain, 2021.
13. Villar, F.; Serrat, R. El envejecimiento como relato: Una invitación a la gerontología narrativa. *Rev. Kairós Gerontol.* **2015**, *18*, 9–29. [CrossRef]
14. Gullette, M. *Aged by Culture*; University of Chicago Press: Chicago, IL, USA, 2004.
15. Hubble, N.; Tew, P. *Ageing, Narrative and Identity: New Qualitative Social Research*; Palgrave Macmillan: London, UK, 2013.
16. Delors, J. Los cuatro pilares de la educación In *La Educación Encierra un Tesoro. Informe a la UNESCO de la Comisión Internacional Sobre la Educación Para el Siglo XXI*; Santillana/UNESCO: Madrid, Spain, 1996; pp. 91–103.
17. Elfert, M. Unesco, the Faure Report, the Delors Report, and the Political Utopia of Lifelong Learning. *Eur. J. Educ.* **2015**, *50*, 88–100. [CrossRef]
18. Ruiz, J. *El Arte de Pensar*; Libros en el Bolsillo: Barcelona, Spain, 2021.
19. Soldevila, A. *Casa Yé-ye*; Desktop Publishing: Lleida, Spain, 2019.
20. Dahlberg, L.; McKee, K.J.; Fritzell, J.; Heap, J.; Lennartsson, C. Trends and gender associations in social exclusion in older adults in Sweden over two decades. *Arch. Gerontol. Geriatr.* **2020**, *89*, 104032. [CrossRef] [PubMed]
21. Fernández-Mayoralas, G.; Pérez-Rojo, F. *Envejecimiento Activo, Calidad de Vida y Género: Una Aproximacion Desde las Experiencias Académica, Institucional y Social*; Tirant lo Blanch: Valencia, Spain, 2021.
22. Van-Dijk, T. Análisis crítico del discurso. *Rev. Austral. Cienc. Soc.* **2016**, *30*, 203–222. [CrossRef]
23. Andrés, S. *La España en la que Nunca Pasa Nada. Periferias, Territorios Intermedios y Ciudades Medias y Pequeña*, 1st ed.; Ediciones Akal: Madrid, Spain, 2021.
24. Fatou, B.; Roldan, E. Reflexiones feministas sobre las mujeres mayores, el envejecimiento y las políticas públicas. *Aproximaciones al caso español. Ex Aequo* **2013**, *28*, 103–117.
25. Lipscomb, V.B.; Marshall, L. *Staging Age: The Performance of Age in Theatre, Dance, and Film*; Palgrave Macmillan: London, UK, 2010.
26. Molina-Luque, F.; Casado Gual, N.; Sanvicén-Torné, P. Mujeres mayores también activas, creativas y fuertes: Modelos para romper estereotipos. *Rev. Prism. Soc.* **2018**, *21*, 43–74. Available online: https://revistaprismasocial.es/article/view/2438 (accessed on 10 January 2022).
27. Butler, R. *The Longevity Revolution: The Benefits and Challenges of Living a Long Life*; Public Affairs: New York, NY, USA, 2008.
28. Sontag, S. The double standard of aging. *Saturday Rev. Lit.* **1972**, *95*, 29–38.
29. Higgs, P.; CBC. IDEAS. Episode Produced by Mary Lynk. 'Fear Ageism, not Aging': How an Ageist Society is Failing its Elders? 2021. Available online: https://www.cbc.ca/radio/ideas/fear-ageism-not-aging-how-an-ageist-society-is-failing-its-elders-1.5978190?__vfz=medium%3Dsharebar&fbclid=IwAR0Ca-D97qfKZw1vbjvV9fReesED-eA6CfV0sEaMVoohhYPF6sA-W4Pz470 (accessed on 11 January 2022).
30. Thone, R. *Women and Aging: Celebrating Ourselves*; Haworth Press: New York, NY, USA, 1992.
31. Bengston, L.V.; Putney, N.M. Future 'conflicts' across generations and cohorts? In *The Futures of Old Age*; Vincent, J.A., Phillipson, C., Downs, M., Eds.; Sage: Londres, UK, 2006; pp. 20–30.
32. Calasanti, T.M.; Slevin, K.F.; King, N. Ageism and feminism: From 'et cetera' to center. *NWSA J.* **2006**, *18*, 13–30. [CrossRef]

33. Agulló-Tomás, M.S.; Zorrilla-Muñoz, V.; Gomez-García, M.V.; Blanco-Ruiz, M. Proposals for better caring ad ageing. In *Handbook of Active Ageing and Quality of Life. From Concepts to Aplications*; Rojo-Pérez, F., Fernandez-Mayoralas, G., Eds.; Springer: New York, NY, USA, 2021; pp. 319–335. Available online: https://link.springer.com/chapter/10.1007/978-3-030-58031-5_19 (accessed on 10 February 2022).

34. Van Gennep, A. *Los Ritos de Paso*; Alianza: Madrid, Spain, 2013.

35. Agulló-Tomás, M.S.; Zorrilla-Muñoz, V.; Gómez-García, M.V.; Criado, B. Liderazgo, envejecimiento y género. In *The Time is Now. Feminism Leadership in a New Era*; Alonso, A., Langle, T., Eds.; UNESCO Chairs in Gender: Buenos Aires, Argentina; ONU Educational Scientific and Cultural Organitzation: París, France, 2019; pp. 112–122. Available online: https://www.catunescomujer.org/globalnetwork/articulos/la-hora-del-liderazgo-feminista-new-publication-of-global-network-unesco-chairs-on-gender/ (accessed on 29 April 2022).

36. Casado Gual, N.; Domínguez Rué, E.; Worsfold, B. *Literary Creativity and the Older Woman Writer: A Collection of Critical Essays*; Peter Lang: Bern, Switzerland, 2016.

37. Lennartsson, C.; Lundberg, O. What's marital status got to do with it?: Gender inequalities in economic resources, health and functional abilities among older adults. In *Health Inequalities and Welfare Resources. Continuity and Change in Sweden*; Fritzekkm, J., Lundberg, O., Eds.; The Policy Press: Bristol, UK, 2007; pp. 179–198.

38. Nerenberg, L. A feminist perspective on gender and elder abuse: A review of the literature. In *National Committee for the Prevention of Elder Abuse*; Whashington, DC, USA, 2002; Available online: https://www.researchgate.net/publication/237400086_A_Feminist_Perspective_on_Gender_and_Elder_Abuse_A_Review_of_the_Literature (accessed on 11 January 2022).

39. Stončikaitė, I.; Mina-Riera, N. A creating writing workshop on sexuality ans ageing: A spanish pilot case study. *Societies* **2020**, *10*, 57. Available online: https://www.academia.edu/43715176/A_Creative_Writing_Workshop_on_Sexuality_and_Ageing_A_Spanish_Pilot_Case_Study (accessed on 23 February 2022). [CrossRef]

40. Hepworth, M. *Stories of Ageing*; Open University Press: Maidenhead, UK, 2000.

41. Hernández, F.; Villar, A. *Educación y Biografías. Perspectivas Pedagógicas y Sociológicas Actuales*; UOC: Barcelona, Spain, 2015.

42. García-Canclini, N. *La Globalización Imaginada*; Paidós: Buenos Aires, Argentina, 2012.

 *land*

Article

The Digitization of Seniors: Analyzing the Multiple Confluence of Social and Spatial Divides

Millán Arroyo-Menéndez [1], Noelia Gutiérrez-Láiz [2,†] and Blanca Criado-Quesada [1,3,*]

[1] Department of Sociology: Theory and Methodology, Faculty of Political Science and Sociology, Member of TRANSOC Research Institute, Complutense University of Madrid, 28223 Madrid, Spain; millan@cps.ucm.es
[2] Department of Marketing, Faculty of Business, Complutense University of Madrid, 28223 Madrid, Spain
[3] Sociology and Anthropology PhD Program, Faculty of Political Science and Sociology, Complutense University of Madrid, 28223 Madrid, Spain
* Correspondence: blancacquesada@gmail.com
† Deceased.

Abstract: The lower digitization among seniors must be understood in the context of the coming together of multiple digital divides. In addition to the obvious generation divide (age is one of the factors most determining digital uses), others also have an influence, such as a lower education or income level, which is characteristic of this group and also strongly correlated with lower use of new technologies. We also find gender differences in the digital uses of seniors (more pronounced than in the population as a whole) and a significant geospatial inequality in several variables. The latter is important due to both the rapid aging of the rural population, greater than that seen in the urban population, and the fact that the geographical areas with a lower income level, where the aging population tends to be concentrated to a greater extent, are also the areas where digitization reaches the least, in terms of both infrastructures and uses. This article addresses the multiconfluence of the aforementioned "digital divides in older people" (or "seniors"), trying to determine the effects and degree of importance of each, identify the main groups at risk of digital exclusion, and to characterize the technological uses of seniors and their main segments. To do this, we have used the microdata from the "Survey on Equipment and Use of Information and Communication Technologies in homes", produced by the Spanish Statistical Office (INE) for the year 2020.

Keywords: ICTs; digitization; digital divides; seniors; gender; socio-spatial inequalities

Citation: Arroyo-Menéndez, M.; Gutiérrez-Láiz, N.; Criado-Quesada, B. The Digitization of Seniors: Analyzing the Multiple Confluence of Social and Spatial Divides. *Land* **2022**, *11*, 953. https://doi.org/10.3390/land11060953

Academic Editor: Hossein Azadi

Received: 29 April 2022
Accepted: 14 June 2022
Published: 20 June 2022

1. Introduction

This analysis looks in greater depth at the causes and factors that explain the low digitization of the elderly (65 years and over), using data from Spanish society. Older people will henceforth be called seniors. Seniors is a euphemism we employ to stress the digital/technological limitations of the elderly. The digital limitations of the elderly are primarily due to their lack of participation in preferences regarding design development, ergonomic, and usability adaptations, among other characteristics where the elderly are considered [1,2].

It is well known that seniors are a group with a very low level of digitization, and that age is one of the most discriminating variables in access to the Internet and information and communication technologies. However, there are not many studies that focus on this group. The most obvious thing is to think that what explains the low digitization of this group is age. However, what we apply and try to discover here is to what extent variables other than this are relevant. When we designed this study, we already knew that seniors tend to have less education, lower income, a smaller male presence, and tend to live in more rural areas than younger people. And important social inequalities can also be seen in these variables. Therefore, a better understanding of the low digitization of seniors implies addressing the perspective of the analysis of the confluence of multiple inequalities, at least those of the aforementioned variables.

Methodologically, this implies resorting to multivariate causal analysis as the main technique. Taking into account the interrelation of the independent variables, a suitable alternative is path analysis through modeling with structural equations.

Important differences can also be seen by territory, province, and autonomous region, which require a special approach.

Once the influence of the multiple divides that affect the digitization of seniors is known, it will be possible to see the main segments of seniors, taking into account that previous studies warn that the group is not homogeneous.

To finish, we carry out a complementary analysis of the causes of their total digital exclusion, which covers more than half of all seniors.

The aims of our analysis are as follows, referring to Spanish society:

1. To quantify the explanatory importance of the variables involved in the seniors' digitization level (uses and skills) with special attention to socio-spatial variables, comparing the older people's results with those for the general population.
2. To study the possible digitization differences between seniors in terms of the main segments in this group.
3. To measure and explain the digital exclusion of this group, describing the sociological profile of those excluded.

2. Background

The concept of the digital divide was created to refer to the difference between people who access or do not access the Internet [3] that is, to refer to the differences between different social groups according to the percentage with Internet access. Although the first studies of the digital divide focused on access, it began to be evident that access did not imply use [4], so priority attention stemmed from access to use. Some authors who claimed to focus on use preferred the concept of digital inequalities [5] to differentiate their approach from the previous one. This was applied to analyze the differences between different Internet users due to different uses.

Other authors preferred to continue using the term digital divide, but differentiating between a first access gap versus a second uses gap [6,7] and also digital skills and competencies [4,8]. The thesis of this line of studies is that inequalities have not been reduced with the increase in digitization, but have diversified as digital uses are increasingly present in daily life, benefiting more people with a better socioeconomic position [9].

Our analysis is framed within the concept of the second digital divide or digital inequalities and argues the special digital vulnerability of seniors in two dimensions of digitization: uses and skills. A vulnerability that, according to this theoretical framework, would not only be digital but would also lead to social disadvantages and a more general social vulnerability. Following Spark's thesis [9,10], digital exclusion-a very frequent phenomenon among seniors-could lead to social exclusion since more and more access to social resources occurs through Information and Communication Technologies (ICTs).

In accordance with this theoretical framework, we present below a bibliographic review aimed at seeing the variables and factors that influence the levels of digitization (in uses and skills) or the digital exclusion of Spanish seniors in particular and also in other countries.

There is extensive empirical evidence of the limited digitization of older people. The following are just a few examples: Friemel finds that while Internet use stands at 80–90% of the population in Western societies, those over 65 years of age achieve much lower levels of use [6]. Other authors also find striking differences between the older persons' digitization and the rest of the population [11,12]. In addition, older persons make up the greatest proportion of the information-weak classes [13].

In Spanish society, we find that Internet access among older people is growing and at a faster rate than in society as a whole, albeit still at a very low level and far behind all other age groups [9,14].

Despite this overwhelming evidence, there are few studies that explore the position of this group and the causes and factors behind it [15]. Studies on older persons by Spanish authors are also considered very limited [16]. Moreover, there are practically no studies that address the role of socio-spatial variables in the low digitization of the senior population. In this aspect, the contribution of this work is original.

Most studies are restricted to describing the situation, presenting the group as disconnected or poorly connected, but do not look further at the causes, with the dominant shared stereotype being that older persons are not involved in technology [17,18].

Specifically in Spain, the scarcity and limitations of studies on the causes of the older people's great digital inequality provide strong justification for this work. Previous studies conducted in Spain are small in number, partial, and limited, although they have pointed to the following causes, factors, and explanations [19].

In 2012, a study tried to identify the technological resources, their use, and types of uses among seniors, although with an excessively small sample (n = 215) that did not allow the authors to find significant differences by gender or even by age, although it did find a difference by education level [20]. It is the main reference of rural seniors and the main contribution from the geospatial perspective.

In 2015, Pino Juste and Rodríguez López [16] conducted a survey on a small sample of fifty-two people who had taken courses for seniors, the results of which are not considered in this study due to the precariousness of the sample.

In 2016, there was a mixed methodology, qualitative and quantitative, study on rural seniors in the autonomous community of Castile and Leon, one of the most rural parts of Spain [20]. This study showed the multi-exclusion of seniors, finding that it is not due to a single cause and highlighting that the highest exclusion rates are found in the smallest municipalities and among the oldest seniors, although there are other variables that also have an influence, such as the distance to provincial capitals or larger municipalities. It also found that the role played by the 'support generation' (the children of the seniors, who are more digitized) is key in the digitization level of seniors.

The main Spanish reference for understanding the causes of exclusion or lower digitization of seniors is from 2017 [17]. The use of the Internet and its applications was studied on the basis of socio-demographic variables. This study found noticeable differences between seniors aged from 65 to 74 and those aged 75 and over, finding that younger seniors use the Internet much more and confirming the existence of seniors who are fully integrated into the use of the Internet. In this respect, the authors criticize the shared stereotype that seniors are cut off from technology and conclude that there is a need for a more detailed segmentation of the group in order to overcome the single uniform vision that studies have offered up to now. They also concluded that age is not sufficient to understand the group and point to other factors, such as psychological variables, that really explain the differences in the use of online banking and social networks.

Outside of Spain, other studies have contributed to the background on the digitization of seniors. One of the pioneering works is that of Wicks [21], which established the differences between old and young seniors that other subsequent studies have also highlighted and that will also be examined here. The study suggested that access barriers are not the only problem and pointed to reluctance as another significant barrier observed among seniors.

Another study from 2016 conducted in Switzerland on a sample of 1105 seniors [6] highlighted the influence of education, income, interest in technology, having worked with computers before retiring, and marital status. It also confirmed that gender-based differences disappear when controlling for the aforementioned variables.

Another noteworthy study is that conducted in the United Kingdom in 2018 [15], which studied the ownership and use of smartphones. This study highlighted the importance of variables such as social influence, facilitation conditions, service expectations, effort, and enjoyment.

The argument that the lower use and acceptance of ICT among seniors is caused by a lack of interest and the absence of social pressure has also been highlighted as a reason why seniors have lower digitization [11,18,22,23]. The importance of a self-perception of low skills has also been highlighted [23]. There are also less innovative [24] and have less need to search for information. [13,25]. Geospatial variables have been little used to date in order to explain the low digitization of seniors. In the bibliography in English and Spanish, we have not detected any relevant studies.

Other studies on digitization that are not specific to this age group are also useful for determining the relevant variables, at least in the digitization levels of the general population. In Spain, it is worth highlighting a study from 2011 [26] that verified, through a logistic regression model, the high explanatory capacity of education, age, and employment status. In contrast, gender and the characteristics of the geographical environment made minimal explanatory contributions. The study emphasized the explanatory capacity of social variables over geographic ones. The rural-urban dichotomy was also emphasized as an important variable in explaining the digital divide in another Spanish study [27]. Outside of Spain, Warf [28] found that Internet providers tend to provide less coverage in rural areas and highlighted the variable as one component of digital inequality, among others (age, education, income, gender, etc.) The contribution of this variable also appears in other studies [6]. However, the studies by Hindman [29] at the start of this century concluded that the differences between metropolitan and non-metropolitan areas were small and more limited than those for variables such as income, age, and education, which these authors demonstrated in Spain [26].

The first studies on digitization at the end of the 20th century and the beginning of the 21st put the emphasis on the great geographical inequalities that were fundamentally appreciated between countries [30]. The economistic interpretation prevailed in the explanation of the differences in infrastructure and access devices, so that digitization was largely explained by the level of territorial development [9,26,31]. In the early days, what was mediated was access to the Internet and these differences in access were reduced as the use of the Internet expanded, but big inequalities linked to the quality and modes of access available to users continued to be evident [9].

Since the first decade of the 21st century, the perspective of analysis has been changing towards a second level of the digital gap, beyond the access gap, in which it was important to examine the levels of competence and forms of use [9]. This change in perspective was associated with a lesser role in the analyzes of geographical aspects and a greater interest in other relevant variables of the social structure, such as age, sex, studies, or family income. Underlying the theoretical approach is that inequalities in the social structure determine digital inequalities. Digital inequality is considered to reproduce the old classical inequalities in the digital society [32].

Among the various authors who highlight this new overview, Dimaggio stands out [4,26]. He criticized the perspective of geographic politics and the "connected/non-connected" dichotomization that prevailed until then. Dimaggio argued that the expansion of the Internet among the population, as well as its services and infrastructures, did not guarantee the reduction of the digital gap, demonstrating that in those territories in which there was universal access to the Internet, important inequalities persisted. This causes academic interest to shift from having or not having access to the differences in digital uses and skills. From the new perspective, the explanatory interest of social variables increases compared to geographic ones, evidencing the great influence of variables such as race, gender, education, age, etc.

However, although in terms of Internet access, the geographical and spatial differences have decreased a lot with the expansion of the Internet, this has also decreased a lot in other social variables. And yet, although great differences are also appreciated in the uses and competencies in social variables, important geographical differences continue to be appreciated (in competencies and uses) also in the geographical variables. On the one hand, this is due to the theoretical interest having shifted from the geographical to the social. On the other hand, causal analyzes in surveys of digital uses are giving a more secondary role to geographical variables than to social ones.

If the causal studies indicate (despite the importance of these differences) a lower explanatory capacity of the socio-spatial variables compared to other social variables, it could be due to inequalities in the social structure are not only determining digital inequalities, but also the special segregation of the population, further contributing to the digital exclusion of the seniors. This is our main hypothesis, and it will have to be examined by studying the interactions of social variables with geographical ones.

In accordance with this bibliography and also considering the possibilities and limitations of the available data source, we have selected the variables or factors that are worth exploring in our research to look in more detail at the reasons behind the lower digitization of seniors in Spain:

Age. The importance of this variable is sufficiently demonstrated in the specialized literature [5,9,11,12,14,24,33], etc. This is the first variable to consider since, as we have seen, significant differences between older people and younger age groups have been identified. Differences have also been found between young and old seniors. There are two ways in which age can have an impact and cause lower digitization: one as a cohort effect and another as a life-cycle effect. In the first, it is explained by differences in the socialization and learning of the different generational groups. The second relates to physical limitations or adaptations to age-appropriate lifestyles. According to Friemel [6], most authors tend to interpret the age effect as a temporal cohort effect. While the author confirmed this effect, he found that those over 85 years of age do not use the Internet due to vision and hearing limitations. The age variable is also correlated to other variables, such as education or income level, which can be reflected in differences between age groups if the intervening effects of these variables are not eliminated.

Education. The numerous studies of Spanish society demonstrate its great explanatory power, which is also seen in most previous studies [9,27,33,34].

Income. With regard to the household income variable, we find that this is one of the determining variables in Spain, along with age, education, and gender [35].

Sex/Gender. There are many studies that address the gender digital divide. Although differences have been narrowing since the start of this century [27], they still exist between men and women in terms of uses and skills [7,27]. As a counterpoint to the importance of the differences always found through bivariate analyzes, in which women have lower digitization, we find that multivariate causal approaches minimize or eliminate the explanatory importance of this variable, on being controlled for by others or by including independent variables with greater explanatory power in the models [6,19]. Alternatively, this can be largely explained as a reflection and reinforcement of educational inequalities [7,27].

Geospatial variables. There are also many bibliographical references to territorial or geospatial inequalities [14,29,35]. In Spain, for example, big differences can be seen among autonomous regions, and also others not so great by the size of the municipality [26]. These variables, however, have a lower explanatory power than such social variables as age, education, and income. This is true in both Spain and other developed countries [29]. We have explored different variables: the size of the municipality, demographic density, and territorial division by province.

Other variables of explanatory interest that will not be considered due to the limitations of the source are the following: psychological, attitudes, the influence of the social environment, and socio-cultural variables. Given the limitations of the available source, we can only consider those mentioned above. These are variables relating to social structure commonly used in other studies analyzing the second digital divide or social inequalities [16,27], and which have served to confirm that the inequalities seen in the social structure prior to digitization also have an impact on it [9,33]. According to some authors, the risk of digital and social exclusion among the most disadvantaged groups, largely made up of seniors, increases as a greater diversity of activities and services move over to the digital world [9,10].

Employment occupation has often been considered as another explanatory variable, although we have not taken it into account because the vast majority of seniors are retired and in the available data there is no information on occupation before retirement.

The group of seniors is not only affected by inequality as an age group but is also affected by the confluence of inequalities that can be seen in the variables mentioned above and that constitute a multiple digital divide: older age, lower education level, lower income level, greater presence of women and rural inhabitants.

The digitization variables or dimensions still need to be explained and substantiated and this is what we will explore. We will focus on what has been called the second digital divide, or digital inequalities, placing less emphasis on Internet access, which has reached high levels of coverage, and instead focusing on two factors. First, the skills and competencies required in the digital world [4,8,36]. Second, is what the Internet is used for [6,7,36]. These are important aspects of the analysis of the so-called second digital divide or digital inequalities [5]. More specifically, we will make operational the variables that define the concepts of digital skills and the intensity and diversity of uses.

Ageism, understood as negative age stereotypes and negative self-perceptions about aging [37], would be another hypothetical factor to explain the lower digitization of seniors. There is evidence that the mechanisms of websites limit the participation of older people, which is explained by the homophily of corporate teams, mechanisms that deprioritize, neglect, or exclude older people [38]. Also, the lack of awareness and understanding of the needs and difficulties of older people by web designers has been proven [39]. But this negative prejudice not only stems from younger people but also from seniors when they internalize their limitations and lack of aptitude, based on the dominant social perception [37]. In this way, it is closely related to other previously mentioned limitations-psychological, attitudinal, or the influence of the sociocultural environment.

3. Materials and Methods

We analyzed the 2020 microdata file, from the Survey on Equipment and Use of Information and Communication Technologies in Households, conducting an original statistical study of this data (Survey of the Spanish Statistical Office (INE) and coordinated at the European level with Eurostat). The fieldwork was conducted between 2 March and 15 September, concentrated in recent months due to the lockdowns resulting from the pandemic during the months of March to June. This survey partially reflects the progress in digitization that took place in Spain following the lockdowns during the COVID-19 pandemic, and also partially reflects the progress in public policies to support the digitization of rural areas. It is a self-weighted sample of 15,343 cases for the total population and 5247 for those over 65 years of age (older people population). These sample sizes allow segmentations to be made that are appropriate to the research objectives (The random error limit for all older people is $+/- 1.38\%$ for $p = q = 50\%$ at a confidence level of the order of two-sigma).

The older people population has been identified as those aged sixty-five and over, an age very close to the real retirement age in Spain and after which the proportion of employed people becomes insignificant.

In order to empirically make the two proposed digitization dimensions operational (uses and skills), the following procedure was performed: to obtain a digital skills variable, the INE methodology was followed, approved by Eurostat, based on four levels among those who have used the Internet with a frequency equal to or greater than the last three months: No skills, Low, Basic and Advanced. In addition to preparing the indicator following the indications from the INE, we have added the 'cannot be assessed' category due to the importance of including all those who do not use the Internet or do so less frequently than in the last three months in order to detect digital exclusion. The result is a variable with five categories coded from 0 to 4, in ascending order of skills. For further information please consult Annex 4 in the study methodology in: https://www.ine.es/metodologia/t25/t25304506620.pdf Also can be seen: https://digital-strategy.ec.europa.eu/en/library/new-comprehensive-digital-skills-indicator (accessed on 7 June 2022).

In order to obtain a digital uses variable, which allows us to calibrate both: The frequency of use and the variety of types of digital uses (aspects highly correlated with each other), an index was created through a factor analysis of principal components, forcing the extraction of a single factor, because we wanted to create a single index of variety and intensity of uses. This was applied to a representative set of twenty-one items available in the questionnaire recoded for the analysis with: code 1 for mentions and 0 for no-mentions and non-responses (non-responses were very scarce and attributable to non-use). The result was a standardized variable with a mean of 0 and standard deviation of 1, with negative values for values below the mean for the Spanish population as a whole and positive values for values above this mean. The index obtained was reviewed and validated with excellent results, through the analysis of Cronbach's Alpha statistic (which measures the consistency of the index based on the degree of correlation between the items), finding a value for the final selection of items of A = 0.943 (very high consistency). The goodness of fit for the factor in the principal component analysis was very high: Kaiser Meyer Olkin (KMO) = 0.962 (Excellent). For some tabulations, the factor scores were grouped into quintiles, taking the total sample as a reference for obtaining the quintiles. In this way, each quintile in the sub-sample of seniors obtained a similar skills level to the total sample.

In order to identify the influence of the explanatory variables on the two digitization dimensions (uses and skills), we prepared causal models through a path analysis (using AMOS, software for modeling through structural equations, in the statistical package for social sciences SPSS). A model was prepared to explain uses and another to explain skills, both of these being dependent variables in each model. Moreover, to better understand the specifics of the older people population, for each explained variable, the results of the models for the senior population were compared to the total population. The explanatory variables in all the models were the same, although their parameters logically varied. These were: age, education, income, rural, and gender. Parameters and covariances were estimated using the maximum likelihood method, adjusting the saturated and independence model. In all the models, we obtained an optimum fit of the data to the modeling proposed (Chi-Square = 0) in all the models presented.

The choice of modeling method reflects the interest in identifying the effects of all the variables that identify the coming together of various digital inequalities in the seniors' group while considering collinearity between the explanatory variables, which is sometimes very strong. Path analysis fits this situation well, and estimation by the maximum likelihood method is reasonably robust when working with variables for which the criteria of univariate and multivariate normality are not met, without this significantly affecting the estimators.

The characteristics and transformations of the explanatory variables are explained below. There was no need for any processing of the age variable, a metric variable with no missing or unclassified cases. Education was recorded as follows: 0 = No education or incomplete primary education. 1 = Complete primary education. 1.5 = Not classifiable (there were only 36 in the total sample). This value was given after verifying, through a simple correspondence analysis, that they were between primary and secondary education). 2 = Secondary (first or second stage of secondary education and similar). 3 = Vocational Training (VT: VT1 and VT2). 4 = Higher Education (University). Income means net family income, divided into fourteen payments per year, which is common in Spain. Starting from the original intervals in the questionnaire, in order to transform the variable into a scale we have recalculated the means of each income interval, or an estimation by approximation of the highest and lowest (taking other national surveys in 2020 as a reference). Non-response for income, which is high, has been estimated through a simple correspondence analysis based on the distance with the closest intervals. The estimation is needed to run the models with the AMOS software because it does allow the introduction of variables with missing cases. The values for each interval were as follows: 800 = Less than €900. 1250 = From 900 to less than 1650. 1675 = Don't know. 2050 = From 1600 to less than 2500. 2750 = From 2500 to less than 3000. 3300 = From €3000. The gender variable was recoded as follows: 1 = Men, 0 = Women. The rural variable was subject to various considerations before being made operational, considering the limitations of the survey. The use of the habitat size variable was ruled out because it grouped municipalities with fewer than 10,000 inhabitants into the lowest population interval. Instead, the demographic density variable was used, whose category 'sparsely populated areas' (or scattered) corresponds mainly to the smallest municipalities and has a much closer relationship with the digital divide, precisely because these are the areas with less Internet coverage (lower coverage of ADSL, fiber, and mobile telephony). It also corresponds better to the 'rural' concept of municipalities with fewer than 10,000 inhabitants (91.2% of these areas are located in municipalities with fewer than 10,000 inhabitants). Therefore, sparsely populated areas, which we will call 'rural' here, were coded with 1 and the rest with 0 (densely populated areas and intermediate areas).

As a complement to the causal analysis of survey data, a causal analysis of the territorial differences between provinces is also carried out, to determine to what extent the social variables explain the geographical differences. It is carried out from the perspective of aggregate data, considering as cases or records (52) each of the 50 provinces and the 2 autonomous cities (Ceuta and Melilla) and combining potentially explanatory variables of the differences in the data file and in other external sources, such as GPD, GPD per capita, and various Internet access indicators: Long Term Evolution (LTE) coverage for mobile phone data and fiber optic coverage Fiber To The Home (FHTT). The source of information for the economic data is the National Institute of Statistics (2020 data). The Source on Internet coverage infrastructures, a report from the Ministry of Economic Affairs and Digital Transformation, referring to 2020 [40].

It has been decided to present the provincial perspective instead of the grouping in 17 autonomous communities because more differences are observed by provinces than by communities and the large size of the sample offers statistically significant results.

Once the effects of the different digital divides on older people had been identified, we performed a segmentation of the older people group and adapted it to differentiate between the various digitization levels. To do this, we used the relevant variables coming out from the previous analysis and the CHAID segmentation method (Chi-Square Automatic Interaction Detector). This technique allowed us to create a segmentation tree of the dependent variable (digitization), starting from a set of variables that will be segmented according to the interactions detected for the best description of the dependent variable.

Although this technique allows us to work with non-categorical variables, in order to simplify the solutions and seek a better fit, we first dichotomized the segmentation variables trying to dichotomize each variable with the most discriminating category or interval possible according to previous explorations: Age: from 65 to 75 years vs. 76 and over. (The age categories respond to two criteria: make better use of the possibilities of the sample and distinguish between younger and older seniors). Education: low vs. not low (low included no education, incomplete and primary education, the rest: not low). Income: Low vs. not low (low = less than €900, the rest were not low). The urban and gender variables were already dichotomous. For the recordings, the values of the digitization variables in each category have been used to achieve the most discriminating dichotomization possible.

As criteria for configuring the segmentation, based on the possibilities of the sample for n = 5247 older people, we have limited the segmentation to three branching levels and a minimum of 200 cases for each final node. The result was a classification of seniors into eight final segments. Two segmentations were performed, one for digital uses and another for digital skills. The resulting classification was exactly the same in each case.

4. Results

4.1. Differences between Seniors and the General Population

Although previous studies have identified differences between seniors and the general population [4–6,14,15,26,28,33,34], we consider it important to make at least a brief reference to the differences identified using our information source. This is in the form of a summary and allows us to discuss these differences and introduce the main variables we are going to analyze. We have considered all variables for which previous studies have detected digital divides, with the sole exception of occupation, as the older people group is mostly retired. Shown below is a table that compares seniors to the rest of the population and the total population. Table 1 shows the variables accounting for the main digital divides and the digitization dimensions studied here.

Table 1. Differences between the seniors and the general population.

		Age		Total
		16–64	+65	
STUDIES	Incomplete	2%	22%	9%
	Primary	8%	35%	18%
	Secondary	47%	28%	41%
	Professional	13%	3%	10%
	University	28%	11%	22%
Total		100%	100%	100%
INCOME	Low	13%	27%	18%
	Medium-low	18%	19%	18%
	Medium	20%	15%	18%
	Medium-high	8%	4%	7%
	Highest	11%	4%	9%
	DK	29%	31%	30%
Total		100%	100%	100%
RURAL	Urban	84%	80%	83%
	Rural	16%	20%	17%
Total		100%	100%	100%
SEX	Female	52%	60%	55%
	Male	48%	40%	45%
Total		100%	100%	100%
DIGITAL SKILLS	Not users	4%	54%	21%
	None	1%	5%	2%
	Low	31%	29%	31%
	Basic	19%	8%	15%
	Advanced	44%	5%	31%
Total		100%	100%	100%
DIGITAL USES	Low	5%	55%	22%
	Med-Low	15%	25%	18%
	Medium	24%	12%	20%
	Med-High	27%	6%	20%
	Highest	29%	2%	20%
Total		100%	100%	100%
n–		10,096	5247	15,343

The data confirm that seniors have much lower education and income levels than the rest of the population and the population as a whole, with clearly higher percentages for women and inhabitants in rural areas. (Studies Chi-square: 4078, fd 5 $p = 0.000$. Income Chisquare: 706 df 5 $p = 0.000$).

We also confirm that their digital skills and digital uses are much lower than those of the rest of the population and the population as a whole. (Digital Uses Chi-square: 6533 df 4 $p = 0.000$. Digital Skills Chi-square: 6187 df 4 $p = 0.000$). These variables for the two digital dimensions presented here will be explained in the next section. With the data presented, we illustrate that the seniors profile includes the characteristic features of the various and most important digital divides.

4.2. Digitization Variables and Their Relationship with Digital Divides

Following the criteria set forth in Section 2 *Materials and methods*, the two digitization dimensions we set out to study have been made operational: uses (intensity and variety of uses) and skills, creating two variables that can be examined empirically to establish a relationship with the variables causing the digital divide. Below (see Table 2) is a table with the descriptive statistics for the resulting variables, in both the general population (as they were constructed with reference to the total sample) and the senior group. The digital skills variable is presented in both a standardized and non-standardized way (both ways have been used in the analyzes).

Table 2. Descriptive statistics of the digital variables.

			Descriptive Statistics. Total Population		
	N	Minimum	Maximum	Average	Stand. Dev.
Digital Uses	1543	−1.3859	1.6741	0.0000	1.0000
Digital Skills (Stand.)	1543	−1.5784	1.1501	0.0000	1.0000
Digital Skills (Not Stand.)	1543	0.0000	4.0000	2.3139	1.4660
			Descriptive Statistics. Seniors		
	N	Minimum	Maximum	Average	Stand. Dev.
Digital Uses	5247	−1.3859	1.6741	−0.8773	0.7280
Digital Skills (Stand.)	5247	−1.5784	1.1501	−0.8634	0.8517
Digital Skills (Not Stand.)	5247	0.000	4.000	1.048	1.249

It can be seen that digital uses (a standardized variable from a factor analysis of twenty-one types of uses) have a mean of 0 and a standard deviation of 1 in the total population, whereas, in the seniors' group, there is a much lower mean of −0.8773. This already indicates that seniors score much lower than the population as a whole in intensity and variety of digital uses. The result is very similar for digital skills if we consider the standardized version of the variable, whose zero mean in the population as a whole becomes a mean of −0.8624. The non-standardized digital skills variable adopts four scale values (from 0 to 4) according to the following categories: not applicable (does not use the Internet), none (no skills), low, basic, and advanced. While the average for the population is 2.3 (between low and basic), that of seniors is 1 (no skills on average for the group).

In order to cross-correlate the variables constituting the digital divides, the non-standardized scale of digital skills has been used, and, in addition, the digital uses variable has been recoded into five quintiles, taking the total sample as a reference. Shown below is a table of results (Table 3) that compares seniors with the sample as a whole.

Table 3. Digital variables: seniors and total population.

Digital Uses Typology		
	Total Pop.	Seniors
Low	21.7	54.9
Med-Low	18.3	25.0
Medium	20.0	12.0
Med-High	20.0	5.8
Highest	20.0	2.3
n=	**15,343**	**5247**
Digital Skills Classification		
	Total Pop	Seniors
Not Internet Users	21.2	53.8
None	2.4	4.6
Low	30.5	29.2
Basic	15.3	7.7
Advanced	30.5	4.7
n=	**15,343**	**5247**

It can be seen that most seniors are concentrated in the lower digital uses intervals (74.9%) and that they completely lack digital skills (58.4%). However, there is a small minority who scored high in uses (2.3%, or 9.1% if we add together the high and medium-high scores) or profess advanced skills (4.7%). The differences between these figures and those for the entire sample are huge.

Shown below is a table listing all items that have been considered when preparing the digital uses variable. The results for the seniors' group are shown, although it should be remembered that the variable was obtained for the entire sample. In addition to presenting the items and examining the mentions in the group, data cross-referenced by the regrouping of factor scores are presented, which will allow us to make some interesting observations.

The main uses have a minor presence within the group. These are: using instant messaging (39%), phoning or making video calls over the Internet (30%), reading the news online (29%), searching for information on goods and services (26%), or on health issues (25%). The average number of uses is 3.4, mainly those mentioned. However, the majority (54.9%) do not use the Internet and do not mention its uses. For this reason, the level of medium-low uses (second quintile in the sample as a whole) is higher than the average: 3.9. The intermediate level is 9.6 uses, the medium-high level is fourteen and the high level is 17.9 (out of a total of 21 uses). The table (Table 4) shows the correspondence between the means of uses and the means of the factor scores, giving a less abstract idea of what the factor scores represent in terms of the diversity of uses. The results support the construct validity of the variable, in terms of diversity of uses. In terms of intensity, this is demonstrated by correlating it with the frequency of use of the Internet, which is $r = 0.797$ among seniors and $r = 0.787$ in the total sample.

Now that we have identified the digitization variables, we can study the relationships between these and the digital divides in the seniors group. We obtain a first approximation from the following cross tables (See Table 5).

Table 4. Items comprising the digital uses variable.

Activities Performed on the Internet in the Last 3 Months	Digital Uses Typology					Total
	Low	Med-Low	Med	Med-High	Highest	
Emailing	0%	21%	77%	97%	100%	22%
Phoning or making video calls over the Internet (WhatsApp, Messenger, etc.)	0%	50%	79%	92%	99%	30%
Social networks	0%	17%	42%	61%	88%	15%
Using instant messaging (via WhatsApp, Skype, Messenger, etc.)	0%	81%	94%	99%	99%	39%
Searching for information on goods and services	0%	33%	85%	97%	100%	26%
Reading news, newspapers, magazines online.	0%	43%	89%	95%	99%	29%
Listening to music (YouTube, Spotify, etc.)	0%	14%	45%	74%	92%	15%
Watching programs streamed over the Internet by TV channels	0%	10%	34%	53%	91%	12%
Watching video content on sharing sites. (YouTube, etc.)	0%	24%	55%	77%	98%	19%
Watching films or videos through on-demand companies (Netflix, HBO, etc.)	0%	10%	33%	56%	86%	12%
Searching for information on health issues	0%	34%	76%	86%	92%	25%
Making a medical appointment via a website or app	0%	13%	41%	59%	67%	13%
Using electronic banking	0%	16%	62%	90%	96%	19%
Posting own content to be shared	0%	5%	18%	38%	69%	7%
Using storage spaces on the Internet to store files	0%	2%	18%	50%	84%	8%
Taking a course online	0%	0%	3%	13%	43%	2%
Using learning material online	0%	2%	8%	31%	70%	5%
Communicating with teachers or students using educational websites	0%	1%	3%	13%	37%	2%
Obtaining information from websites	0%	7%	40%	81%	97%	13%
Sending completed forms	0%	7%	34%	72%	89%	12%
Downloading official forms	0%	3%	24%	65%	90%	10%
Average of uses (about 21)	**0**	**3.9**	**9.6**	**14.0**	**17.9**	**3.4**
Factor Score Average	−1.39	−0.78	0.07	0.70	1.25	−0.88
Senior Population % (n = 5247)	**54.9**	**25**	**12**	**5.8**	**2.3**	**100**

The table shows the correspondence between low levels of digital uses in the senior group and: low educational levels, older age, low income, rural residences, and female gender. And vice-versa, there is correspondence between high levels of digital uses and high education levels, lower age, high income, and non-rural residence. (Cramer's V: 0.443 for age, 0.573 for studies, 0.463 for income, 0.128 for rural and 0.114 for sex. Always $p = 0.000$). All the variables examined show a close relationship with digital uses, particularly studies, income, and age.

Table 5. Digital uses typology by digital divides.

		Digital Uses Typology					Total
		Low	Med-Low	Medium	Med-Hight	Hight	
AGE	65–75	33%	71%	79%	88%	92%	53%
	76+	67%	29%	21%	12%	8%	47%
	Total	100%	100%	100%	100%	100%	100%
STUDIES	Incomplete	34%	12%	4%	1%	2%	22%
	Primary	43%	35%	19%	9%	3%	35%
	Secondary	18%	38%	42%	40%	35%	28%
	Professional	1%	5%	7%	8%	7%	3%
	Universitary	3%	10%	28%	42%	53%	11%
	Total	100%	100%	100%	100%	100%	100%
INCOME	Low	37%	20%	11%	7%	7%	27%
	Medium-low	21%	20%	18%	13%	6%	19%
	Medium	8%	19%	28%	28%	33%	15%
	Medium-hight	1%	4%	9%	11%	11%	4%
	Hight	1%	3%	9%	18%	30%	4%
	DK	32%	35%	25%	23%	14%	31%
	Total	100%	100%	100%	100%	100%	100%
RURAL	Urban	75%	84%	88%	85%	88%	80%
	Rural	25%	16%	12%	15%	12%	20%
	Total	100%	100%	100%	100%	100%	100%
SEX	Female	63%	62%	53%	47%	41%	60%
	Male	37%	38%	47%	53%	59%	40%
	Total	100%	100%	100%	100%	100%	100%
Seniors	**(n = 5247)**	**2879**	**1314**	**628**	**303**	**123**	**5247**

Table 6 shows the correspondence between low levels of digital skills and: low educational levels, older age, low income, rural residences, and female gender. And vice-versa, there is correspondence between high levels of digital skills and high education levels, lower age, high income, and non-rural residence. (Cramer's V: 0.443 for age, 0.590 for studies, 0.469 for income, 0.128 for rural and 0.161 for sex. Always $p = 0.000$). All the variables examined show a close relationship with digital skills, particularly studies, income, and age, as it happens with digital uses.

The data examined confirmed that the five variables constituting digital divides clearly differentiate the digitization levels of the seniors group. We have also verified that in the sample examined (and in keeping with the existing literature), notable differences are seen in these variables if we compare seniors to the total sample. In the next section, we will analyze the explanatory weight or contribution of these five variables to the low digitization levels of seniors.

Table 6. Digital skills classification by digital divides.

		Digital Skills Classification					Total
		Not Users	**None**	**Low**	**Basic**	**Advanced**	
AGE	65–75	32%	65%	75%	82%	86%	53%
	76+	68%	35%	25%	18%	14%	47%
	Total	100%	100%	100%	100%	100%	100%
STUDIES	Incomplete	35%	14%	10%	3%	0%	22%
	Primary	43%	44%	31%	10%	5%	35%
	Secondary	18%	37%	40%	39%	34%	28%
	Professional	1%	3%	5%	8%	7%	3%
	University	3%	2%	14%	40%	53%	11%
	Total	100%	100%	100%	100%	100%	100%
INCOME	Low	37%	25%	18%	7%	5%	27%
	Medium-low	21%	24%	20%	13%	9%	19%
	Medium	8%	11%	21%	31%	32%	15%
	Medium-high	1%	1%	5%	11%	12%	4%
	Highest	1%	2%	4%	16%	25%	4%
	DK	32%	37%	32%	22%	17%	31%
	Total	100%	100%	100%	100%	100%	100%
RURAL	Non rural	75%	81%	85%	87%	84%	80%
	Rural	25%	19%	15%	13%	16%	20%
	Total	100%	100%	100%	100%	100%	100%
SEX	Female	63%	67%	63%	42%	36%	60%
	Male	37%	33%	37%	58%	64%	40%
	Total	100%	100%	100%	100%	100%	100%
Seniors	**(n = 5247)**	**2825**	**241**	**1531**	**403**	**247**	**5247**

The spatial variables available in the questionnaire were the following: habitat size, demographic density, and territorial division by autonomous communities and provinces. Table 7 presents data on the senior population according to habitat size and demographic density. Disaggregated data at the provincial level is also available in Appendix B.

Table 7. Factor score's means by geographic variables.

Habitat Size	Digital Uses	Digital Skills
500 thousand & +	−0.67471	−0.63737
Capitals-500	−0.80883	−0.78517
10 a 49 thousand	−0.86033	−0.86222
50 a 99 thousand	−0.81826	−0.77171
20 a 49 thousand	−0.95258	−0.95016
10 a 19 thousand	−0.93739	−0.91998
−10 thousand	−1.02600	−1.03833
Total	−0.87732	−0.86335
Demographic Density Areas	**Digital Uses**	**Digital Skills**
Densely populated	−0.77954	−0.75469
Intermediate	−0.94760	−0.93158
Sparsely populated	−1.04633	−1.06359
Total	−0.87732	−0.86335

Digital uses and skills decrease as habitat size and population density decrease. The differences in these means are significant both in the senior population and in the general population according to the Kruskall-Wallis test ($p < 0.001$). With the same test, the differences in digital uses and skills are also significant when compared by province. By provinces, moreover, the differences are greater. This information is interesting because, as we will explain later in the causal analysis, the geographical variables do not stand out. If the relationship of geographic variables with digitalization is much stronger in a bivariate analysis than in a multivariate causal analysis, it could be due to the unequal spatial distribution of social variables, such as age, studies, or income. This is our hypothesis that we will prove later.

We have paid special attention to the rural-non-rural dichotomy, within the possibilities of the survey. The category of less than 10,000 inhabitants as an indicator for rural areas was ruled out. The 'sparsely populated' category of the demographic density variable was more consistent with the concept and more discriminating in the causal analysis, after verifying that these areas are far from urban centers and 93% belong to municipalities with fewer than 10,000 inhabitants. They are rural areas far from representative services of what is popularly called 'España vaciada' (empty Spain). So, the variable called 'Rural' responds to the dichotomy of living or not in these scattered areas.

4.3. The Explanatory Weight of Digital Divides

To identify the explanatory weight of the different digital divides, four explanatory models have been prepared using path analysis, allowing us to examine the interrelationships between the independent variables and the direct effects of the explanatory variables on the variable being explained. The explanatory models for each of the digitization dimensions, uses, and digital skills, are represented below. In turn, for each of the two dimensions, the results obtained for the population as a whole (16 and over) and for the senior population (65 and over) are compared. The comparison between seniors and the general population allows us to better understand the specific characteristics of seniors. The models present standardized estimates, which measure the strength of the relationships between variables. The direct effects of the explanatory variables on the dependent variable are standardized regression coefficients, whereas the relationships between the explanatory variables are covariances (to see the values of the estimates and statistical significances of the path analysis models, see Appendix A).

A larger number of variables than those finally presented in the models have been explored. Those that were not significant have been eliminated, with the exception of the sex/gender variable, to which we pay special attention ($p > 0.05$ for digital uses models, see Appendix A). Habitat size was not significant after having incorporated the 'rural-non-rural' variable into the model, which is more discriminating and with which it is closely correlated (Spearman's Rho = 0.657 for Senior Population). The close correlation between both variables explains why habitat size is not significant in the model when we include 'rural-non-rural'. On the other hand, the territorial variables were not included in these models and were subject to a special treatment that is presented in the following Section 4.4.

When we examine digital uses in the total population (Figure 1), we see that the variable with the greatest explanatory power is age. This is the most significant digital divide. It is followed by education and, to a lesser extent, income. The rural/urban dichotomy has a low but statistically significant explanatory power. However, the influence of the gender variable, with a standardized coefficient close to zero, is not statistically significant, so its explanatory contribution is not shown. The most digitized are: the youngest, those with more education and more income, and those in non-rural areas. Vice versa, the least digitized are the oldest, those who have more studies, less income, and those who live in rural areas.

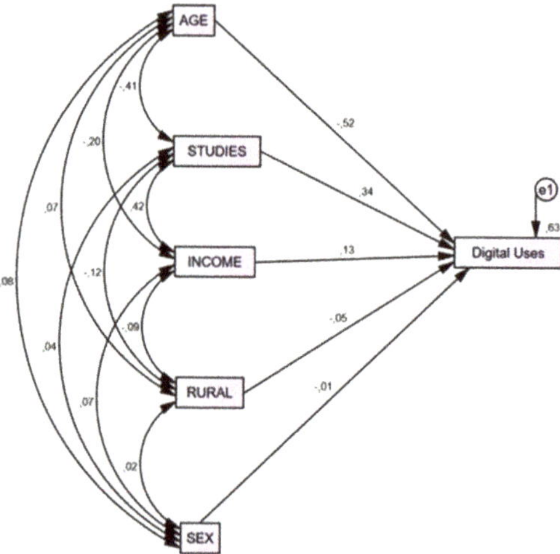

Figure 1. Digital Uses Total Population Model.

As can be seen, if we look at the covariances between gender and other variables, gender differences in digital uses are influenced by interactions with other variables, mainly age (women live longer and are older in Spain, according to the INE time series), income (they live in households with a lower family income) and education (they are less educated than men).

Another interesting observation in the comparison between seniors and the entire sample is that the residual error of the model for seniors is lower than that for the entire sample, which we interpret as seniors being less affected by other variables not included in the model, perhaps because they are a more homogeneous group than the population as a whole.

The main interrelationships between explanatory variables can be seen to be between education and age, education, and income, and, to a lesser extent, income and age. The others are less important. The higher age, the lower education. The higher education, the higher income, and the higher income, the lower age.

In the model that explains digital uses in the population of seniors (Figure 2), we find that the variable with the greatest weight is education level, followed by age, and then income, whereas the explanatory weight of gender and rural remain low, at levels similar to those for the total population. The gender variable is not statistically significant for seniors either, so we cannot state that it explains digital uses, in either the total population or among seniors. However, the effect of the rural variable, while small, is statistically significant (see Appendix A for more details). Therefore, the profile of those who score low in digital uses is: low education, high age, low income, and rural residence.

The fact that age ceases to be the main digital divide in the seniors' group is because the range for the variable is much more concentrated than it is in the population as a whole, so it is less discriminating of digital uses. However, despite this, age is the second divide, above income level. The differences in the three main divides are smaller in the population of seniors and they are more similar to each other.

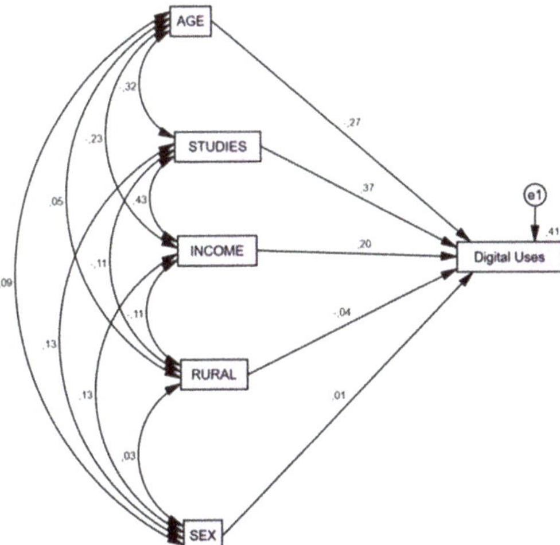

Figure 2. Digital Uses Senior Population Model.

Examining the models that explain differences in digital skills among the total population and among seniors, we reach very similar conclusions, with small nuances. This strong similarity in the results is due to the high correlation between the two indices (Spearman's rho = 0.881 in the seniors' group and rho = 0.964 in the total population. Spearman's correlation is presented instead of Pearson's because the digitization variables do not fit a normal distribution).

They are two closely related digitization dimensions: the greater the intensity and variety of digital uses, the more skills are developed, and vice-versa, the more skills, the greater the likelihood of use.

In the total sample (Figure 3), the main explanatory variable for digital skills is by far age, followed in second place by education, and in third place by income. Being rural appears but has low explanatory power and gender is very low. Despite this, the weak explanatory power of gender in terms of differences in digital skills in the total population is apparently significant ($p = 0.002$: in the digital skills models, it was not significant). Although the lower digital skills of women seem to be explained mainly by their lower income, lower education, and older age, a very weak but significant explanatory power persists that is independent of the interactions between gender and other variables considered in the models. We can confirm that, although no significant gender differences are observed in digital uses (as measured), they are seen, albeit very weakly, in the lower digital skills of women (see Appendix A for the statistical significance).

If we examine the model that explains digital skills among the seniors' group (Figure 4), we find that the first digital divide is education, followed by age, and in third place income. With a much lower impact, we continue to find being rural and gender. The nuances between the results among seniors for skills and for uses consist mainly of the fact that the influence of age is somewhat more important and income somewhat less important. Therefore, the profile of those who score low in digital skills is the same as those who score low in digital uses: low education, high age, low income, and rural residence.

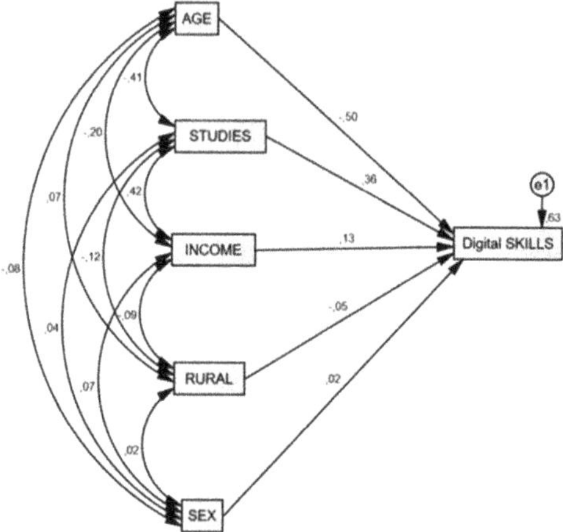

Figure 3. Digital Skills Total Population Model.

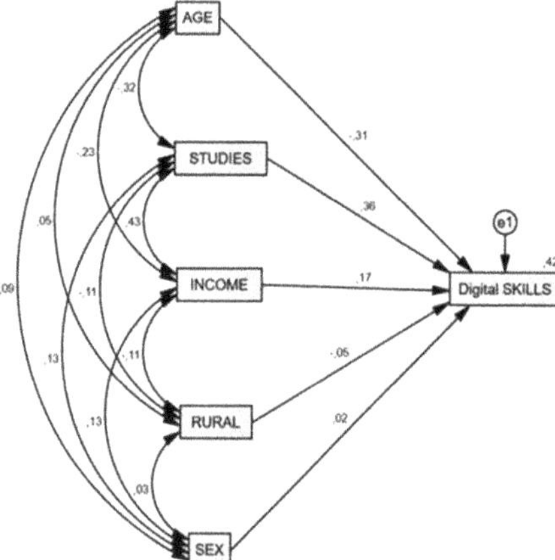

Figure 4. Digital Skills Senior Population Model.

An additional analysis through partial correlations allows us to verify, through the triangulation of methods, the interpretive hypothesis suggested by the path analysis models, that the gender or rural residence differences could be due to these variables interacting with other variables, included mainly age, education, and income. (See Table 8).

Table 8. Partial correlation effects between sex, rural and digitalization.

		Total Correlations		Partial Correlations */	
General Population		**Sex**	**Rural**	**Sex**	**Rural**
Digital Skills	Pearson	0.078 **	−0.136 **	0.023 *	−0.074 **
	Sig.	0.000	0.000	0.004	0.000
	N	15.343	15.343	15.338	15.338
Digital Uses	Pearson	0.058 **	−0.142 **	−0.010	−0.085 **
	Sig.	0.000	0.000	0.213	0.000
	N	15,343	15,343	15,338	15,338
Senior Population		**Sex**	**Rural**	**Sex**	**Rural**
Digital Skills	Pearson	0.114 **	−0.119 **	0.018	−0.059 **
	Sig.	0.000	0.000	0.181	0.000
	N	5247	5247	5242	5242
Digital Uses	Pearson	0.115 **	−0.115 **	0.019	−0.050 **
	Sig.	0.000	0.000	0.165	0.000
	N	5247	5247	5242	5242

*/ Controlling by: Age, Studies and Income. **. Correlation is significative at level 0.01. *. Correlation is significative at level 0.05.

The initial positive correlations between the digitization dimensions and the gender and rural variables decrease considerably when controlled for simultaneously by age, education, and income. However, they do not disappear completely. They continue to be small but statistically significant for the rural variable, both in the sample as a whole and among seniors. However, for the gender variable, the correlation with digital skills is only statistically significant in the total population, not among seniors, which is fully consistent with what the path analysis models reveal.

These results confirm that the differences we see between women and men in terms of digital uses are almost fully explained by the fact that women are older and have lower education and income levels. However, these differences, albeit important, do not fully explain the lesser digital habits of women in the population as a whole, without the same being true when we analyze the seniors' group.

4.4. Explaining Differences between Provinces

From previous studies and from the data examined here, we already know that there are important territorial variations in Internet use by autonomous regions, greater than those observed in the variable size of the municipality [20]. However, a logistic regression analysis of the aforementioned study revealed that the territorial differences by autonomous region had less explanatory power than social variables, such as age or education. The differences by province are greater than by autonomous region, examining digital uses and skills and having obtained statistically-significant results, we have preferred to address the territorial differences between provinces (in Spain, most autonomous regions are made up of several provinces and the level of territorial disaggregation is greater in the provinces).

Table 9 presents a summary of the variability by the province of digital uses and skills. The most interesting statistic is the range, which oscillates around 0.7 standard deviations between the maximum and minimum values observed in the 52 territories. This range of variations is quite high, which indicates that the provinces are a fairly discriminating variable of digital uses and skills.

Table 9. Variations in 52 provinces.

Variations in 52 Provinces. Descriptive Statistics.						
Total Population						
	N	Range	Mín.	Max.	Average	Std. Dev
Digital Uses	52	0.725	−0.497	0.229	−0.054	0.144
Digital Skills	52	0.699	−0.497	0.202	−0.048	0.141
Senior Population						
	N	Range	Mín.	Max.	Average	Std. Dev
Digital Uses	52	0.622	−1.205	−0.58	−0.92	0.135
Digital Skills	52	0.774	−1.251	−0.47	−0.903	0.161

The provinces with the highest scores in digital uses coincide with being the most urban and also often the richest (Barcelona, Madrid, Melilla, Valencia . . .) while those that score the lowest are the less urban and often less affluent provinces (Lugo, León, Ourense, Zamora . . .). For more details, consult Appendix B where detailed data are offered at the provincial level of these and other variables.

The objective of the analysis in this section is to find out to what extent territorial variations in digitization are due to differences in the composition of the social structure of the territories (in non-spatial variables such as sex, age, studies, or family income) or whether they are due to other causes that we have managed to operationalize for this analysis: weight of the rural population, GPD and GPD per capita as indicators of economic development in the territory, degree of implementation of fiber optics (FTTH) and degree of implementation of quality data in mobile telephony (LTE) as indicators of quality Internet access infrastructures [35]. Underlying the hypothesis that territorial differences would be explained by differences in access to infrastructure and economic development as well as in part by the differences in the sociological and sociodemographic composition of the territories.

The prominence of the so-called social variables compared with spatial variables suggests an important weight of these in the territorial variations. However, the variables available in the surveys do not allow the causal weight of hypothetically explanatory variables such as those mentioned on economic wealth and Internet access infrastructure to be adequately explored.

To integrate and be able to establish the explanatory weights of these two sets of variables, we have adopted the perspective of aggregate data, so that the analysis matrix prepared consists of 52 records (50 provinces and the two autonomous cities of Ceuta and Melilla). We have introduced as variables the average values in each province of some survey variables as well as the data available at a provincial level from other secondary sources. For more details, Appendix B presents the analysis matrix, with detailed scores by province for all the variables explored.

In the explanatory models, we will only present results on digital uses but not on skills. The final models are included below, in which only the statistically-significant variables ($p \leq 0.05$) are included and have been developed following criteria of parsimony and the greatest possible explanatory adjustment.

In the model of the total population (Figure 5), the main variables that explain the differences between provinces in digital uses are age and studies. These two variables explain as much as the others: fiber optic rollout (FHTT) and GDP.

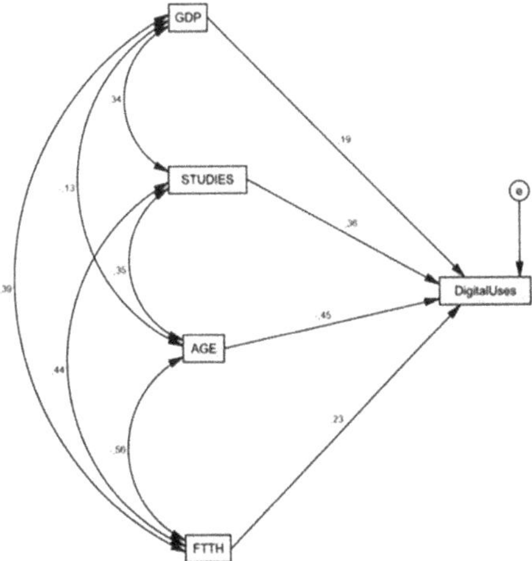

Figure 5. Differences between provinces. Total population digital uses.

These data confirm the suspicion that an important cause of the territorial differences must be sought in the spatial segregation of the sociological and sociodemographic characteristics of the population. This hypothesis is at least confirmed by the explanatory weight of studies and age, since they are the variables that best explain the differences in digital uses between provinces. On the other hand, the results confirm that economic wealth and access to infrastructure are also important explanatory factors, although minor, of the territorial differences.

Significant covariances are also seen between the explanatory variables, especially between fiber optics for people of a younger age, higher education, and higher GDP. Our interpretation is that the spatial segregation favors the younger and more educated population to concentrate in well-connected areas with greater wealth and vice versa, those who are older and with less education tend to be concentrated more in territories with a lower GPD and less Internet connection. This interpretation is supported by the strong covariation of the fiber optic with the variables of age, studies, and GPD.

The model of the senior population (Figure 6) is somewhat different. Partly for methodological reasons, since the smaller sample has conditioned that only three variables have been statistically significant. But it is also due to the differential specificities of the group.

The variable that explains the most by far is studies, the explanatory effect of this variable being greater than the sum of the other two. The absence in the model of economic variables is striking, since they seem to be less determinant of the territorial differences in the group. However, we already know that the studies variable is correlated with income, so the study implicitly indicates personal status, although this is much less decisive in explaining digital uses than qualification. The weight of the rural population is also important to a lesser extent, given that in the provinces with a greater relative weight of the rural population, less Internet use can be very clearly appreciated. This variable is not significant in the general population model, but it is in the senior population. The indicator of connection infrastructure that best territorially differentiates the uses of seniors is not fiber but the LTE connection (which allows access to 4G data in mobile telephony). This is due to the fact that the elderly mainly connect to the Internet through their smartphones and much less through PCs. Specifically, 37% of seniors indicate that there is a personal computer in their homes, while ownership of mobile phones is 90%. On the other hand, the proportion of the rural population in each territory is another significant variable in this model. Its weight in explaining territorial variations is important, and it is closely related to the availability of LTE connections, which are less accessible in rural areas with a low population density.

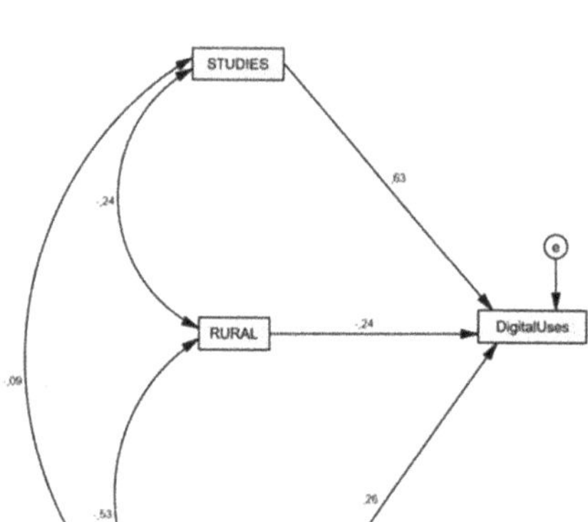

Figure 6. Differences between provinces. Senior population digital uses.

Among the seniors, the data also confirm the suspicion that an important cause of the territorial differences must be sought in the spatial segregation of at least the level of studies, although the results are not as resounding as in the whole of the population. Its greater presence in provinces with a greater rural component and the differences in access to mobile data suggest that in this group the geospatial dimension has a more direct weight than in the general population.

4.5. The Senior Population's Segmentation

Once the weights of the different digital divides had been identified, our objective was to examine the digitization differences between the main segments of the seniors' group. To identify these segmentation criteria and determine the segments, it was appropriate to use the CHAID (Chi-Square Automatic Interaction Detection) segmentation technique. Presented below are the results of this segmentation. Remember that it has been limited to three branching levels and a minimum of 200 cases for each final node. Two segmentations were performed, one with the uses variable as a dependent variable and another with the digital skills variable as a dependent variable. In both cases, the segments are identical in the three-level segmentation tree. The digital skills variable has been standardized so that the results are comparable with those for digital uses, also a standardized variable. The methodological procedure used for this segmentation ensures that we can configure the relevant segments in order to analyze the different degrees of digitization among seniors, differentiating between them on the basis of relevance. See the results in Table 10 and Figure 7.

Firstly, we see that the values for uses and digital skills are negative in all segments. This means that they remain below the population mean, which is equal to zero. On the one hand, this is due to the strong impact of the confluence of various inequalities in the group, and on the other, it is because we have prioritized the less-digitized categories over the higher ones, focusing our approach on lower digitization. For this reason, there are categories above the mean, such as university education and income over €3000.

Table 10. Chaid segmentation summary.

Node	Studies	Age		N	%	USES	SKILLS
7	Not Low	76+	Income −	131	2.5%	−1.1510	−1.1618
8	Not Low	−75	Income +	543	10.3%	−0.81676	−0.74047
9	Not Low	76+	Income −	203	3.9%	−0.70258	−0.65766
10	Not Low	−75	Income +	1326	25.3%	−0.21141	−0.1123
11	Low	76+	Income −	756	14.4%	−1.3303	−1.4602
12	Low	76+	Income +	1052	20.0%	−1.2732	−1.3650
13	Low	−75	Urban	959	18.3%	−0.96806	−0.89126
14	Low	−75	Rural	277	5.3%	−1.1286	−1.0785
0	**All**	**65+**	**Seniors**	**5247**	**100.0%**	**−0.8773**	**−0.8634**

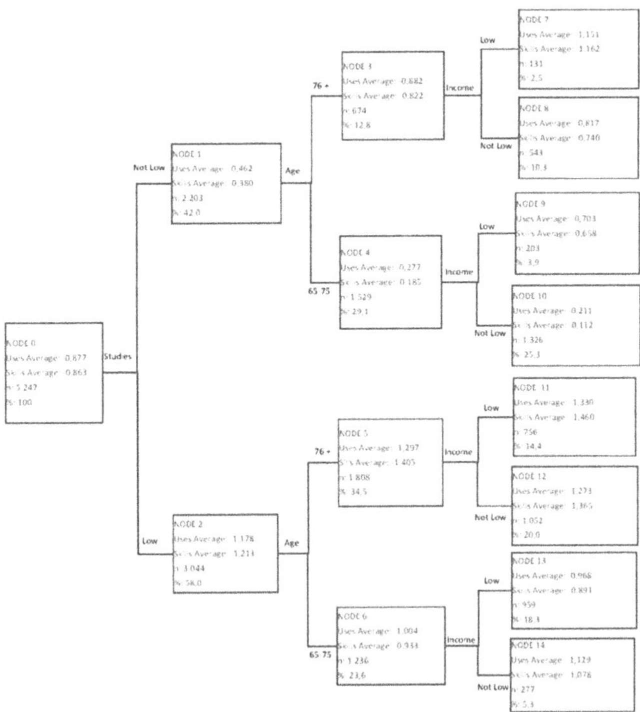

Figure 7. Segmentation Tree.

To compensate for possible biases from the recoding performed in the CHAID segmentation, we have conducted a simple segmentation of the means for uses and digital skills, with the five variables examined, without regrouping the response categories. The results of this analysis are outlined below. The mean factor score for digital skills in seniors who went to university is +0.1145, a value that is slightly above the mean, but not by much. The mean score for seniors with incomes over €3000 is +0.0841, that is, they are at the population mean. In digital uses, those who have higher education scored 0.01 and those with incomes over €3000 scored 0.05. These segments with non-negative values are very small, accounting for 11% and 4% of seniors, respectively. When we jointly segment through education and income, by age we see that among those aged under 75, those with incomes over €2500 achieve non-negative mean scores in both digitization variables, but all those who do not have university education continue to have negative means. In contrast, when we segment education with income, we see that university graduates with incomes over €1600 have non-negative means (also in uses and digital skills). No other segmentation of the five variables considered allows us to obtain non-negative means, in either uses or skills (with two segmentation levels, exhausting the possibilities of the sample when contemplating all the response categories for the variables).

Secondly, returning to the CHAID segmentation, we see that the scores for digital uses and skills do not differ much. Any differences are nuances that still allow us to reach the same overall conclusions. The main difference is a slightly higher variability in skill levels than in uses.

It is also interesting to note that the first segmentation level is determined by the education variable, the second level by the age variable, and in the third level, income differences dominate, although the rural/urban dichotomy also appears. This is highly consistent with the conclusions on the explanatory weights for digital divides examined in the previous section. Note that the rural variable has a modest presence in the segmentation and the gender variable does not appear as a discriminant variable. The results of both analyzes provide support for each other. The perspective of the CHAID segmentation serves to confirm, through the triangulation of methods, the results of the causal analysis obtained through path analysis, in terms of ranking the importance of the explanatory variables.

We find the lowest scores in all low education segments, which have negative scores below one standard deviation, with the sole exception of those under 76 years in urban locations, with a somewhat lower score, similar to all seniors. Those with low education and over 76 years of age are particularly low, regardless of their income levels. This combination of education and age is the most determinant of low levels of digitization and accounts for just over a third of the group (34.4%).

We find the highest digitization levels in the segment of high education, under 75 years of age, and high income, which accounts for 25.3% of seniors.

The following segments are clearly below the mean for seniors and appear in order: 11 (low education, over 76, low income), 12 (low education, over 76, high income), 7 (high education, over 76, high income), 14 (low education, under 75, rural). Around the mean are those who have low education, aged under 75 and in an urban location, and those who have a high education, aged under 75 and have a high income. Those who have a high education, aged over 76, and a high income are clearly above the mean, along with those who have a high education, aged under 75, and a low income.

Although the results of the segmentation are presented for a segmentation level of three variables, we have explored up to a fourth level, restricting the configuration of final nodes to 100 cases. When doing this, we have found some additional segmentations with significant differences in means, which are the four below (graph and tables not presented due to space issues):

- Differences between women and men, with high income, young seniors with a high education (Uses and skills).
- Differences between rural and urban, with high income, old seniors with a low education (Uses and skills).
- Differences between women and men, urban, young seniors with a low education (Uses and skills).
- Differences between women and men, rural, young, low education (Only skills, in uses, the differences are not significant).

In the fourth segmentation level, there are differences by age and residence. Bear in mind that the differences in means do not necessarily imply causality. In this segmentation level, the differences are small (albeit significant) and they always have negative scores, lower than the mean of the population as a whole. Skills generally differ more than uses.

4.6. Profiles and Divide Weights of the Digitally Excluded

A complementary perspective to that outlined so far consists of characterizing the profile and dimensioning the digital divides of the digitally excluded, who are mostly seniors. The clearest criterion for differentiating between them is found in the 'low' category of the digital uses variable, which includes all those who do not use the Internet or do so less frequently than in the last three months and therefore do not mention any Internet use. Moreover, 100% of these completely lack digital skills. While in the general population they represent the lowest quintile (21.7%) and in those aged under 65, this falls to a minority of 4.5%, among seniors they account for 54.9%. In addition, a vast majority of the digitally excluded in the general population are seniors: 86.3%.

According to the tables already presented (see Tables 3, 5 and 6), the most distinctive features of the profile of the digitally excluded are as follows:

- 77% have primary or incomplete education.
- 67% are aged over 76.
- 63% are women.
- 58% have a low or medium-low income.
- 25% are in a rural location.

The measure of the over-representation of those excluded compared to the mean for seniors in each of these groups of variables was calculated using index numbers over 100, in which 100 is equal to the mean for seniors, above 100 is higher, and below this is lower. These are, from the highest to the lowest figures:

- Index 141 Over 76 years.
- Index 136 Primary or incomplete education
- Index 126 Low or medium-low income
- Index 125 Rural
- Index 105 Women.

The indices allow us to identify the categories where we find the main differences, with age being the most prominent, followed by education, income, rural, and finally women, with the latter category differing little from the mean. In all these categories or groups of variables, those excluded stand out as being above the mean for seniors.

The indices are not sufficient to estimate the impact of the variables that represent the categories examined for digital exclusion. To identify the effects with greater accuracy, a new path analysis model has been developed with the five explanatory variables and one dependent variable, digital exclusion, with a value of 1 for those excluded and 0 for the rest. The results are shown in Figure 8 (standardized estimates).

The variables that most explain digital exclusion among seniors are age and education, much more similar in terms of direct effects on the dependent than in previous models, although age has a little more weight. Income appears at a clearly secondary level, and in the third level of importance, with low weights, are the rural and gender variables. Examining the statistical significance of the regression coefficients, we verify that the contribution of the rural variable to the dependent variable is statistically significant ($p = 0.000$), whereas the almost zero contribution of the gender variable is not ($p = 0.136$). The other higher contributions are all highly significant (see significances in Appendix A).

This means that digital exclusion is explained mainly and with much difference from other variables by age and studies. In addition, the explanatory weight of both is very similar. Income appears as the third variable, more distant and to a lesser extent, the rural habitat also provides explanatory power. The results of the exclusion analysis are partly similar to those that explain digital uses and skills, as is logical. But they differ in the greater explanatory importance of age and studies.

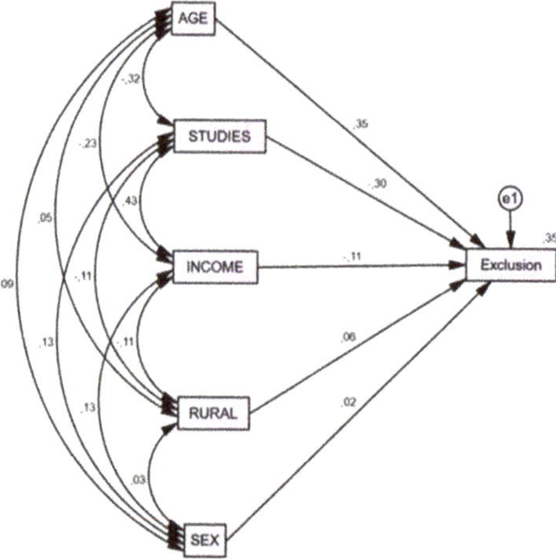

Figure 8. Model Seniors Digital Exclusion.

5. Discussion

In relation to the digital skills and digital uses variables, we see a great similarity in results and very small differences, just nuances. We interpret this as a consequence of the close relationship between uses and skills that is reflected in the high correlation between these variables. Other studies had already found very similar correlations between the various digitization variables and suggested that there is a latent dimension when we study digitization [41].

Whether we look at uses or skills, the variables that most explain the digitization levels of Spanish seniors are firstly education, followed closely by age. The explanatory weight of these two variables is very high. In third place is income, and in a distant fourth place is rural habitat. The age variable has insignificant or zero explanatory power.

What differentiates seniors from the general population both in uses and in digital skills is mainly the lower explanatory importance of age, and the greater importance of educational level. Income level is also somewhat more important among seniors than in the general population. However, there are hardly any differences in relation to the explanatory power of habitat and gender, despite the fact that before conducting the statistical tests, we thought that these variables would differentiate seniors more.

Therefore, those who use digital media more and have more digital skills are younger seniors, with a higher level of education, high incomes, and residing in non-rural areas. Vice versa, those who use the least and have the least skills are the older seniors, with low education, low income, and residing in rural areas.

The scarce or zero explanatory power of the gender variable contrasts with the extensive literature on the gender divide. However, we had seen previous evidence that the relationship between gender and certain digitization variables disappears when controlled for by other more explanatory variables, in studies on Spanish society [17–19] and other countries [6]. This is exactly what we find in this study. But it would not be correct to think that there are no gender differences. These differences are evident. Our study finds that digitization differences between Spanish men and women, seniors, and non-seniors (in the digitization variables we studied), are explained almost entirely by the lower education level of women, lower income, and older age. There is practically no difference between the general population and the senior population, despite the fact that among seniors, the differences are greater than in the general population. We found that this is due to the fact that there are greater differences in education, income, and age among seniors.

However, we also find that while the rural-urban variable has a small explanatory role, the explanatory power of this variable does not disappear (although it logically decreases) when controlled for by other explanatory variables, meaning that there are specific causes (spatial or geographic) that explain lower digitization in rural areas, both in the population as a whole and among seniors (to a similar extent and not greater among seniors than in the population as a whole). Other studies had previously confirmed the weak effect of geographic variables, with less impact than other social variables [14,29,30]. Therefore, the Spanish case follows this pattern.

It is striking that the important differences that can be seen in the geographical variables become secondary from the multivariate causal perspective, both in our research and in the previous ones. The analysis of the differences in digital uses by provinces in the population as a whole allows us to verify that it is mainly due to the fact that the territories segregate the populations unequally based on other variables, such as age, educational level, family income, and sex. The different provinces have different sociodemographic and socioeconomic compositions of their populations, and these variations are what mainly explain the territorial differences. The causal models 'punish' the implicit redundancy in the geographical variables with lower regression coefficients, sometimes not significant. Another second reason is the limitation of the survey method to incorporate other variables that explain territorial differences, such as those related to connection infrastructure, economic wealth, or public policies, with a more evident spatial projection. However, it has been proven that the territorial differences in the levels of Internet use have much more to do with variations in age and inter-territorial study than with other variables such as connection coverage or the GPD. The spatial segregation of population is the main explanation.

This explanation is also valid in general for the seniors, although less strongly than the total population and with some nuances. The variable that mainly explains the territorial differences in the use of the Internet between the seniors is the level of studies, with a resounding role with respect to other explanatory variables. To a lesser extent, variations in rural population and mobile data coverage (LTE) are also relevant variables. This last discovery is important because it confirms that the most relevant infrastructure to favor digital uses among seniors is the development of access to quality data in mobile telephony, much more than the expansion of fiber and ADSL, which are the ones that have prioritized public policies to universalize digital connection.

In summary, it is evident that the main variables that explain the low digitization of seniors are education, age, and income. The geographical differences largely depend on the spatial distribution of the senior population and its characteristics in terms of age, studies, and income. That is primarily why they are more present in rural areas and provinces with a low demographic density and less urbanized, in which there is an older population, with less income and a lower level of education.

If the main reason behind the differences found by age is differences in socialization between generational groups [6], it is likely that these differences will diminish in the future. It is highly likely that a lack of interest and absence of a social presence [11,17,23] is related to generational socialization, along with the self-perception of low skills [5].

The lower education level of seniors is a generational characteristic, which could perhaps also ease over time to the extent that access to education has been more widespread for younger generations. However, the education variable, despite weighing less in the population as a whole, is very important in it, establishing differences in digital uses and skills among the younger generations [42]. We must also consider the great persistence of the effect of variables such as age and income [43].

What we do not find in this study, despite the large sample size and our exploiting of the advantages of the extensive sample size, is confirmation of the diversity within the seniors' group, in terms of digitization, as referred to by other authors. [17,20]. There are hardly any segments with digitization levels above the mean for the Spanish population as a whole, and these segments are among the seniors who are very much in the minority: seniors with a university education or with family incomes of over €3000 per month, basically. We can confirm in this regard that there is a relative diversity, but digitization levels clearly below the population mean generally predominate. The main segmentation shows differences between low and very low digitization levels, and these differences relate to the most explanatory variables: education, age, and income. What can be seen is the very low digitization of the low education level segments in general and those over 75 years of age, also in general, with some exceptions. This incredibly low digitization does contrast with other segments that are close to the mean but do not reach it.

The digitally excluded (do not use the Internet and do not have digital skills) among seniors represent 54.9%. Seniors comprise a large majority of all the excluded in the population as a whole since they account for 86.3% of the total number of excluded people in Spain. With these figures, we can confirm that in Spain, digital exclusion as a phenomenon is almost entirely exclusive to seniors, affecting more than half of this group. The determining variables of this exclusion are mainly education and age, followed at a lower level by income, and rural residence. The impacts of education and age are very similar when explaining exclusion. Income and rural residence emerge as secondary variables, both also significant.

We do not know to what extent this situation of low digitization of the elderly will ease over time when future cohorts of the seniors have more studies, or whether it will continue or increase when future technological applications require new learning. But it is very worrying because the Internet and digital media are increasingly the gateway to services and benefits and not using them places seniors at a disadvantage in terms of their social status. A greater awareness of ageism and homophilic prejudices in the design of applications and devices, a greater knowledge and concern for the difficulties of seniors, and the promotion of intergenerational learning strategies could contribute to reducing or closing this digital divide. As regards this, studies conducted on intergenerational digital learning programs, from grandchildren to grandparents, demonstrate the success of such programs [44–46]. They all demonstrate success in the digital learning of the elderly and some significant changes of perception and attitude toward the elderly by younger people, such as children [46] and students [47] are also appreciated.

It is necessary to act to increase the digitization of seniors, which is excessively low and unequal.

Author Contributions: M.A.-M. has managed and coordinated the research and was in charge of the final drafting of the article, also developing the path analysis models, the segmentation analysis, and producing the digital uses variable. N.G.-L. has been partially responsible for compiling and working on the bibliographic references and the background section, and has dealt with various complementary statistical operations, among which it is worth highlighting the production of the digital skills variable and the contingency tables. B.C.-Q. has proposed the research theme, has been partially responsible for compiling and working on the bibliographical references and the background section, has processed and prepared the database for its use with SPSS statistical software, and has collaborated in the preparation of statistical tables. All authors have read and agreed to the published version of the manuscript.

Funding: This research received no external funding.

Institutional Review Board Statement: The study did not require ethical approval.

Informed Consent Statement: Not applicable.

Data Availability Statement: The data used for the study are available on the following website: https://www.ine.es/dyngs/INEbase/es/operacion.htm?c=Estadistica_C&cid=125473617674 1&menu=resultados&idp=1254735976608#!tabs-1254736194579 (accessed on 7 June 2022).

Acknowledgments: We are grateful for the support of Vanessa Zorrilla-Muñoz and María Silveria Agulló-Tomás. Members of the Institute on Gender Studies University Carlos III of Madrid (IEG-UC3M) and the ENCAGEn-CM program (https://encage-cm.es/ (accessed on 10 March 2022)).

Conflicts of Interest: The authors declare no conflict of interest.

Appendix A. Path Analysis Models (Estimates)

Appendix A.1. Digital Uses Total Population Model

Table A1. Regression Weights: (Group number 1-Default model).

			Estimate	S.E.	C.R.	*p* */	Label
D_USE	<—	AGE	−0.028	0.000	−95.373	***	par_10
D_USE	<—	STUDIES	0.279	0.005	58.496	***	par_11
D_USE	<—	INCOME	0.000	0.000	24.750	***	par_12
D_USE	<—	RURAL	−0.136	0.013	−10.290	***	par_13
D_USE	<—	SEX	−0.019	0.010	−1.882	0.060	par_14

*/ *p* = *** means *p* = 0.000 (high statistical significance), in this table and subsequent.

Table A2. Standardized Regression Weights: (Group number 1-Default model).

			Estimate
D_USE	<—	AGE	−0.518
D_USE	<—	STUDIES	0.343
D_USE	<—	INCOME	0.135
D_USE	<—	RURAL	−0.051
D_USE	<—	SEX	−0.009

Table A3. Covariances: (Group number 1-Default model).

			Estimate	S.E.	C.R.	*p*	Label
RURAL	<–>	SEX	0.003	0.002	2.191	0.028	par_1
SEX	<–>	INCOME	26.428	2.928	9.026	***	par_2
SEX	<–>	STUDIES	0.022	0.005	4.503	***	par_3
SEX	<–>	AGE	−0.753	0.076	−9.955	***	par_4
RURAL	<–>	INCOME	−25.215	2.224	−11.340	***	par_5
RURAL	<–>	STUDIES	−0.055	0.004	−14.735	***	par_6
RURAL	<–>	AGE	0.490	0.057	8.546	***	par_7
INCOME	<–>	STUDIES	373.165	7.803	47.826	***	par_8
INCOME	<–>	AGE	−2704.644	112.270	−24.091	***	par_9
STUDIES	<–>	AGE	−9.487	0.201	−47.176	***	par_15

Appendix A.2. Digital Uses Senior Population Model

Table A4. Regression Weights: (Group number 1-Default model).

			Estimate	S.E.	C.R.	*p*	Label
D_USE	<—	AGE	−0.025	0.001	−24.271	***	par_10
D_USE	<—	STUDIES	0.224	0.007	30.149	***	par_11
D_USE	<—	INCOME	0.000	0.000	16.961	***	par_12
D_USE	<—	RURAL	−0.077	0.019	−4.019	***	par_13
D_USE	<—	SEX	0.022	0.016	1.377	0.169	par_14

Table A5. Standardized Regression Weights: (Group number 1-Default model).

			Estimate
D_USE	<—	AGE	−0.273
D_USE	<—	STUDIES	0.367
D_USE	<—	INCOME	0.201
D_USE	<—	RURAL	−0.043
D_USE	<—	SEX	0.015

Table A6. Covariances: (Group number 1-Default model).

			Estimate	S.E.	C.R.	*p*	Label
RURAL	<->	SEX	0.006	0.003	2.191	0.028	par_1
SEX	<->	INCOME	39.628	4.316	9.181	***	par_2
SEX	<->	STUDIES	0.077	0.008	9.432	***	par_3
SEX	<->	AGE	−0.365	0.054	−6.777	***	par_4
RURAL	<->	INCOME	−27.824	3.545	−7.849	***	par_5
RURAL	<->	STUDIES	−0.051	0.007	−7.688	***	par_6
RURAL	<->	AGE	0.166	0.044	3.747	***	par_7
INCOME	<->	STUDIES	323.690	11.335	28.556	***	par_8
INCOME	<->	AGE	−1134.966	71.035	−15.978	***	par_9
STUDIES	<->	AGE	−3.031	0.137	−22.125	***	par_15

Appendix A.3. Digital Skills Total Population Model

Table A7. Regression Weights: (Group number 1-Default model).

			Estimate	S.E.	C.R.	*p*	Label
DIGSKILLS	<—	AGE	−0.039	0.000	−92.413	***	par_11
DIGSKILLS	<—	STUDIES	0.436	0.007	62.345	***	par_12
DIGSKILLS	<—	INCOME	0.000	0.000	23.991	***	par_13
DIGSKILLS	<—	RURAL	−0.179	0.019	−9.273	***	par_14
DIGSKILLS	<—	SEX	0.046	0.015	3.148	0.002	par_15

Table A8. Standardized Regression Weights: (Group number 1-Default model).

			Estimate
DIGSKILLS	<—	AGE	−0.500
DIGSKILLS	<—	STUDIES	0.364
DIGSKILLS	<—	INCOME	0.130
DIGSKILLS	<—	RURAL	−0.046
DIGSKILLS	<—	SEX	0.016

Table A9. Covariances: (Group number 1-Default model).

			Estimate	S.E.	C.R.	*p*	Label
RURAL	<->	SEX	0.003	0.002	2.191	0.028	par_1
SEX	<->	INCOME	26.428	2.928	9.026	***	par_2
SEX	<->	STUDIES	0.022	0.005	4.503	***	par_3
SEX	<->	AGE	−0.753	0.076	−9.955	***	par_4
RURAL	<->	INCOME	−25.215	2.224	−11.340	***	par_5
RURAL	<->	STUDIES	−0.055	0.004	−14.735	***	par_6
RURAL	<->	AGE	0.490	0.057	8.546	***	par_7
INCOME	<->	STUDIES	373.165	7.803	47.826	***	par_8
INCOME	<->	AGE	−2704.644	112.270	−24.091	***	par_9
STUDIES	<->	AGE	−9.487	0.201	−47.176	***	par_10

Appendix A.4. Digital Skills Senior Population Model

Table A10. Regression Weights: (Group number 1-Default model).

			Estimate	S.E.	C.R.	*p*	Label
DIGSKILLS	<—	AGE	−0.048	0.002	−27.395	***	par_11
DIGSKILLS	<—	STUDIES	0.376	0.013	29.573	***	par_12
DIGSKILLS	<—	INCOME	0.000	0.000	14.760	***	par_13
DIGSKILLS	<—	RURAL	−0.144	0.033	−4.351	***	par_14
DIGSKILLS	<—	SEX	0.043	0.027	1.564	0.118	par_15

Table A11. Standardized Regression Weights: (Group number 1-Default model).

			Estimate
DIGSKILLS	<—	AGE	−0.307
DIGSKILLS	<—	STUDIES	0.359
DIGSKILLS	<—	INCOME	0.174
DIGSKILLS	<—	RURAL	−0.046
DIGSKILLS	<—	SEX	0.017

Table A12. Covariances: (Group number 1-Default model).

			Estimate	S.E.	C.R.	*p*	Label
RURAL	<–>	SEX	0.006	0.003	2.191	0.028	par_1
SEX	<–>	INCOME	39.628	4.316	9.181	***	par_2
SEX	<–>	STUDIES	0.077	0.008	9.432	***	par_3
SEX	<–>	AGE	−0.365	0.054	−6.777	***	par_4
RURAL	<–>	INCOME	−27.824	3.545	−7.849	***	par_5
RURAL	<–>	STUDIES	−0.051	0.007	−7.688	***	par_6
RURAL	<–>	AGE	0.166	0.044	3.747	***	par_7
INCOME	<–>	STUDIES	323.690	11.335	28.556	***	par_8
INCOME	<–>	AGE	−1134.966	71.035	−15.978	***	par_9
STUDIES	<–>	AGE	−3.031	0.137	−22.125	***	par_10

Appendix A.5. Digital Uses. Differences between Provinces Total Population

Table A13. Regression Weights: (Group number 1-Default model).

			Estimate	S.E.	C.R.	*p*	Label
DigitalUses	<—	AGE	−0.023	0.004	−6.222	***	par_7
DigitalUses	<—	STUDIES	0.254	0.048	5.279	***	par_8
DigitalUses	<—	FTTH	0.003	0.001	3.010	0.003	par_9
DigitalUses	<—	GDP	0.000	0.000	2.913	0.004	par_10

Table A14. Standardized Regression Weights: (Group number 1-Default model).

			Estimate
DigitalUses	<—	AGE	−0.448
DigitalUses	<—	STUDIES	0.357
DigitalUses	<—	FTTH	0.235
DigitalUses	<—	GDP	0.191

Table A15. Covariances: (Group number 1-Default model).

			Estimate	S.E.	C.R.	*p*	Label
AGE	<–>	FTTH	−19.149	5.503	−3.479	***	par_1
FTTH	<–>	STUDIES	1.090	0.380	2.867	0.004	par_2
FTTH	<–>	GDP	181,530,906.531	70,525,173.346	2.574	0.010	par_3
AGE	<–>	STUDIES	−0.195	0.082	−2.365	0.018	par_4
AGE	<–>	GDP	−13,690,421.953	14,826,349.802	−.923	0.356	par_5
STUDIES	<–>	GDP	2,620,038.450	1,126,469.983	2.326	0.020	par_6

Appendix A.6. Digital Uses. Differences between Provinces. Senior Population

Table A16. Regression Weights: (Group number 1-Default model).

			Estimate	S.E.	C.R.	*p*	Label
DigitalUses	<—	RURAL	−0.159	0.070	−2.268	0.023	par_4
DigitalUses	<—	STUDIES	0.294	0.043	6.891	***	par_5
DigitalUses	<—	LTE	0.070	0.029	2.444	0.015	par_6

Table A17. Standardized Regression Weights: (Group number 1-Default model).

			Estimate
DigitalUses	<—	RURAL	−0.243
DigitalUses	<—	STUDIES	0.629
DigitalUses	<—	LTE	0.255

Table A18. Covariances: (Group number 1-Default model).

			Estimate	S.E.	C.R.	*p*	Label
RURAL	<->	LTE	−0.052	0.016	−3.334	***	par_1
LTE	<->	STUDIES	−0.013	0.020	−0.674	0.501	par_2
RURAL	<->	STUDIES	−0.014	0.008	−1.651	0.099	par_3

Appendix A.7. Digital Exclusion Senior Population Model

Table A19. Regression Weights: (Group number 1-Default model).

			Estimate	S.E.	C.R.	*p*	Label
DIGSKILLS	<—	AGE	−0.048	0.002	−27.395	***	par_11
DIGSKILLS	<—	STUDIES	0.376	0.013	29.573	***	par_12
DIGSKILLS	<—	INCOME	0.000	0.000	14.760	***	par_13
DIGSKILLS	<—	RURAL	−0.144	0.033	−4.351	***	par_14
DIGSKILLS	<—	SEX	0.043	0.027	1.564	0.118	par_15

Table A20. Standardized Regression Weights: (Group number 1-Default model).

			Estimate
DIGSKILLS	<—	AGE	−0.307
DIGSKILLS	<—	STUDIES	0.359
DIGSKILLS	<—	INCOME	0.174
DIGSKILLS	<—	RURAL	−0.046
DIGSKILLS	<—	SEX	0.017

Table A21. Covariances: (Group number 1-Default model).

			Estimate	S.E.	C.R.	*p*	Label
RURAL	<->	SEX	0.006	0.003	2.191	0.028	par_1
SEX	<->	INCOME	39.628	4.316	9.181	***	par_2
SEX	<->	STUDIES	0.077	0.008	9.432	***	par_3
SEX	<->	AGE	−0.365	0.054	−6.777	***	par_4
RURAL	<->	INCOME	−27.824	3.545	−7.849	***	par_5
RURAL	<->	STUDIES	−0.051	0.007	−7.688	***	par_6
RURAL	<->	AGE	0.166	0.044	3.747	***	par_7
INCOME	<->	STUDIES	323.690	11.335	28.556	***	par_8
INCOME	<->	AGE	−1134.966	71.035	−15.978	***	par_9
STUDIES	<->	AGE	−3.031	0.137	−22.125	***	par_10

Appendix B. Provinces Analysis

Table A22. Total Population.

Province	Digital Uses	Digital Skills	GPD	GPD_PC	INCOME	Age	Studies	Sex	Rural	FTTH	LTE	Sp100M	n
A Coruña	−0.086658	−0.110379	26,682,181	23,816	1728.7	57	2.03226	0.4	0.304	63.89	99.98	75.6	434
Álava	0.0512163	0.0607017	11,882,941	36,404	1734	55	2.39568	0.55	0.173	91.09	99.99	91.7	139
Albacete	0.1156923	0.1397187	8,235,408	21,153	1575.4	51	2.11765	0.48	0.214	87.14	99.89	92.08	187
Alicante	0.0778471	0.0160279	36,521,398	19,757	1552.5	54	2.15198	0.45	0.03	86.38	99.98	90.88	329
Almería	−0.191573	−0.139016	13,979,829	19,919	1421.8	55	1.83486	0.44	0.174	71.03	99.96	71.97	109
Asturias	−0.164979	−0.180417	23,258,673	22,709	1678.2	59	2.02108	0.4	0.19	75.19	99.73	82.42	830
Ávila	−0.00992	0.0045183	3,252,395	20,423	1560.6	57	2.0641	0.51	0.538	57.19	98.63	57.19	78
Badajoz	−0.126553	−0.15712	12,423,261	18,453	1440.1	56	2.00522	0.44	0.446	80.16	99.85	80.16	383
Barcelona	0.2285182	0.1819395	171,350,447	30,947	1813.4	55	2.37347	0.45	0.057	95.54	99.99	95.55	980
Bizkaia	−0.066767	−0.007407	36,085,689	31,792	1803	57	2.36174	0.47	0.059	91.25	99.99	96.93	528
Burgos	0.0594854	0.0149232	10,505,020	29,571	1781.6	55	2.50365	0.48	0.175	72.12	99.02	76.04	137
Cáceres	−0.090851	−0.0657	7,664,977	19,464	1456.3	56	1.98536	0.49	0.381	67.37	99.73	68.88	239
Cádiz	−0.150635	−0.156511	22,535,246	18,050	1578.3	55	1.81556	0.44	0.049	81.51	99.99	86.63	225
Cantabria	−0.074002	−0.032137	13,737,756	23,646	1683.5	56	2.3126	0.46	0.19	72.29	99.82	77	611
Castellón	−0.059952	−0.075635	16,149,473	28,367	1559.8	57	2.04887	0.47	0.158	82.38	99.92	90.31	133
Ceuta	0.0547178	0.1391325	1,720,295	20,251	1713.8	52	2.01786	0.36	0	92.78	99.88	92.78	56
Ciudad Real	−0.122497	−0.092987	10,689,033	21,563	1532.7	55	2.01402	0.43	0.21	85.18	99.94	86.03	214
Córdoba	0.0238531	−0.061983	14,534,325	18,525	1584.7	54	2.05036	0.42	0.137	89.75	99.88	90.55	139
Cuerca	−0.082449	−0.089423	4,536,392	22,691	1603	52	2.18817	0.43	0.398	75.96	99.07	76.88	93
Girona	0.0264724	0.0538664	21,202,782	28,184	1712.5	54	2.14286	0.42	0.293	71.84	99.89	71.84	140
Granada	0.0416022	0.0714295	16,687,601	18,181	1604.3	53	2.27132	0.51	0.085	77.08	99.92	80.82	129
Guadalajara	−0.284492	−0.252009	5,245,815	20,415	1446.2	61	1.80556	0.43	0.556	76.46	99.38	76.46	72
Guipuzkoa	−0.060884	0.0209451	24,060,930	33,851	1889.7	57	2.41723	0.47	0.088	87.35	99.95	96.02	296
Huelva	−0.068298	−0.006147	10,607,333	20,273	1524.1	52	2.07317	0.49	0.171	78.81	99.97	79.73	82
Huesca	0.1090783	0.1025899	6,134,249	28,015	1814.7	55	2.31696	0.43	0.464	78.55	99.55	78.55	112
Islas Baleares	0.0815328	0.0526876	32,767,619	27,870	1736.8	55	2.14617	0.45	0.151	88.2	99.84	90.37	496
Jaén	−0.1336	−0.136891	11,808,429	18,628	1422.6	56	1.5566	0.47	0.226	87.79	99.86	90.83	106
La Rioja	−0.088315	−0.073265	8,593,185	27,482	1664.8	56	2.2407	0.46	0.242	90.72	99.65	90.72	833
Las Palmas	0.0998281	0.0718095	23,553,372	20,813	1469.7	52	2.08383	0.45	0.039	84.97	99.95	85.43	334
León	−0.339573	−0.336961	10,006,588	21,579	1600.6	61	2.00237	0.48	0.545	71.73	98.86	71.82	211
Lleida	−0.054425	−0.029537	12,218,853	28,456	1762.4	56	2.23529	0.47	0.565	74.57	99.6	74.68	85
Lugo	−0.496646	−0.49707	7,692,177	23,320	1549.8	61	1.74815	0.47	0.489	41.82	99.93	60.4	135
Madrid	0.2109524	0.2022956	231,133,592	35,091	1824.3	54	2.43342	0.46	0.021	96.84	99.99	96.85	1607

Table A22. *Cont.*

Province	Digital Uses	Digital Skills	GPD	GPD_PC	INCOME	Age	Studies	Sex	Rural	FTTH	LTE	Sp100M	n
Málaga	−0.01487	0.0410141	31,023,255	18,801	1538.7	55	2.03937	0.51	0.091	92.73	99.99	93.88	254
Melilla	0.1759288	0.1932697	1,582,540	18,700	1950.7	47	2.43056	0.35	0	100	100	100	72
Murcia	0.0627996	0.0381814	31,198,376	21,094	1542.7	52	2.0625	0.47	0.012	84.06	99.94	86.94	584
Navarra	0.0435399	0.0679968	20,047,454	31,026	1834.1	56	2.40653	0.46	0.333	71.25	99.91	79.81	781
Ourense	−0.29578	−0.263187	6,813,831	22,120	1521.8	59	1.85252	0.4	0.403	50.41	99.84	65.95	139
Palencia	−0.11262	−0.081759	4,407,310	27,346	1537.3	59	2.14925	0.45	0.313	71.42	98.92	73.92	67
Pontevedra	−0.176731	−0.149446	21,247,944	22,586	1542.6	55	1.98777	0.41	0.122	65.47	99.99	74.46	327
Salamanca	−0.245509	−0.234175	7,048,640	21,187	1464.7	61	1.92647	0.48	0.353	70.17	98.88	72.65	102
Santa Cruz de T	0.0186333	0.0301631	22,269,949	21,076	1633.6	53	2.09459	0.44	0.071	83.39	99.94	83.57	296
Segovia	−0.059065	−0.110059	3,418,981	22,212	1529.7	57	2.0339	0.53	0.322	70.11	99.51	70.11	59
Sevilla	0.0197862	−0.016871	39,535,345	20,314	1620	55	2.09187	0.43	0.084	91.75	100	92.62	332
Soria	−0.178336	−0.179132	2,380,731	26,626	1681.4	59	1.87179	0.49	0.513	65.64	98.3	65.64	39
Tarragona	−0.030302	0.0081539	24,567,640	30,810	1687.6	58	2.25281	0.44	0.18	73.92	99.89	73.92	89
Teruel	−0.213103	−0.179132	3,367,236	25,262	1464.4	55	1.98718	0.4	0.449	58.75	98.09	58.75	78
Toledo	−0.095853	−0.12354	12,814,575	18,617	1524.8	55	1.84232	0.49	0.32	75.97	99.95	76.71	241
Valencia	0.1265819	0.0873524	59,123,107	23,363	1667.4	54	2.24251	0.46	0.066	86.01	99.94	92.93	534
Valladolid	0.0239926	0.0144926	13,998,460	26,901	1706.4	58	2.34595	0.45	0.124	88.49	99.53	88.49	185
Zamora	−0.285439	−0.305063	3,459,100	19,813	1453.7	61	1.97333	0.48	0.32	52.67	98.41	55.58	75
Zaragoza	0.021271	0.0544684	27,348,811	28,386	1732.8	57	2.25865	0.46	0.15	90.73	99.75	90.73	607
Total	0.0000000	0.0000000	1,204,241,000	25,771	1674.3	56	2.18627	0.45	0.172	84.93	99.87	87.58	15343

Table A23. Senior Population.

Province	Digital Uses	Digital Skills	GPD	GPD_PC	INCOME	AGE	STUDIES	RURAL	SEX	FTTH	LTE	Sp100M	n
A Coruña	−0.973679	−0.979043	26,682,181	23,816	1554.19	76.6	1.20809	0.4046	0.39	63.89	99.98	75.6	173
Álava	−0.865608	−0.789241	11,882,941	36,404	1523.04	76.3	1.76471	0.1373	0.51	91.09	99.99	91.7	51
Albacete	−0.899018	−0.931222	8,235,408	21,153	1387.18	76	1.17949	0.3333	0.51	87.14	99.89	92.08	39
Alicante	−0.813678	−0.803525	36,521,398	19,757	1393.93	75.7	1.38835	0.0291	0.41	86.38	99.98	90.88	103
Almería	−0.972918	−0.932143	13,979,829	19,919	1250.66	75.1	1.05263	0.2895	0.32	71.03	99.96	71.97	38
Asturias	−0.941463	−0.96284	23,258,673	22,709	1578.85	76.6	1.25444	0.1953	0.37	75.19	99.73	82.42	338
Ávila	−1.048728	−1.053656	3,252,395	20,423	1309.62	76.8	1.19231	0.6923	0.35	57.19	98.63	57.19	26
Badajoz	−0.98419	−0.976808	12,423,261	18,453	1256.5	76.4	1.22047	0.5039	0.37	80.16	99.85	80.16	127
Barcelona	−0.68712	−0.681615	171,350,447	30,947	1548.44	76.7	1.56854	0.0467	0.4	95.54	99.99	95.55	321
Bizkaia	−0.864704	−0.82699	36,085,689	31,792	1570.69	75.6	1.67005	0.0508	0.39	91.25	99.99	96.93	197
Burgos	−0.928119	−0.94275	10,505,020	29,571	1561.36	76.7	1.79545	0.1818	0.43	72.12	99.02	76.04	44
Cáceres	−0.947955	−0.934999	7,664,977	19,464	1235.8	75	1.13636	0.4886	0.41	67.37	99.73	68.88	88
Cádiz	−0.976223	−0.966461	22,535,246	18,050	1342.65	75.9	1.08088	0.0588	0.37	81.51	99.99	86.63	68
Cantabria	−0.895654	−0.87647	13,737,756	23,646	1534.3	76.2	1.64734	0.1932	0.4	72.29	99.82	77	207
Castellón	−0.976309	−0.921982	16,149,473	28,367	1418.4	76	1.33019	0.2075	0.38	82.38	99.92	90.31	53
Ceuta	−0.841694	−0.735741	1,720,295	20,251	1577.94	74.6	1.58824	0	0.35	92.78	99.88	92.78	17
Ciudad Real	−1.00039	−0.967509	10,689,033	21,563	1257.46	75.5	1	0.194	0.48	85.18	99.94	86.03	67
Córdoba	−1.01102	−1.102466	14,534,325	18,525	1369.77	76.1	0.97674	0.1628	0.26	89.75	99.88	90.55	43
Cuenca	−1.205496	−1.250948	4,536,392	22,691	1227	74.8	1.18	0.6	0.28	75.96	99.07	76.88	25
Girona	−0.666998	−0.629321	21,202,782	28,184	1425	76.1	1.67391	0.3478	0.3	71.84	99.89	71.84	46
Grarada	−0.883872	−0.840934	16,687,601	18,181	1408.78	76.9	1.56757	0.1081	0.49	77.08	99.92	80.82	37
Guadalajara	−1.20078	−1.237306	5,245,815	20,415	1228.91	77.7	1.1875	0.5	0.34	76.46	99.38	76.46	32
Guipuzkoa	−0.839686	−0.751365	24,060,930	33,851	1619.91	76.4	1.65929	0.0708	0.41	87.35	99.95	96.02	113
Huelva	−1.055769	−0.896241	10,607,333	20,273	1292.05	72.8	1.22727	0.4545	0.27	78.81	99.97	79.73	22
Huesca	−0.681279	−0.555177	6,134,249	28,015	1567.36	74	1.76389	0.6389	0.39	78.55	99.55	78.55	36
Islas Baleares	−0.8661	−0.812545	32,767,619	27,870	1463.34	76.4	1.37423	0.1534	0.37	88.2	99.84	90.37	163
Jaén	−1.014306	−1.001184	11,808,429	18,628	1245.51	76.6	0.69231	0.2564	0.41	87.79	99.86	90.83	39
La Rioja	−0.949345	−0.911352	8,5931,85	27,482	1477.37	76.1	1.59019	0.3291	0.42	90.72	99.65	90.72	316
Las Palmas	−0.874818	−0.896241	23,553,372	20,813	1296.88	74.8	1.36364	0.0455	0.42	84.97	99.95	85.43	88
León	−0.890206	−0.958886	10,006,588	21,579	1545.92	77.7	1.65306	0.5	0.39	71.73	98.86	71.82	98
Lleida	−0.921853	−0.974199	12,218,853	28,456	1651.43	75	1.51429	0.6286	0.49	74.57	99.6	74.68	35
Lugo	−1.105812	−1.162698	7,692,177	23,320	1399.22	77.4	1.29688	0.5625	0.42	41.82	99.93	60.4	64
Madrid	−0.697609	−0.690941	231,133,592	35,091	1565.24	75.9	1.70097	0.0097	0.36	96.84	99.99	96.85	515

Table A23. *Cont.*

Province	Digital Uses	Digital Skills	GPD	GPD_PC	INCOME	AGE	STUDIES	RURAL	SEX	FTTH	LTE	Sp100M	n
Málaga	−0.7175	−0.642426	31,023,255	18,801	1429.07	76	1.43023	0.093	0.48	92.73	99.99	93.88	86
Melilla	−0.583406	−0.47647	1,582,540	18,700	1573.08	74.6	2.23077	0	0.31	100	100	100	13
Murcia	−0.961825	−0.984258	31,198,376	21,094	1234.52	76.1	1.07742	0	0.41	84.06	99.94	86.94	155
Navarra	−0.835077	−0.842791	20,047,454	31,026	1572.67	76.6	1.66791	0.3582	0.41	71.25	99.91	79.81	268
Ourense	−1.019081	−0.985701	6,813,831	22,120	1418.03	76.9	1.22131	0.459	0.38	50.41	99.84	65.95	61
Palencia	−0.745896	−0.759816	4,407,310	27,346	1415	76.4	1.44	0.28	0.36	71.42	98.92	73.92	25
Pontevedra	−0.98622	−1.014064	21,247,944	22,586	1442.95	75.1	1.21818	0.1364	0.41	65.47	99.99	74.46	110
Salamanca	−1.032965	−0.975559	7,048,640	21,187	1329.07	76.8	1.24419	0.4651	0.51	70.17	98.88	72.65	43
Santa Cruz de T	−0.964645	−0.981507	22,269,949	21,076	1514.06	75.6	1.2125	0.0625	0.36	83.39	99.94	83.57	80
Segovia	−1.12927	−1.047826	3,418,981	22,212	1341.67	79.4	1.27778	0.6111	0.56	70.11	99.51	70.11	18
Sevilla	−0.937435	−0.87598	39,535,345	20,314	1484.9	75.9	1.26238	0.0891	0.41	91.75	100	92.62	101
Soria	−1.086951	−1.177118	2,380,731	26,626	1622.06	78.1	1.64706	0.3529	0.47	65.64	98.3	65.64	17
Tarragona	−0.838028	−0.785626	24,567,640	30,810	1527.7	75.6	1.77027	0.1622	0.46	73.92	99.89	73.92	37
Teruel	−1.080952	−1.066774	3,367,236	25,262	1147.32	75.9	1.28571	0.5714	0.39	58.75	98.09	58.75	28
Toledo	−1.06967	−1.046841	12,814,575	18,617	1203.25	76.6	0.81818	0.4416	0.48	75.97	99.95	76.71	77
Valencia	−0.820961	−0.815239	59,123,107	23,363	1477.66	76.4	1.4625	0.0313	0.41	86.01	99.94	92.93	160
Valladolid	−0.755909	−0.779561	13,998,460	26,901	1457.24	75.8	1.72368	0.1316	0.38	88.49	99.53	88.49	76
Zamora	−0.95353	−0.988421	3,459,100	19,813	1192.57	76.3	1.64865	0.4324	0.49	52.67	98.41	55.58	37
Zaragoza	−0.817581	−0.736273	27,348,811	28,386	1570.46	75.6	1.67478	0.1327	0.44	90.73	99.75	90.73	226
Total	−0.877319	−0.86335	1,201,241,000	25,771	1473.2	76.1	1.44978	0.2043	0.4	84.93	99.87	87.58	5247

References

1. Agulló-Tomás, M.S.; Zorrilla-Muñoz, V. Technologies and Images of Older Women. In *Human Aspects of IT for the Aged Population. Technology Design and User Experience. HCII 2020*; Lecture Notes in Computer Science; Gao, Q., Zhou, J., Eds.; Springer: Cham, Denmark, 2020; Volume 12209. [CrossRef]
2. Zorrilla-Muñoz, V.; Agulló-Tomás, M.S.; Forjaz, M.J.; Fernandez, E.; Rodriguez-Blazquez, C.; Ayala, A.; Fernandez-Mayoralas, G. Technology, Gender and COVID-19. Analysis of Perceived Health in Adults and Older People. In *Human Aspects of IT for the Aged Population. Supporting Everyday Life Activities. HCII 2021*; Lecture Notes in Computer Science; Gao, Q., Zhou, J., Eds.; Springer: Cham, Denmark, 2021; Volume 12787. [CrossRef]
3. Van Dijk, J. Digital divide research, achievements and shortcomings. *Poetics* **2006**, *34*, 221–235. [CrossRef]
4. DiMaggio, P.; Hargittai, E. From the "Digital Divide" to "Digital Inequality": Studying Internet use as Penetration Increases. Princeton University, Woodrow Wilson School of Public and International Affairs, Center for Arts and Cultural Policy Studies. Princeton. Working Paper Series 15. 2001. Available online: https://www.academia.edu/2802657/From_the_Digital_Divide_toDigital_Inequality_Studying_internet_use_as_penetration_increases (accessed on 7 June 2022).
5. Van Deursen, A.J.A.M.; Van Dijk, J.A.G.M. Measuring Internet Skills. *Int. J. Hum–Comput. Interact.* **2010**, *26*, 891–916. [CrossRef]
6. Friemel, T.N. The digital divide has grown old: Determinants of a digital divide among seniors. *New Media Soc.* **2014**, *18*, 313–331. [CrossRef]
7. Castaño, C. Nuevas tecnologías y Género. La segunda brecha digital y las mujeres. *Telos* **2008**, *75*, 183–200. Available online: http://www.telos.es/articuloAutorInvitado.asp?idarticulo=2&rev=75 (accessed on 7 June 2022).
8. Hargittai, E. Second-Level Digital Divide: Differences in People's Online Skills. *First Monday* **2002**, *7*, 1–14. Available online: http://firstmonday.org/article/view/942/864#author (accessed on 7 June 2022). [CrossRef]
9. Calderón-Gómez, D. Una aproximación a la evolución de la brecha digital entre la población joven en España (2006–2015). *Rev. Española Sociol.* **2019**, *28*, 27–44. [CrossRef]
10. Sparks, C. What is the "Digital Divide" and why is it Important? *Javn. Public* **2013**, *20*, 27–46. [CrossRef]
11. Reisenwitz, T.; Iyer, R. A Comparison of Younger and Older Baby Boomers: Investigating the Viability of Cohort Segmentation. *J. Consum. Mark.* **2007**, *24*, 202–213. [CrossRef]
12. Chua, S.L.; Chen, D.-T.; Wong, A.F. Computer Anxiety and its Correlates: A Meta-analysis. *Comput. Hum. Behav.* **1999**, *15*, 609–623. [CrossRef]
13. Jun, W. A Study on the Current Status and Improvement of the Digital Divide among Older People in Korea. *Int. J. Environ. Res. Public Health* **2020**, *17*, 3917. [CrossRef]
14. Torres-Albero, C. Sociedad de la información y brecha digital en España. *Panor. Soc.* **2017**, *25*, 18–33. Available online: https://www.funcas.es/wp-content/uploads/Migracion/Articulos/FUNCAS_PS/025art03.pdf (accessed on 7 June 2022).
15. Choudrie, J.; Pheeraphuttranghkoon, S.; Davari, S. The Digital Divide and Older Adult Population Adoption, Use and Diffusion of Mobile Phones: A Quantitative Study. *Inf. Syst. Front.* **2020**, *22*, 673–695. [CrossRef]
16. Pino-Juste, M.R.; Soto-Carballo, J.G.; Rodríguez-López, B. Las personas mayores y las TIC. Un compromiso para reducir la brecha digital. Pedagogía Social. Revista Interuniversitaria. *Pedagogía Social, Revista Interuniversitaria.* **2015**, *26*, 337–359. [CrossRef]
17. Peral-Peral, B.; Arenas-Gaitán, J.; Villarejo-Ramos, A.F. From digital divide to psycho-digital divide: Elders and online social networks. *Comunicar* **2015**, *45*, 57–64. [CrossRef]
18. Criado-Quesada, B.; Zorrilla-Muñoz, V.; Agullo-Tomas, M.S. El uso de tecnologías de asistencia sanitaria digital por parte de la población mayor desde una perspectiva de género e intrageneracional. *Teknokultura* **2021**, *18*, 103–113. [CrossRef]
19. Agudo-Prado, S.; Pascual-Sevillano, Á.; Fombona, J. Uses of Digital Tools among the Elderly. *Comunicar* **2012**, *20*, 193–201. [CrossRef]
20. Morales Romo, N. El reto de la brecha digital y las personas mayores en el medio rural español: El caso de Castilla y León. *FJC* **2016**, *13*, 165–185. [CrossRef]
21. Wicks, D.A. Building Bridges for Seniors: Older Adults and the Digital Divide. In *Proceedings of the Annual Conference of CAIS/Actes du congrès annuel de l'ACSI*; University of Alberta: Edmonton, AB, Canada, 2013; *Preprint.* [CrossRef]
22. Eastman, J.K.; Iyer, R. The Impact of Cognitive Age on Internet Use of the Elderly: An Introduction to the Public Policy Implications. *Int. J. Consum. Stud.* **2005**, *29*, 125–136. [CrossRef]
23. Iyer, R.; Eastman, J.K. The Elderly and Their Attitudes Toward the Internet: The Impact on Internet Use, Purchase, and Comparison Shopping. *J. Mark. Theory Pract.* **2006**, *14*, 57–67. [CrossRef]
24. Dean, D.H. Shopper Age and the Use of Self-service Technologies. *Manag. Serv. Qual.* **2008**, *18*, 225–238. [CrossRef]
25. Kim, H.J.; Lee, J.M. Determinants of Mobile Digital Information Usage among Senior Consumers: Focusing on secondary digital divide. *Family Envorn. Res.* **2018**, *56*, 493–506. [CrossRef]
26. Robles-Morales, J.M.; Torres-Albero, C.; Molina, Ó.M. La Brecha Digital. Un Análisis de las Desigualdades Tecnológicas en España. *ICTLogy.* 2010. Available online: https://ictlogy.net/bibliography/reports/projects.php?idp=1722 (accessed on 7 June 2022).
27. Navarro Beltrá, M. La brecha digital de género en España: Cambios y persistencias. *Feminismos* **2009**, 183–199. [CrossRef]
28. Warf, B. Teaching Digital Divides. *J. Geogr.* **2019**, *118*, 77–87. [CrossRef]
29. Hindman, D.B. The Rural-Urban Digital Divide. *J. Mass Commun. Q.* **2000**, *77*, 549–560. [CrossRef]
30. Norris, P. Information Poverty and the Wired World. *Int. J. Press Politics* **2000**, *5*, 1–6. [CrossRef]
31. Ragnedda, M. *The Third Digital Divide: A Weberian Approach to Digital Inequalities*; Routledge: New York, NY, USA, 2017; 128p.

32. Lupac, P. *Beyond the Digital Divide. Contextualizing the Information Society*; Emerald Publishing Limited: Bingley, UK, 2018; ISBN 978-1-78756-548-7.
33. Bouza, F. Tendencias a la desigualdad en Internet: La brecha digital (digital divide) en España. In *Tendencias en Desvertebración social y en Políticas de Solidaridad*; Tezanos, J.F., Tortosa, A., Eds.; Ed. Sistema: Madrid, Spain, 2003; pp. 93–121.
34. Robles, J.M.; Torres-Albero, C.; Villarino, G. Inequalities in digital welfare take-up: Lessons from e-government in Spain. *Policy Stud.* **2021**. [CrossRef]
35. Giner-Pérez, J.M.; Tolosa-Bailén, M.C. La brecha digital local en España: Factores explicativos. Working Paper: En: XXXIV Reunión de Estudios Regionales, Baeza-Jaén, Spain, 27–29 November 2008. Available online: http://rua.ua.es/dspace/handle/10045/23143 (accessed on 7 June 2022).
36. Van Deursen, A.; Van Dijk, J. Toward a Multifaceted Model of Internet Access for Understanding Digital Divides: An Empirical Investigation. *Inf. Soc.* **2015**, *31*, 379–391. [CrossRef]
37. Chang, E.-S.; Kannoth, S.; Levy, S.; Wang, S.-Y.; Lee, J.E.; Levy, B.R. Global reach of ageism on older persons' health: A systematic review. *PLoS ONE* **2020**, *15*, e0220857. [CrossRef]
38. Rosales, A.; Fernández-Ardevol, M. Ageism in the era of digital plataforms. *Int. J. Res. New Media Technol.* **2020**, *26*, 1074–1087. [CrossRef]
39. Manor, S.; Herscovici, A. Digital ageism: A new kind of discrimination. *Hum. Behav. Emerg. Technol.* **2021**, *3*, 1084–1093. [CrossRef]
40. Report: Cobertura de Banda Ancha en España en el año 2020. Secretaría de Estado de Telecomunicaciones en Infraestructuras Digitales. 2021. Available online: https://avancedigital.mineco.gob.es/banda-ancha/cobertura/Documents/Cobertura-BA-2020.pdf?csf=1&e=lVCXmu (accessed on 7 June 2022).
41. Várallyai, L.; Herdon, M.; Botos, S. Statistical analyses of digital divide factors. *Procedia Econ. Financ.* **2015**, *19*, 364–372. [CrossRef]
42. Granado Palma, M. Educación y exclusión digital: Los falsos nativos digitales. Revista De Estudios Socioeducativos. *ReSed* **2019**, *7*, 27–41. [CrossRef]
43. Shah, D.; de Zúñiga, H.G.; Rojas, H.; Cho, J. Beyond Access: The Digital Divide and Internet Uses and Gratifications. IT & Society. 2003. Available online: https://www.academia.edu/808155/Beyond_access_The_digital_divide_and_Internet_uses_and_gratifications?auto=citations&from=cover_page (accessed on 7 June 2022).
44. Drigas, A.; Papagerasimou, Y. Intergenerational Learning for the E-Inclusion of Senior Citizens through the prism of the GRANKIT Project. *Int. J. Comput. Sci. Issues (IJCSI)* **2015**, *12*, 64. Available online: http://ijcsi.org/papers/IJCSI-12-6-64-71.pdf (accessed on 7 June 2022).
45. Orzea, I.; Bratianu, C. Intergenerational learning in ageing societies. In Proceedings of the 9th International Conference on Intellectual Capital, Knowledge Management and Organizational Learning-ICICKM, Bogotá, Colombia, 18–19 October 2012; pp. 193–200. Available online: https://www.academia.edu/20895868/Intergenerational_Learning_in_Ageing_Societies (accessed on 7 June 2022).
46. Cummings, S.M.; Williams, M.M.; Ellis, R.A. Impact of an Intergenerational Program on 4th Graders' Attitudes Toward Elders and School Behaviors. *J. Hum. Behav. Soc. Environ.* **2002**, *6*, 91–107. [CrossRef]
47. Slaght, E.; Stampley, C. Promoting intergenerational practice. *J. Intergener. Relatsh.* **2006**, *4*, 73–86. [CrossRef]

 *land*

Article

A System of Indicators for Socio-Economic Evaluation and Monitoring of Global Change: An Approach Based on the Picos de Europa National Park

Iván López [1], Rodrigo Suarez [2] and Mercedes Pardo [3,*]

[1] Department of Psychology and Sociology, University of Zaragoza, Pedro Berbuna, 12, 50009 Zaragoza, Spain; ivalopez@unizar.es

[2] Management Department, Picos de Europa National Park, Arquitecto Reguera, 13, 33004 Oviedo, Spain; rsuarez@pnpeu.es

[3] Department of Social Analysis, University Carlos III of Madrid, 126, 28903 Getafe, Spain

* Correspondence: mercedes.pardo@uc3m.es

Abstract: National Parks are spaces that are of great interest for evaluating and monitoring global environmental change as these parks encompass natural, cultural, and rural features, along with ecological processes, which are subject to social or economic changes that are much more difficult to track outside of these spaces. To do this, it is necessary to have a sufficient set of data and indicators to monitor the effects of global change in the short, mid, and long term. The majority of indicators have been developed to monitor the bio-geophysical environment; socio-economic indicators of global change for National Parks are much more limited. The aim of this paper is to present a system of indicators for socio-economic evaluation and monitoring of global change for the Picos de Europa National Park. This park has two unique features: it has one of the two systems of socio-economic indicators developed for the Spanish National Parks, and it is practically the only one of Spain's 16 National Parks with human populations living within its boundaries. Many of the indicators specifically developed for this park can be used for other national parks that have similar characteristics.

Keywords: protected areas; climate change; socioeconomic indicators; Spain

Citation: López, I.; Suarez, R.; Pardo, M. A System of Indicators for Socio-Economic Evaluation and Monitoring of Global Change: An Approach Based on the Picos de Europa National Park. *Land* **2022**, *11*, 741. https://doi.org/10.3390/land11050741

Academic Editors: Vanessa Zorrilla-Muñoz, Maria Silveria Agulló-Tomás, Eduardo Fernandez, Blanca Criado Quesada, Sonia De Lucas Santos, Jesus Cuadrado Rojo and Hossein Azadi

Received: 20 March 2022
Accepted: 8 May 2022
Published: 16 May 2022

Publisher's Note: MDPI stays neutral with regard to jurisdictional claims in published maps and institutional affiliations.

1. Introduction

Global change, and particularly climate change, is one of the gravest problems facing humanity [1,2]. Monitoring this threat in the short, mid, and long term is a crucial task in order to obtain greater knowledge and understanding, and thereby help societies to better mitigate and adapt to its effects. Preparing for and adapting to climate change is as much a cultural and intellectual challenge as it is an ecological one [3].

Within this context, National Parks (NPs) are relevant places for monitoring global change [4] due to the need to protect their natural resources. Additionally, these protected areas encompass natural, cultural, and rural elements, and ecological processes subject to social or economic changes, which are much more difficult to track outside of their boundaries. They are likewise important spaces to monitor because these changes are global in nature, and as such, these protected areas are not only of intrinsic interest, but they are also valuable for comparative purposes. Although global change in the NPs has been monitored, it has been mainly for the biophysical setting [5]. Monitoring the socio-economic environment in NPs has been much more limited, and socio-economic indicators designed for their monitoring are even more lacking [6].

The importance of socio-economic monitoring derives from the fact that national park features and changes need to be analyzed and interpreted as a socio-ecological system, since human systems and ecosystems are inextricably linked [7]. Within the context of climate change, this task is even more relevant, since its speed, together with the uncertainty it entails, increases the potential for more intense and frequent stresses, shocks, and resultant protracted crises [1].

In Spain, a system of indicators with these characteristics has only been developed for two of its 16 National Parks: the Sierra de Guadarrama National Park [8] and the Picos de Europa National Park [9]. The latter is the object of analysis in this paper, contributing to knowledge that is currently inexistent in the case of the National Parks of Spain, as well as other parks with similar biophysical and socio-economic features, which, in any event, would require adaptation [6]

2. Literature Review

A corpus of knowledge on social and economic indicators to measure development and social well-being in protected areas already exists [10,11]; although, the task of defining the concepts of development and social well-being, and determining which quantitative and qualitative variables to monitor, is not always straightforward [11]. In the global review by these authors regarding the use of indicators to assess the management of protected areas, they found that although social analysis in conservation is [11] (p. 1) "increasingly recognized as important for successful environmental outcomes . . . there was limited diversity and representation of important well-being dimensions".

For this study, this issue becomes even more complex. Its aim is to develop useful social and economic indicators, not only regarding social well-being, but also in connection with features of the National Parks' biophysical environment and the main goal of their conservation. In addition, there is the threat of climate change, a grave problem, with all the uncertainty it entails [1]. At the same time, research from a social science perspective on the matter is still limited [12], with even less consideration of global change in the NPs' socioeconomic indicators [13,14].

Two Challenges: Climate Change and Conservation of the National Parks

Global change, and especially climate change, has already been included in research, political, economic, and social agendas. It is thus of essence to increase the corpus of theoretical and empirical knowledge on global change and its consequences for our societies [1], still quite limited [1,2], as well as strengthen the political, legal, technological, economic, and sociological instruments for mitigation and adaptation by societies.

It should also be kept in mind that the interaction within the biophysical system itself, and between this system and social systems, amplifying or attenuating the effects, is a key feature of global change, making its evolution difficult to predict [2,15].

Additionally, one must take into account the socio-economic characteristics and the evolution of populations living in the high mountain natural areas under extreme meteorological conditions, especially exposed to climate change [16], as is the case of the national park under study. Throughout history, these populations have not only been able to develop an economy and way of life in adverse locations, but they have also contributed to its conservation and even improved upon some aspects [17]. The ecological importance of the high mountain pastures, for example, would not be sustained without the traditional livestock raising and vice versa [18]. Natural and human systems are closely interconnected; it is an interrelationship that goes beyond a static, deterministic environmental vision.

Nevertheless, at present, many of these traditional mountain economies are undergoing processes of economic and demographic decline. Their survival depends on a transformation towards higher added-value production (for example, traditional foods with national park denomination of origin) and an economy linked to the controlled tourist attraction of a national park [19].

Future evaluation, after monitoring, must be based on [7,20] holistic approaches that take into account the principal forces of change identified through monitoring, the dynamics of growth and decline, as well as the recognition of uncertainty.

This research lies along all these lines. Accordingly, any progress in this direction will be tentative and preliminary, more like a hypothesis for a system of indicators that will require empirical verification over time.

Hence, after the literature review, this paper first presents the challenge of conserving NPs in the face of global change; it then describes the research objectives and methodology used; the proposed indicators follow; and finally, the discussion and conclusions are presented.

3. Characteristics of National Parks and the Picos de Europa National Park

To study the consequences of global change in Spain's National Parks, the Global Change Monitoring Program in the National Parks Network was established. This monitoring program includes sociological monitoring as one of its three lines, as stated in the following [21] (p. 1).

> *"The NP network is at the service of society. Expressly included among the Network objectives is the contribution to society's environmental awareness and the implementation of models for sustainable development within the parks' environment. Sociological monitoring of the NP Network seeks to acquire further knowledge about the social role of the Network, assessing its projection, its presence and its repercussion on society, beyond the physical limits of the protected areas. This materializes in monitoring the interaction between the Network and society at different levels: visitors to the NP, the population in the NP socioeconomic area of influence, educators, the scientific community, etc.".*

The National Parks Network of Spain [22] (p. 1) states that *"a Park is a natural area with great ecological and cultural value, little changed by human activity, where, on account of its exceptional natural assets, its representative nature, the singularity of its flora, fauna, and its geomorphological formations, its conservation warrants the utmost attention, and it is declared to be in the general interest of the Nation as it is representative of Spain's natural patrimony. For a territory to be declared a National Park, it must be representative of its natural system, have a sufficiently large area to enable natural evolution and ecological processes, with natural conditions clearly predominating, display limited human intervention in its natural resources, territorial continuity, and, as a rule, not have inhabited areas within its limits (with exceptions), and be surrounded by an area that could be deemed a peripheral protection zone".*

The basic objective of a National Park is thus to ensure the conservation of its natural resources. It represents the highest degree of legal protection for a large tract of territory to guarantee its conservation. National Parks are singular and unique areas, and, clearly, they are few and far between. They are places where "nonintervention" is prioritized and their principle is to enable the free evolution of natural processes. The second objective of the National Park Network is to reconcile conservation with public use and enjoyment of the parks' natural assets. Third, the NPs are at the service of research and at furthering scientific knowledge.

Beyond that, the NPs can be a resource for economic development in the regions where they are located, as long as it is sustainable development. In short, the challenge is to make conservation compatible with social justice and economic development for those communities affected by these parks' restrictive legislation.

The Physical and Social Setting of the Picos de Europa National Park

The Picos de Europa National Park [23] is located in the north of the Iberian Peninsula, between Asturias, Cantabria, and Castile-Leon (the province of Leon). It extends over three mountains and the north-facing flank of the Cantabria Cordillera Range (where the park's densest forests are located).

These peaks, formed by limestone rock that surged up from the seabed due to orogenic activity, make up a terrain of high summits alternating with deep gorges and canyons. Hence, in the park, there are 200 elevations of over 2000 m, with differences in relief of more than 2300 m. The limestone has produced interesting karstic processes with slopes of over 1000 m, intense glacier erosion processes, and glacier lakes.

The park has a surface area of 67,455 hectares, making it the second most extensive one in Spain after Sierra Nevada National Park. Its location coordinates are Latitude 43°18′58″ N, 5°07′15″ O; and Longitude 43°04′28″ N, 4°37′03″ O. The National Park was created by Law 16/1995, which was later expanded in 2015. However, as it is heir to the first NP created in Spain, a part of the current park was already officially protected as of 1918 (Montaña de Covadonga National Park), one of Europe's first as well.

The park has received international recognition for its state of conservation and planning, and for the comprehensive management of its natural resources. Part of its territory is a Biosphere Reserve, Special Area of Conservation (SAC), and Special Protection Area (SPA) for Birds.

The mesotrophic limestone soil and the variability in altitude (almost 2500 m in an area of relatively small size (674.55 km^2), as well as the influence from the nearby sea, converge in a wide variety of ecosystems, with an ample diversity of flora: around 1900 species and subspecies, with some taxa specific to the park. The fauna associated with these ecosystems includes such emblematic species as the Cantabrian chamois, the wolf, the woodland grouse, and the brown bear, among others, some in danger of extinction. The unique geomorphological formations have shaped a landscape of great variety and esthetic value. The park encompasses deep canyons carved out by rivers, striking grooves, and gullies that drop down 1500 m or more, beech and oak forests, and the extraordinary mixed Cantabrian forest (in some areas residual Tertiary laurel forest), with meadows that are the result of centuries of human presence and livestock raising. Presiding over it all are the distinctive white limestone cliffs, in a permanent process of change due to the erosive action of water, ice, and wind, creating such unusual formations as the *lapiaces* (limestone grooves and hollows). Obviously, this variability of ecosystems and the fragility of many of them condition their detailed management and constrain the adoption of any decision that could affect their conservation.

Another unique feature of the park is that it is practically the only Spanish National Park with human populations living within its boundaries. Picos de Europa was occupied very early on in history, with a permanent human presence dating back to at least the Neolithic Period. Currently, the number of people dwelling within the park is approximately 981 (2021), distributed among 11 municipalities.

Its inhabitants have continued to make use of its territory up to the present day, shaping its landscapes to a considerable extent, at least in the less mountainous areas of the park, mainly through raising livestock and small-scale farming, along with wood gathering in the forest cover. In a large part of its extension, Picos de Europe is a humanized park.

In addition, the park's socioeconomic area of influence is inhabited by 14,164 people (2020) and includes an extension of 133,682.56 hectares pertaining to the 11 municipalities distributed among the three Autonomous Communities of Asturias, Cantabria, and Castile-Leon.

There has been an intense interrelation throughout history between the territory and its human populations, whose manifestations are so evident that a symbiosis seems to have existed between the two elements [23]. Indeed, this remarkable landscape is as much a result of the natural terrain and the morphological composition as it is the human activity that has taken place over centuries in its valleys and around the higher altitude *majadas* (temporary settlements used during the grazing season in the mountains).

All of this makes it one of the most visited NPs in Spain, with nearly two million visitors a year (1,620,739 visitors in 2021 despite COVID-19 restrictions on mobility), which at the same time presents a challenge for its management.

4. System for Evaluation and Monitoring the Socio-Economic Impact of Global Change in the Picos de Europa National Park

These biophysical and social characteristics of the park make assessing and monitoring the impact of global change an even more urgent task.

As mentioned, in Spain there is a Global Change Monitoring Program in the NP Network [24], dependent on the Ministry of Ecological Transition and Demographic Challenge. The scope of its monitoring is broad; although, it exclusively targets the biophysical environment, concretely in the placement of automatic weather stations (the Picos de Europa National Park has eight installations) along with specific studies on certain species and areas [24]. Analysis aimed at socio-economic monitoring of global change is non-existent, beyond its annual report (number of visitors, livestock using pastures, etc.) and, as noted earlier, the system of indicators for Sierra de Guadarrama and Picos de Europa National Parks [9].

In this context, the objective of this research is to design and operationalize a system for evaluating and monitoring the socio-economic impact of global change for the Picos de Europa National Park. It seeks to respond to the need for a sufficient set of data to monitor the effects of global change in the short, mid, and long term in the social and economic settings of the park. The research focus is on the Picos de Europa National Park due to its distinctive feature of being an inhabited national park, constituting, hence, the most complex case, while at the same time, looking to draw recommendations that can be applied to other national parks.

This general objective has been specified into the following goals:

-   Creating and fine-tuning an updated database for all the indicators of social and economic change that are able to be monitored.
-   Defining a system of indicators for monitoring (and eventually evaluating) global change in the social and economic setting of the Picos de Europa National Park, specifying what can be extrapolated to other national parks with similar features.

5. Theoretical Framework for the Indicators and Justification

The indicators were selected to obtain a balance between the general use of protected natural areas and those developed for the particular case of the Picos de Europa National Park. Utilizing indicators of general use is always helpful, since they allow for comparison between different protected areas. In addition, by incorporating broader monitoring, it is possible to obtain a longer time series.

There are diverse systems of indicators that use different frames of reference. One of them, the most simple, consists of structuring indicators by topic and subtopic so that the results appear grouped together and ordered. Another more structured framework is the 'Pressure-State-Response' (PSR) model, which was originally developed and recommended by the OECD [25]. In this model, indicators are first identified for 'Pressure', which corresponds to the causes for the phenomena under study (global change in this case). The second is the 'State' of the environment receiving the impacts from the pressure, and the third is the 'Response' of society to the problems presented. Through this framework of reference, cause–effect relationships are explicit in the monitoring system.

There are other models, such as the one later extended by the OECD—the DPSIR, Driving force-Pressure-State-Impact-Response framework. As Mosaffaie et al. [26] (p. 1) point out, "the DPSIR model depicts 'why it happened' through P and D terms. After knowing why it happened, we can and should further analyze 'how to deal with it'. There are many ways to achieve this goal, either directly through S or I, or acting on P or D".

For national parks, where there is a lack of disaggregated socio-economic data for the specific park area, the simpler model—PSR—could be more useful as a framework of reference, to provide coherence from a systematic perspective, while at the same time being a more viable model. This is an approach based on the objective of an evaluation and monitoring system, formed as such by a series of interrelated elements from different processes. The indicators should be variables (or indexes) that provide information about the trends of change in these elements and processes and that explain the global functioning of the system and its deviation from or its approximation to greater sustainability. Pressure indicators (global change) are beyond the scope of this study, as a planetary process is scarcely affected by the activities taking place in the park, and the focus of our system of indicators is on impact and adaptation.

The first block of indicators is the frame of reference on the STATE of the situation in the PSR model; hence, it is called the *'Receptor environment'* (RE) in this study. The variation in its indicators, as a result of the pressure due to global environmental change and that of the mitigation and adaptation measures, is what will reflect the social and economic impacts of global change. The impact can be evaluated once the monitoring system is developed, along with its evolution over time. Specifically, the following categories in the levels of differentiation for group and subgroup (in parenthesis) are included in the system of socio-economic indicators:

- Natural resources (land use: water consumption; water treated by the purification system; energy consumption; agricultural resources, environmental resources; waste treatment).
- Demographic base (population and characteristics; activity, occupation; unemployment).
- Economic base (production of goods and services; employment in productive activities; tourism activity; income and transfer; public investments).
- Social, political, and cultural base (health; political and social organization; social cohesion; well-being; poverty; security; culture).

The second frame of reference corresponds to RESPONSE in the scheme, called *'Mitigation and Adaptation'* (M&A), which includes two levels of group differentiation and others for the subgroup (in parenthesis):

- Governance (official; unofficial; national park management; legislation).
- Social and research instruments (information and communication; perception; education; training and participation; socio-economic research).

This is the framework of reference that defines the selection of indicators for this project.

6. Materials and Methods

The area of study entails the eleven municipalities where the Picos de Europa National Park is located along with its socio-economic area of influence, as it is defined legally [27] (Figure 1).

The methodology used has been quantitative and qualitative in a combined way [28]. The materials and methods utilized are the following:

(1) Bibliographic search and analysis, including academic databases and public administration reports. Given the nature of the Spanish National Parks, in addition to scientific literature, it is necessary to examine studies and documentation from public administrations and responsible agencies [29], as well as from other similar experiences in other national parks. Although each park has its own distinctive features, all of them have their objective of conservation and their complex management in common. Thus, there are many elements that are highly useful for comparison and learning purposes.

(2) Collecting and analyzing national, regional, and local legislation, and programs and plans affecting the park as well as specific park legislation and planning. The foregoing must be taken into account as they represent the context that will determine the characteristics for this system of indicators.

(3) Collecting and analyzing the statistical information necessary to create a database for monitoring purposes for the short, mid, and long term, distinguishing between official sources and other reliable sources from grey literature [29]. This combination of information will not only allow data to be compared and contrasted, but most importantly, lead to critical analysis.

(4) Holding a workshop with academic experts and specialists in national park management, aimed at discussion and consensus regarding the focus for a system of indicators of this nature, inexistent in the other Spanish National Parks. Specifically, a workshop was held with eight participants chosen from different fields of study, including Soci-

ology and Economics, and from National Park management. This interdisciplinary academic and management focus has enriched the proposed approaches by combining scientific knowledge with the empirical knowledge of day-to-day management [30].

(5) Designing and developing indicators for state, impact, and response. The indicators contain current information (chiefly, although not limited to, statistical data) but also reveal the deficits and shortcomings to resolve in future monitoring. This design was corroborated by park managers for several reasons: much of the required information came necessarily from the managing institution; their empirical experience added a greater degree of practicality to the indicators selected.

(6) Identifying the interrelations between indicators. This is key in creating a system of indicators, and not simply drawing up a lengthy list [31]. This is one of the most complex tasks, since global change is a multidimensional issue with impacts that are likewise multidimensional. Our research presents some of these interrelations, as a hypothesis that will need to be revised over time for adaptive monitoring [32].

Figure 1. Location of the Picos de Europa National Park. Source: PNPE.

Methodological Conceptualization

To achieve these objectives, the starting point must be a conceptualization that allows us to orient the scope and content of the indicators to select.

Hutto and Belote [33] (p. 1) point out that the type of monitoring to be performed will differ according to the questions previously posed. They establish four types of monitoring—surveillance, implementation, effectiveness, and ecological effects "that are designed to answer very different questions and achieve very different goals. Surveillance monitoring is designed to uncover changes in target variables over space and time; implementation monitoring is designed to record whether management actions were applied as prescribed; effectiveness monitoring is designed to evaluate whether a given management action was effective in meeting a stated management objective; and ecological effects monitoring is designed to uncover unintended ecological consequences of management actions".

The research presented here represents the first type, surveillance; although, in some ways, it takes into account the second type, implementation, since it includes indicators not only on the state, but also on impacts, mitigation, and adaptation. However, for this same type, the differentiation made by Hellawell [34] is pertinent, as this author noted that monitoring could only be referred to when trying to determine the degree of deviation of the indicator values from the reference values established beforehand. When the indicator value is not compared with the reference value, the indicator acts as a type of surveillance.

Therefore, monitoring should not be referred to when doing only a diagnostic evaluation, that is to say, evaluating the state of a system or some of its components at a moment in time. However, if repetition of this diagnostic assessment were set up, as is the case of this system of indicators, it would be close to the concept of monitoring, establishing reference thresholds to compare the evolution of the system with the initial state.

Designing a system of evaluation and monitoring, as is the goal of this study, is to operationalize an observation process repeated in space and time of the variables to be studied, which requires establishing methods to obtain comparable data.

It is necessary to define the objectives of the evaluation and monitoring system before identifying the set of indicators for the system and make it possible to detect changes in their processes, establishing continuation of protocols to measure these indicators.

As the theoretical basis and method for defining indicators can be common to the socio-economic system and to the bio-geophysical ecosystems and landscapes (which are the result of the interrelation between the bio-geophysical and the social environment), when identifying the objectives of the evaluation and monitoring process it is possible to generically refer to the systems under study, with the understanding that they can be either natural or social.

For the case of evaluating and monitoring the socio-economic impact of global change on the Picos de Europa National Park, the specific objectives for the indicators have been defined as follows:

- To record the dynamics of the systems under study, analyzing the trends of change, whether due to natural or social (anthropic) causes.
- To improve knowledge about the systems studied, through the collection or generation of new information related to the social and economic impact of global change on the park.
- To predict specific and/or global changes in the systems studied, especially alterations or damage from unforeseen events.
- To identify effects on the dynamics of social systems resulting from management practices, and to detect undesired effects.

Indicators are the instruments that support any evaluation or monitoring system. Although there are many definitions for the concept of indicators for natural protected areas, for this study, they are understood as a variable or the relationship between variables (indexes) whose measurement can reveal certain references about the evolution of the system in which they are immersed. The variables and indexes that have this indicator characteristic are those that are sensitive to changes and trends of a natural or human origin.

The indicators must be significant and meaningful, in this case, from the point of view of sustainability [35]. Social and economic variables lacking meaning or significance linked to sustainable development will be less useful as indicators in this system. As such, the evaluation and monitoring system should never be understood as a mere list of indicators, but rather always in its connection with the goal of sustainable development, as posed by Agenda 2030 [36], to which Spain has subscribed.

Clearly, the highly complex interlinked natural and social systems in protected natural areas make it virtually unfeasible to measure all possible indicators. A limited selection is thereby needed in order to maximize information and minimize cost.

A first filter to reduce the initial list could be verifying if the preselected indicators satisfy certain basic requisites. These could include, in addition to their rigor and reliability, that they are sufficiently sensitive so as to provide early warning about change; be able to be distributed over a large geographical area or be easy to apply; be capable of providing continuous estimates under a wide range of stress; be relatively independent of the sample size; be easy to measure or interpret; be easy to differentiate between natural cycles and changes brought about by anthropogenic influence; and be relevant to important socioeconomic and ecological phenomena (global and climate change in this case).

Finally, the indicators were selected to obtain a balance between the general use of protected natural areas and those developed for the particular case of the Picos de Europa National Park. Utilizing indicators of general use is always helpful, since they allow for comparison between different protected areas. In addition, by incorporating broader monitoring, it is possible to obtain a longer time series.

For gathering data for the set of indicators, only published statistical sources have been used. These include the Municipal Register of Inhabitants; the Population and Housing Census; the Economic and Social Data Report from the Territorial Units of Spain (Caja España-Caja Duero); the Institute of Statistics from the Autonomous Communities; the Public Employment Service; the Ministry of the Interior; Ecoembes and Ecovidrio; the Ministry of the Economy and Competitiveness; annual reports from the Picos de Europa National Park (visitor statistics, natural resources databases, investments, management actions, etc.); and the National Institute of Statistics.

For a detailed description of each municipality in the Picos de Europa National Park, a data sheet has been drawn up for every one of them with their main socio-economic characteristics (Table 1). Thus, each municipality has its own data sheet according to the following areas and sections.

Table 1. Data sheet municipalities.

Area		Sections
01.	Territorial data	- Total population - Population density - Area included in the park - Distance to the capital of the province
02.	Population and demography	- Population evolution - Structure and movement - Immigrants
03.	Productive structure	- Economy and productive sectors - Services and tourist activity - Employment and unemployment - Housing
04.	Living conditions	- Facilities - Motorization - Associations present
05.	Environment	- Waste
06.	Municipal income and budget	- Municipal budget
07.	Participation in elections	- Voter participation

Source: own elaboration.

Territorial data include figures on the total population of the municipality, population density, surface area forming part of the park, and distance from the municipality to the capital of the province of the Autonomous Community.

The section on population and demography synthesizes the evolution of the population (gender and year-on-year growth rate), the structure and movement of the population (population pyramid, demographic dependency ratio, rate of aging, trends, and replacement rate), as well as the immigrant population (by gender and evolution).

Information on the productive structure of the municipalities in the park includes the economic and productive sector (number of workers and companies by activity sector or land use), services and tourism activities (commercial establishments, tourist capacity, and basic facilities), employment and unemployment (evolution of the unemployment rate), and housing (type of housing and type of ownership for principal dwellings).

Living conditions in the park municipalities include data on educational facilities (non-university educational centers), healthcare facilities (primary healthcare centers, clinics, and pharmacies), motorization (registered vehicles), and associations, sports entities, and other collectives located within the administrative limits.

The section on environmental information refers here to selective waste collection (plastics and containers, paper and cardboard, and glass).

The data sheet also includes data on municipal budgets (municipal financial expenditures and real expenses)—there are no published data on income—and voter participation (abstention in local elections).

In short, these data sheets provide a general overview of the socio-economic situation and evolution for each municipality located within the Picos de Europa National Park.

7. Results

A data sheet (Annex Table 2) has been elaborated with information for each of the indicators. The indicators developed (Table 3) vary in detail for diverse reasons, among which is the lack of disaggregated statistical information for the park area; this has been the case, for example, for the indicators on the estimation of income generated by private individuals and by companies.

Table 2. Example of Indicator data sheet of each indicator.

System of Indicators for Socio-Economic Evaluation and Monitoring of Global Change for the Picos de Europa National Park Indicator Data Sheet		
Indicator Name	Agriculture and Livestock Area Index	
Frame of Reference	Receptor environment	*Reference number* MR-02
Group of indicators	Natural resources	
Subgroup	Land use	
CHARACTERISTICS OF SELECTED INDICATOR		
Objective, definition, and justification of the indicator	The agricultural and livestock exploitations within the territory, which include the strata of agricultural crops, scrub, pasture, and grassland. It seeks to reflect land use by those that do not entail an irreversible transformation of the National Park.	
Measurement parameters	Percentage of agricultural and livestock area with respect to the total park area.	
Calculation formula	Agricultural area multiplied by 100, divided by the total area.	
Unit of measurement	Percentage rate, result of dividing hectares by hectares.	
Data disaggregation	By park municipalities.	
Source of information	National Forest Inventory.	
Scope	The territory included within the delimitation of the National Park.	
Data availability	Upon request from the National Park Management Office.	
Measurement periodicity	Corresponding to the update from the National Forest Inventory.	
Responsibility for the accuracy of data	Ministry of Ecological Transition and Demographic Challenge (current name)	
Indicators to which it is related	RE-01, RE-02, RE-04. (Table 3)	
Reference values	Other National Parks.	

Receptor Environment Indicators (RE)	
:ator Name	Related Indicators (A Hypothesis That Must be Empirically Tested in the Future)
ce economy rate	RE-26
ist accommodation capacity	RE-29, RE-32
visits	RE-28
or to resident ratio	RE-19, RE-29
onality of park visits	RE-28, RE-29
ndary uses of housing for tourism	RE-28, RE-29, RE-30
ic investment per capita	RE-25, RE-26
icipal investment per capita	RE-35, RE-29
icipal indebtedness per capita	RE-34, RE-29
versity graduate rate	RE-19
th infrastructure ratio	RE-19, RE-29, RE-33, RE-34
ne equipment (heating)	RE-21, RE-39
erly population living alone	RE-19, RE-21
Mitigation and Adaptation (M&A)	
tings held by the governing and social icipatory bodies	M&A-02
al participation agreements	M&A-01
islation	To all the M&A indicators
islation compliance	M&A-03, M&A-05
rrent agreements with institutions	M&A-03, M&A-04
ministrative sanctions	M&A-07
ministrative authorizations	M&A-06
n-regulated governance activities	M&A-01, M&A-05
ared area for fire protection and improvement of zing use	M&A-10
ea for controlled burning to prevent fires and prove grazing Use	M&A-09
lf damage to livestock	M&A-12, M&A-03, M&A-04, M&A-07
ld boar damage to the productive capacity grasslands	M&A-03, M&A-07
rticipants in the environmental ucation program	M&A-14, M&A-28, M&A-29, M&A-32
rticipants in the park's volunteer program	M&A-13, M&A-28, M&A-32
ste removed from the park	M&A-18
idence of forest fires	M&A -17
vestment in prevention and extinction of est fires	M&A-16, M&A-18
blic investments in the park	M&A-16, M&A-17, M&A-19
bsidies in the municipalities of the park	M&A-05, M&A-18
mpensation for cessation of activity compatible th the park	M&A-19

Table 2. *Cont.*

System of Indicators for Socio-Economic Evaluation and Monitoring of Globa
Indicator Data Sheet

	INDICATOR VALUES FOR THE DIFFERENT ARE.	
Year	Municipalities of the NP in Asturias	Municipalities of the NP in Ca & Castile-Leon
	(*) Comments	

(*) Values above 100 imply the predominance of agricu
livestock raising since there are few agricultural uses. Soι

Table 3. Indicators for the Socio-Economic Monitori
the Picos de Europa National Park.

	Receptor Environment Indicators (RΙ	
N°	**Indicator Name**	**Related Indi Empirically**
RE-01	Wooded forest index	RE-02, RE-03,
RE-02	Agricultural and livestock area index	RE-01, RE-03,
RE-03	Agriculture and livestock forest index	RE-01, RE-02-
RE-04	Scrubland index	RE-01, RE-02-
RE-05	Livestock use	RE-06
RE-06	Cattle, goats, and sheep	RE-05
RE-07	Water consumption from the supply network	RE-08
RE-08	Water treated by purification systems	RE-07
RE-09	Energy consumption	RE-10
RE-10	Energy production	RE-09
RE-11	Electrical energy balance	RE-09, RE-10
RE-12	Gasoline consumption	R-13
RE-13	Diesel consumption	RE-12
RE-14	Urban waste	RE-15, RE-16, R
RE-15	Waste paper collected to recycle	RE-14, RE-16, R
RE-16	Cans and plastic collected to recycle	RE-14, RE-15, RΙ
RE-17	Glass collected to recycle	RE-14, RE-15, RΙ
RE-18	Waste selection containers	RE-14, RE-15, RΙ
RE-19	Demographic population pyramid	RE-20, RE-21, RΕ
RE-20	Demographic dependency rate	RE-19, RE-21
RE-21	Aging rate	RE-19, RE-20
RE-22	Immigration rate	RE-19, RE-20, RE
RE-23	Active population rate	RE-24, RE-19
RE-24	Occupied population rate	RE-23, RE-25
RE-25	Registered unemployment	RE-19, RE-23
RE-26	Agrarian workers rate	RE-19, RE-27

N°	**Ind**
RE-27	Ser
RE-28	Tou
RE-29	Par
RE-30	Vis
RE-31	Sea
RE-32	See
RE-33	Pu
RE-34	Mu
RE-35	Mu
RE-36	Un
RE-37	He
RE-38	Ho
RE-39	Ele
M&A-01	Me pa
M&A-02	Sc
M&A-03	Le
M&A-04	Le
M&A-05	Cu
M&A-06	A
M&A-07	A
M&A-08	N
M&A-09	C g
M&A-10	A iı
M&A-11	V
M&A-12	V c
M&A-13	F e
M&A-14	I
M&A-15	V
M&A-16	I
M&A-17	I
M&A-18	I
M&A-19	
M&A-20	

Table 2. *Cont.*

System of Indicators for Socio-Economic Evaluation and Monitoring of Global Change for the Picos de Europa National Park

Indicator Data Sheet

INDICATOR VALUES FOR THE DIFFERENT AREAS AND PERIODS

Year	Municipalities of the NP in Asturias	Municipalities of the NP in Cantabria & Castile-Leon	Total municipalities of the NP

(*) Comments

(*) Values above 100 imply the predominance of agricultural use of the territory, which is almost exclusively livestock raising since there are few agricultural uses. Source: own elaboration.

Table 3. Indicators for the Socio-Economic Monitoring and Evaluation System of Climate Change for the Picos de Europa National Park.

Receptor Environment Indicators (RE)		
N°	Indicator Name	Related Indicators (A Hypothesis That Must be Empirically Tested in the Future)
RE-01	Wooded forest index	RE-02, RE-03, RE-04
RE-02	Agricultural and livestock area index	RE-01, RE-03, RE-04
RE-03	Agriculture and livestock forest index	RE-01, RE-02-RE-04
RE-04	Scrubland index	RE-01, RE-02-RE-03
RE-05	Livestock use	RE-06
RE-06	Cattle, goats, and sheep	RE-05
RE-07	Water consumption from the supply network	RE-08
RE-08	Water treated by purification systems	RE-07
RE-09	Energy consumption	RE-10
RE-10	Energy production	RE-09
RE-11	Electrical energy balance	RE-09, RE-10
RE-12	Gasoline consumption	R-13
RE-13	Diesel consumption	RE-12
RE-14	Urban waste	RE-15, RE-16, RE-17, RE-18
RE-15	Waste paper collected to recycle	RE-14, RE-16, RE-17, RE-18
RE-16	Cans and plastic collected to recycle	RE-14, RE-15, RE-17, RE-18
RE-17	Glass collected to recycle	RE-14, RE-15, RE-16, RE-18
RE-18	Waste selection containers	RE-14, RE-15, RE-16, RE-17
RE-19	Demographic population pyramid	RE-20, RE-21, RE-22
RE-20	Demographic dependency rate	RE-19, RE-21
RE-21	Aging rate	RE-19, RE-20
RE-22	Immigration rate	RE-19, RE-20, RE-21, RE-23, RE-24, RE-25, RE-26
RE-23	Active population rate	RE-24, RE-19
RE-24	Occupied population rate	RE-23, RE-25
RE-25	Registered unemployment	RE-19, RE-23
RE-26	Agrarian workers rate	RE-19, RE-27

Table 3. *Cont.*

Receptor Environment Indicators (RE)		
N°	**Indicator Name**	**Related Indicators (A Hypothesis That Must be Empirically Tested in the Future)**
RE-27	Service economy rate	RE-26
RE-28	Tourist accommodation capacity	RE-29, RE-32
RE-29	Park visits	RE-28
RE-30	Visitor to resident ratio	RE-19, RE-29
RE-31	Seasonality of park visits	RE-28, RE-29
RE-32	Secondary uses of housing for tourism	RE-28, RE-29, RE-30
RE-33	Public investment per capita	RE-25, RE-26
RE-34	Municipal investment per capita	RE-35, RE-29
RE-35	Municipal indebtedness per capita	RE-34, RE-29
RE-36	University graduate rate	RE-19
RE-37	Health infrastructure ratio	RE-19, RE-29, RE-33, RE-34
RE-38	Home equipment (heating)	RE-21, RE-39
RE-39	Elderly population living alone	RE-19, RE-21
Mitigation and Adaptation (M&A)		
M&A-01	Meetings held by the governing and social participatory bodies	M&A-02
M&A-02	Social participation agreements	M&A-01
M&A-03	Legislation	To all the M&A indicators
M&A-04	Legislation compliance	M&A-03, M&A-05
M&A-05	Current agreements with institutions	M&A-03, M&A-04
M&A-06	Administrative sanctions	M&A-07
M&A-07	Administrative authorizations	M&A-06
M&A-08	Non-regulated governance activities	M&A-01, M&A-05
M&A-09	Cleared area for fire protection and improvement of grazing use	M&A-10
M&A-10	Area for controlled burning to prevent fires and improve grazing Use	M&A-09
M&A-11	Wolf damage to livestock	M&A-12, M&A-03, M&A-04, M&A-07
M&A-12	Wild boar damage to the productive capacity of grasslands	M&A-03, M&A-07
M&A-13	Participants in the environmental education program	M&A-14, M&A-28, M&A-29, M&A-32
M&A-14	Participants in the park's volunteer program	M&A-13, M&A-28, M&A-32
M&A-15	Waste removed from the park	M&A-18
M&A-16	Incidence of forest fires	M&A -17
M&A-17	Investment in prevention and extinction of forest fires	M&A-16, M&A-18
M&A-18	Public investments in the park	M&A-16, M&A-17, M&A-19
M&A-19	Subsidies in the municipalities of the park	M&A-05, M&A-18
M&A-20	Compensation for cessation of activity compatible with the park	M&A-19

Table 3. *Cont.*

Receptor Environment Indicators (RE)		
N°	Indicator Name	Related Indicators (A Hypothesis That Must be Empirically Tested in the Future)
Mitigation and Adaptation (M&A)		
M&A-21	People served at visitor centers	M&A-22, M&A-23
M&A-22	People served at the information points	M&A-21
M&A-23	School group visits to the park	M&A-24, M&A-25
M&A-24	Non-school group visits to the park	M&A23, M&A-25
M&A-25	Visitors on guided tours of the park	M&A-24, M&A-23
M&A-26	Informative brochures published by the park	M&A-27
M&A-27	Specific publications related to Global Change edited by the park	M&A-26
M&A-28	Social Perception of Global Change	M&A-13, M&A-14, M&A-29, M&A-32
M&A-29	Training actions on global change	M&A-13, M&A-14, M&A-28, M&A-32
M&A-30	Research on the impact of Global Change on the physical environment of the park	M&A-31
M&A-31	Research on the impact of Global Change on the socioeconomic of the park	M&A-30

Source: own elaboration.

8. Discussion

This pilot model for the System of Indicators for the Socio-Economic Evaluation and Monitoring of Global Change for the Picos de Europa National Park has been designed to take into account its utility not only for the park itself, but for other national parks as well. Thus, among its contributions is minimizing the lack of homogenization in indicators for monitoring global change and climate change in Spain [24] and in other protected areas throughout the world [37].

Despite the foregoing, it should be taken into account that this System of Indicators is designed based on the specific reality of the Picos de Europa National Park, and on its evolution and unique features. Accordingly, it is always necessary to adapt it to each park's bio-geophysical and socio-economic characteristics.

Therefore, park management and the development of indicators in the context of global and climate change must incorporate human beings as part of the protected space and accordingly safeguard them as well [20]. This human factor has a two-fold facet for its potential impact on the natural setting and for playing a key role in the park protection and conservation [23]. This is in line with the results of the analysis [38] of indicators from 180 countries, which revealed limited diversity and representation of important well-being dimensions such as health and governance.

Nevertheless, including social factors in the management of protected areas, such as national parks, [39] can lead to doubts about the weight given to protection with respect to the economic development of the zone. Accordingly, Hummel [39] (p. 1) calls for "a balanced and inclusive combination of the societal-focussed approach and the traditional view of conservation, protecting nature, and biodiversity, in order to become adopted in current management strategies." However, difficulties abound in the social factors that are to be taken into account, and how they are considered, as is the case for commitment to equitably management of protected areas assumed by 194 countries worldwide in the Aichi Target 11, Convention, whose results are not comparable across sites [38]. This is because among other reasons, no adequate standardized metrics to assess equitably management exists [40].

At the same time, as it is a protected area managed by three public regional administrations with different political criteria, Picos de Europa co-management, government, and governance all require special attention, due to the obstacles and difficulties entailed [41], while taking advantage of opportunities presented for understanding and cooperation. The co-management of protected areas in virtually all parts of the world, whether the responsibility of various public administrations or through the intervention of other social actors, is an issue of political and scientific relevance still requiring further theoretical and empirical knowledge so that it can be improved [42].

Hence, there are indicators that are especially relevant. This is the case for public participation, which includes information, communication, training, and tracking the social perception of global and climate change in national parks; that is, the social instruments of environmental management, which are still limited in their application to the management of protected areas [43].

Indicators on tourism are also especially important, given that there are almost two million visitors a year in the Picos de Europa National Park. There is a risk of negative impact on this protected area if this tourism is not sufficiently limited and controlled [44]. Nevertheless, tourism has the potential to be used for increasing social awareness about the importance of sustainable development for the national parks and all types of protected natural areas, and advantage can even be taken from income generated by tourism by reinvesting it in park sustainability policies [44,45]. Both indicators have been included in this system of indicators; although, they require strict monitoring.

At the same time, this system can be a reference for the design of other models in other protected areas, considering the aforementioned lack of both empirical and theoretical knowledge on socio-economic indicators of global change. This would mean including some social indicators in the system that are rarely examined for national parks, such as the social and political base, governance, political and social organization, well-being, poverty, culture, and people's social perception [10].

Adaptation of this System of Indicators as such would require previous research into each park, and thus, an ad hoc study. In any case, further work is needed on this model, especially for its empirical testing, in order to improve it for future editions.

However, above all, as pointed out by Peterson et al. [46] (p. 1) in their review of the knowledge on climate change adaptation in national forests, "adaptation to climate change will be successful only if it can be fully implemented in established planning processes and other operational aspects of national forest management". A critical gap still remains between the synthesis of scientific information on climate change vulnerability and adaptation and the actual integration of these ideas into management plans and practices [47]. This is a task that has not yet been carried out in either of the two national parks in Spain for which the system of indicators monitoring climate change has been developed. Nor has it been undertaken in the majority of national parks worldwide [3]. This represents a significant limitation in determining the effectiveness of a system of indicators and the feedback required.

9. Conclusions

The first relevant conclusion is that, as the literature review revealed, there is still very limited research into the socioeconomic impact of global change on national parks. However, taking into account the grave nature of this issue, its assessment and monitoring, in this case of the socioeconomic environment, is a key activity in the fight against global change and in particular climate change.

Despite the advancement in knowledge regarding this problem, designing a system of indicators as proposed here presents numerous difficulties. These include the still insufficient disaggregated quantitative and qualitative information on the national parks; the complexity of interpreting the socioecological processes that are produced within them, and of those, the pressure of global change and climate change in particular; and the continuity in monitoring necessary to obtain sufficient time-series data. The aforementioned factors have been the main limitations in the development of this project.

Nevertheless, the system of socio-economic indicators presented in this paper provides a framework for monitoring and interpreting changes in the Picos of Europe National Park, which admittedly need to be improved over time, according to the results and data viability. The approach taken here could also be applied in other similar national parks and be useful for their management.

Author Contributions: Conceptualization, I.L. and R.S.; methodology, I.L. and M.P.; investigation, I.L., R.S. and M.P.; writing—original draft preparation, I.L.; writing—review and editing, M.P.; funding acquisition, I.L. and M.P. All authors have read and agreed to the published version of the manuscript.

Funding: This research was funded by Fundación Biodiversidad 2010/00171/001. No funds were received to cover the costs of publishing in open access.

Informed Consent Statement: Not applicable.

Data Availability Statement: The socio-economic data of the municipalities are available online at: https://portal.uc3m.es/portal/page/portal/grupos_investigacion/sociologia_cambio_climatico/Datos%20Municipales%201%20AA%20parte.pdf (accessed on 13 May 2022). https://portal.uc3m.es/portal/page/portal/grupos_investigacion/sociologia_cambio_climatico/Datos%20Municipales%202%20AA%20parte.pdf (accessed on 13 May 2022). The following supporting information can be downloaded at: https://portal.uc3m.es/portal/page/portal/grupos_investigacion/sociologia_cambio_climatico/PN%20PICOS%20EUROPA.INDICADORES%20SOCIOECONOMICOS_0.pdf (accessed on 13 May 2022).

Conflicts of Interest: The authors declare no conflict of interest.

References

1. IPCC. Summary for Policymakers. In *Climate Change 2021: The Physical Science Basis. Contribution of Working Group I to the Sixth Assessment Report of the Intergovernmental Panel on Climate Change*; Masson-Delmotte, V.P., Zhai, A., Pirani, S.L., Connors, C., Péan, S., Berger, N., Caud, Y., Chen, L., Goldfarb, M.I., Gomis, M., et al., Eds.; Cambridge University Press: Cambridge, UK, 2021; *in press*.
2. Future Earth. *Sustainability in the Digital Age, and International Science Council. Global Risks Perceptions Report 2021*; Future Earth Canada Hub. Available online: https://doi.org/10.5281/zenodo.5764288 (accessed on 1 September 2021).
3. Baron, J.S.; Allen, C.D.; Fleishman, E.; Gunderson, L.; McKenzie, D.; Meyerson, L.; Stephenson, N. National parks. In *Adaptation Options for Climate-Sensitive Ecosystems and Resources*; The US Climate Change Science Program: Washington, DC, USA, 2008.
4. Franz, H.; Vogel, M. National Parks as outdoor laboratories for climate change impact. In Proceedings of the European Conference Climate Change and Nature Conservation in Europe—An Ecological, Policy and Economic Perspective, Bonn, Germany, 25–27 June 2013; p. 87.
5. Zamora, R.; Pérez-Luque, A.J.; Bonet, F.J. Monitoring global change in high mountains. In *High Mountain Conservation in a Changing World*; Catalan, J., Ninot, J., Aniz, M., Eds.; Springer: Berlin/Heidelberg, Germany, 2017; pp. 385–413.
6. Machlis, G.E.; Tichnell, D.L.; Eidsvik, H.K. *The State of the World's Parks: An International Assessment for Resource Management, Policy, and Research*; Routledge: London, UK, 2019.
7. Berkes, F.; Folke, C.; Colding, J. (Eds.) *Linking Social and Ecological Systems: Management Practices and Social Mechanisms for Building Resilience*; Cambridge University Press: Cambridge, UK, 2000.
8. López, I.; Pardo, M. Socioeconomic indicators for the evaluation and monitoring of climate change in national parks: An analysis of the Sierra de Guadarrama National Park (Spain). *Environments* **2018**, *5*, 25. [CrossRef]
9. López, I.; Pardo, M. Tourism versus nature conservation: Reconciliation of common interests and objectives—An analysis through Picos de Europa National Park. *J. Mt. Sci.* **2018**, *15*, 2505–2516. [CrossRef]
10. Corrigan, C.; Robinson, C.J.; Burgess, B.D.; Kingston, N.; Hockings, M. Global review of social indicators used in protected area management evaluation. *Conserv. Lett.* **2018**, *11*, e12397. [CrossRef]
11. Loveridge, R.; Sallu, S.M.; Pesha, I.J.; Marshall, A.R. Measuring human wellbeing: A protocol for selecting local indicators. *Environ. Sci. Policy* **2020**, *114*, 461–469. [CrossRef]
12. Callaghan, M.W.; Minx, J.C.; Forster, P.M. A topography of climate change research. *Nat. Clim. Change* **2020**, *10*, 118–123. [CrossRef]
13. Carter, R.W.; Walsh, S.J.; Jacobson, C.; Miller, M.L. Global change and human impact challenges in managing iconic national parks. *Georg. Wright Forum* **2014**, *31*, 245 255.

14. Miller, M.L.; Carter, R.W.; Walsh, S.J.; Peake, S. A conceptual framework for studying global change, tourism, and the sustainability of iconic national parks. *Georg. Wright Forum* **2014**, *31*, 256–269.

15. Duarte, C.M.; Alonso, S.; Benito, G.; Dachs, J.; Montes, C.; Pardo Buendía, M.; Valladares, F. *Cambio Global. Impacto de la Actividad Humana Sobre el Sistema Tierra*; CSIC: Madrid, Spain, 2006.

16. Palomo, I. Climate change impacts on ecosystem services in high mountain areas: A literature review. *Mt. Res. Dev.* **2017**, *37*, 179–187. [CrossRef]

17. Martin-Lopez, B.; Leister, I.; Lorenzo Cruz, P.; Palomo, I.; Grêt-Regamey, A.; Harrison, P.A.; Walz, A. Nature's contributions to people in mountains: A review. *PLoS ONE* **2019**, *14*, e0217847. [CrossRef]

18. Durán, M.; Canals, R.M.; Sáez, J.L.; Ferrer, V.; Lera-López, F. Disruption of traditional land use regimes causes an economic loss of provisioning services in high-mountain grasslands. *Ecosyst. Serv.* **2020**, *46*, 101200. [CrossRef]

19. Hugo, M.; Carvalho, P.; Almeida, N. Destination Brand Experience: A Study Case in Touristic Context of the Peneda-Gerês National Park. *Sustainability* **2021**, *13*, 11569.

20. Funnell, D.; Parish, R. *Mountain Environments and Communities*; Routledge: London, UK, 2005.

21. MITECO. Red de Parques Nacionales. Seguimiento y Evaluación—Seguimiento Sociológico. Available online: https://www.miteco.gob.es/es/parques-nacionales-oapn/plan-seguimiento-evaluacion/sociologico.aspx (accessed on 15 September 2021).

22. MITECO. Red de Parques Nacionales. Características de los Parques Nacionales. Available online: https://www.miteco.gob.es/es/red-parques-nacionales/red-seguimiento/rscg.aspx (accessed on 15 September 2021).

23. PNPE—Parque Nacional Picos de Europa. Presentación. Available online: https://parquenacionalpicoseuropa.es/el-parque/presentacion/ (accessed on 1 September 2021).

24. MITECO. Red de Seguimiento del Cambio Global en la Red de Parques Nacionales. Available online: https://www.miteco.gob.es/es/red-parques-nacionales/red-seguimiento/rscg.aspx (accessed on 1 September 2021).

25. OECD. *Environmental Indicators. OECD Core Sets*; OECD: Paris, France, 1994.

26. Mosaffaie, J.; Jam, A.S.; Tabatabaei, M.R.; Kousari, M.R. Trend assessment of the watershed health based on DPSIR framework. *Land Use Policy* **2021**, *100*, 104911. [CrossRef]

27. MITECO. Parque Nacional Picos de Europa. Area de Influencia Socioeconómica. Available online: https://www.miteco.gob.es/es/red-parques-nacionales/nuestros-parques/picos-europa/area-influencia/default.aspx (accessed on 1 September 2021).

28. Bryman, A. *Social Research Methods*; Oxford University Press: Oxford, UK, 2016.

29. Paez, A. Gray literature: An important resource in systematic reviews. *J. Evid. Based Med.* **2017**, *10*, 233–240. [CrossRef]

30. Nelson, E.; Mathieu, E.; Thomas, J.; Harrop Archibald, H.; Ta, H.; Scarlett, D.; Tambalo, D. Parks Canada's adaptation framework and workshop approach: Lessons learned across a diverse series of adaptation workshops. *Parks Steward. Forum* **2021**, *36*, 1. [CrossRef]

31. Diamantopoulos, A.; Winklhofer, H.M. Index construction with formative indicators: An alternative to scale development. *J. Mark. Res.* **2001**, *38*, 269–277. [CrossRef]

32. Lindenmayer, D.B.; Likens, G.E. Adaptive monitoring: A new paradigm for long-term research and monitoring. *Trends Ecol. Evol.* **2009**, *24*, 482–486. [CrossRef]

33. Hutto, R.L.; Belote, R.T. Distinguishing four types of monitoring based on the questions they address. *For. Ecol. Manag.* **2013**, *289*, 183–189. [CrossRef]

34. Hellawell, J.M. Development of a rationale objective for monitoring. In *Monitoring for Conservation and Ecology*; Goldsmith, F.B., Ed.; Chapman and Hall: London, UK, 1991; pp. 1–14.

35. Dudley, N.; Ali, N.; Kettunen, M.; MacKinnon, K. Editorial essay: Protected areas and the sustainable development goals. *Parks* **2017**, *23*, 10–12. [CrossRef]

36. UN. *Transforming Our World: The 2030 Agenda for Sustainable Development*; UN: New York, NY, USA, 2015.

37. Hill, S.L.; Harfoot, M.; Purvis, A.; Purves, D.W.; Collen, B.; Newbold, T.; Mace, G.M. Reconciling biodiversity indicators to guide understanding and action. *Conserv. Lett.* **2016**, *9*, 405–412. [CrossRef]

38. Moreaux, C.; Zafra-Calvo, N.; Vansteelant, N.G.; Wicander, S.; Burgess, N.D. Can existing assessment tools be used to track equity in protected area management under Aichi Target 11? *Biol. Conserv.* **2018**, *224*, 242–247. [CrossRef]

39. Hummel, C.; Poursanidis, D.; Orenstein, D.; Elliott, M.; Adamescu, M.C.; Cazacu, C.; Hummel, H. Protected Area management: Fusion and confusion with the ecosystem services approach. *Sci. Total Environ.* **2019**, *651*, 2432–2443. [CrossRef]

40. Zafra-Calvo, N.; Pascual, U.; Brockington, D.; Coolsaet, B.; Cortes-Vazquez, J.A.; Gross-Camp, N.; Burgess, N.D. Towards an indicator system to assess equitable management in protected areas. *Biol. Conserv.* **2017**, *211*, 134–141. [CrossRef]

41. Fabricius, C.; Collins, S. Community-based natural resource management: Governing the commons. *Water Policy* **2007**, *9*, 83–97. [CrossRef]

42. Soliku, O.; Schraml, U. Making sense of protected area conflicts and management approaches: A review of causes, contexts and conflict management strategies. *Biol. Conserv.* **2018**, *222*, 136–145. [CrossRef]

43. Nita, A.; Ciocanea, C.M.; Manolache, S.; Rozylowicz, L. A network approach for understanding opportunities and barriers to efftive public participation in the management of protected areas. *Soc. Netw. Anal. Min.* **2018**, *8*, 1–11. [CrossRef]

44. Leung, Y.F.; Spenceley, A.; Hvenegaard, G.; Buckley, R. *Tourism and Visitor Management in Protected Areas: Guidelines for Sustainability*; IUCN: Gland, Switzerland, 2018.

45. Dharmaratne, G.S.; Sang, F.Y.; Walling, L.J. Tourism potentials for financing protected areas. *Ann. Tour. Res.* **2000**, *27*, 590–610. [CrossRef]
46. Peterson, D.L.; Millar, C.I.; Joyce, L.A.; Furniss, M.J.; Halofsky, J.E.; Neilson, R.P.; Morelli, T.L. *Responding to Climate Change in National Forests: A Guidebook for Developing Adaptation Options*; Gen. Tech. Rep. PNW-GTR-855; US Department of Agriculture, Forest Service, Pacific Northwest Research Station: Portland, OR, USA, 2011.
47. Carlton, J.S.; Angel, J.R.; Fei, S.; Huber, M.; Koontz, T.M.; MacGowan, B.J.; Prokopy, L.S. State service foresters' attitudes toward using climate and weather information when advising forest landowners. *J. For.* **2014**, *112*, 9–14. [CrossRef]

Article

Ageing Perception as a Key Predictor of Self-Rated Health by Rural Older People—A Study with Gender and Inclusive Perspectives

Vanessa Zorrilla-Muñoz [1,2,*], María Silveria Agulló-Tomás [1,3], Carmen Rodríguez-Blázquez [4], Alba Ayala [5], Gloria Fernandez-Mayoralas [6] and Maria João Forjaz [7]

[1] University Institute on Gender Studies, University Carlos III of Madrid, Getafe, 28903 Madrid, Spain; msat@polsoc.uc3m.es

[2] Department of Mechanical Engineering, University Carlos III of Madrid, Leganes, 28911 Madrid, Spain

[3] Department of Social Analysis, University Carlos III of Madrid, Getafe, 28903 Madrid, Spain

[4] National Center of Epidemiology and CIBERNED, Institute of Health Carlos III and CIBERNED, 28029 Madrid, Spain; crodb@isciii.es

[5] Department of Statistics, University Carlos III of Madrid, Getafe, 28903 Madrid, Spain; aayala@est-econ.uc3m.es

[6] Institute of Economics, Geography and Demography (IEGD), Centre for Human and Social Sciences (CCHS), Spanish National Research Council (CSIC), 28037 Madrid, Spain; gloria.fernandezmayoralas@cchs.csic.es

[7] National School of Public Health, Institute of Health Carlos III and REDISSEC, 28029 Madrid, Spain; jforjaz@isciii.es

* Correspondence: vzorrill@ing.uc3m.es

Citation: Zorrilla-Muñoz, V.; Agulló-Tomás, M.S.; Rodríguez-Blázquez, C.; Ayala, A.; Fernandez-Mayoralas, G.; Forjaz, M.J. Ageing Perception as a Key Predictor of Self-Rated Health by Rural Older People—A Study with Gender and Inclusive Perspectives. *Land* **2022**, *11*, 323. https://doi.org/10.3390/land11030323

Academic Editor: Claudia A. Radel

Received: 31 December 2021
Accepted: 14 February 2022
Published: 23 February 2022

Publisher's Note: MDPI stays neutral with regard to jurisdictional claims in published maps and institutional affiliations.

Abstract: This paper investigates positive perceptions of ageing in rural people aged 65 and over as a key predictor of the self-assessment of one's health. Method: The sample covers a total of 3389 people from the 'Survey of Health, Ageing and Retirement' (SHARE), wave 6 (W6, 2015). This research analyses men and women who live in a rural environment. A linear regression model is proposed to consider the dependent variable 'self-rated health' and independent variables based on measures of quality of life in older adults. This study confirms that rural women perceive their health on the basis of factors different to those of their male contemporaries. The variable 'How often do you feel/think that you can do the things that you want to do?' is associated with women's self-perceived health. In men, a high relationship (with $p < 0.001$) is obtained for the variables 'How often do you feel/think look back on your life with a sense of happiness?' and 'How often do you feel/think that family responsibilities prevent you from doing what you want to do?' Certain daily activities (e.g., leisure or care), along with a positive perception of life, influence one's perceptions of one's own health, especially in the case of women. In sum, rural older women make a positive evaluation of their own health and ageing, while rural older men relate self-rated health to passivity and reminiscing. There is a need for further research on psycho-social and socio-spatial issues from an intergenerational, technological and gender perspective for rural and territorial influences to attain better health and quality of life for rural older people in comparison to urban people.

Keywords: rural; self-perceived health; older people; gender; activity; socio-spatial

1. Introduction

The age of the world population has increased in recent decades. According to data from the World Bank (2020), 9.321% of the total global population are aged 65 years and above, which indicates an increase of 3.069% since the year 1990 [1]. Global projections have estimated an increase in life expectancy for 2050, when two-thirds of the population in low- and middle-income countries will be over 60 [2]. Moreover, the current pandemic has highlighted the need to consider challenges in policies and actions to face demographic change in order to maintain well-being, quality of life and active ageing. Within all this, the

World Health Organization (WHO) is leading an international commitment to improve the lives of older people, their families and communities by declaring 'The Decade of Healthy Ageing', stating: 'A decade of concerted global action on healthy ageing is urgently needed to ensure that older people can fulfil their potential in dignity and equality and in a healthy environment'. The action areas are centred on four themes: (1) 'Age-friendly environments' and better places to promote non-discriminatory opportunities free of physical and social barriers; (2) 'Combating Ageism', an effort to reduce and eliminate negative behaviours, perceptions, prejudices and stigma that have negative effects on the well-being and health of older people; (3) 'Integrated care', with non-discriminatory access to health services; (4) 'Long-term care', which addresses issues relating to more accessible resources, support, spaces and technologies to contribute to improving the situations of dependency and disability, providing dignity to people living in these situations [3].

Given that healthy old age is linked to well-being during the ageing process, socio-gerontological studies have shown that certain activities of daily life (such as leisure and voluntary activities, among others), along with a favourable perception of life itself, have a positive influence on quality of life ([4–6], among others). In addition, classic authors (for example, Havighurst [7], Spreitzer and Snyder [8] and Brandtst [9], among others) defend the thesis that activity is central to enjoying a higher quality of life as we become older. The idea that active ageing improves quality of life is also commented on in the WHO's own definition quality of life in the 'World Health Organization Quality of Life Questionnaire' (WHOQOL) in 1994 [10] and, more intensely, from 2000 [11,12] to the present day [13–15]. Moreover, there are gender differences in the leisure activities and voluntary work performed by older people; i.e., older women are caregivers regardless of if they live in rural or urban areas. Moreover, these older caregivers demand more pro-grammes based on family needs assessment and family careers [16,17]. In particular, some examples of programmes demanded by rural women who care for older people are leisure programmes independent of the living environment [16,18] and programmes for reducing loneliness and isolation in rural environments [19]. The improvement of quality of life also addresses aspects that enrich the terms of active or successful ageing, such as health and well-being, among others [11,20]. From this perspective, some resources, such as the WHO-QOL (https://www.who.int/healthinfo/survey/whoqol-qualityoflife/en/, accessed on 20 November 2021) (see, for example, von Steinbüchel [21,22], among others), have been used to assess quality of life in this population and demonstrate differences between women and men in ageing. All of these differences show that quality of life is also associated with health-related indicators, as other authors have previously considered [23,24]. From an-other perspective, the concept of 'ikigai' is an oriental term that refers to the vital meaning of and attitudes about life (see commentaries related to "ikigai" in blogs, such as: https://qaspresearch.wixsite.com/blog/post/ikigai-activity-and-older-european-women, accessed on 20 November 2021), perceived health and other aspects that can be viewed differently by older women and men. Moreover, 'ikigai' is outstanding in the case of living in rural communities [25].

The improvement of quality of life and health determinants has brought success to ageing and the consequent increase in life expectancy. Additionally, the success of ageing entails the need to achieve important challenges in relation to welfare policies, economies and society in countries with high life expectancy, e.g., European countries, including Spain, which is expected to have the world's highest life expectancy in 2040 [26]. The phenomenon of increased life expectancy relates more to women because their quality of life is better, which does not mean that women age better in the sense of having better health.

Some authors, such as Wenger [27], among others, have noted that rural ageing has received less attention than urban ageing in aspects regarding rural lifestyles, the importance of community networks and family support, life satisfaction and access to care services, among others. However, the few studies conducted indicate that rural living, or ageing in rural areas, creates a positive vision of cognitive health and social support [28], behaviour that promotes health and satisfaction with life in rural older women living alone [29] and social support factors associated with self-rated health, such as intergenerational social contact [30]. One negative aspect is the difficulty of accessing primary health services, especially for populations with disabilities [31], which include older people in situations of dependency.

Health is undoubtedly an indicator of quality of life in older people. It can be perceived through variables related to functional capacity, psychological well-being and mental health, but also others related to personal experiences during the ageing process [32], especially when there is a related positive perception of ageing [33]. The perception of ageing is an area of interest in ageing population surveys conducted to acquire important data about health and social variables. Questionnaires about perceptions of ageing make it possible to analyse the emotional representation of the perceptions of ageing [34]. Self-perception of ageing is a factor which affects the success of ageing [35,36]. Furthermore, the self-perception of ageing is a factor which allows us to understand death anxiety [37] and frailty [38] and predict the persistence of depression and anxiety [39] among older people. However, few studies have analysed the perception of ageing as it relates to older adults' health (see, for example, the study of Warmoth et al. [40], Hickey et al. [41] or Sadegh Mohhamad [42]). Several resources for gathering perception about ageing have been validated, for example, the Brief Ageing Perceptions Questionnaire (B-APQ) [34], the Dutch Aging Perceptions Questionnaire [43] and the Brief Ageing Perception Questionnaire in Malaysia [44]. However, there is no resource linked to ageing perception as a key predictor of self-rated health by rural older people. Moreover, positive perceptions of ageing and self-rated health among rural older people are still scarce. It is hypothesized that rural older women have a different perception of health compared with their male contemporaries. For these reasons, the objective of this study is to determine which self-perceived ageing variables are related to the perception of self-rated health in rural older people in Spain. The analysis is conducted on the basis of differences by sex.

2. Materials and Methods

This study utilises data from the Survey of Health, Ageing and Retirement (SHARE: http://www.share-project.org/home0.html, accessed on 15 November 2021), wave 6 (W6, 2015) [45]. Wave 6 contains the latest updated variables regarding perceptions of ageing in older people. Access to the questionnaire data is free for the purpose of scientific use. SHARE data are protected by factual anonymity (as defined by the strict norms of the German Federal Statistics Act and the German Federal Data Protection Law) [46]. The collection of the data is anonymously recorded. The use of the data is subject to the European Union General Data Protection Regulation (GDPR), 'Regulation (EU) 2016/679 of 27 April 2016 on the protection of natural persons related to the processing of personal data and on the free movement of such data, and repealing Directive 95/46/EC'. The SHARE data collection procedures are subject to continuous ethics review. SHARE-ERIC's activities related to human subject research are guided by international research ethics principles, such as the Respect Code of Practice for Socio Economic Research (http://www.share-project.org/fileadmin/pdf_documentation/respect_code_socio_economic_research.pdf, accessed on 10 January 2022) (professional and ethical guidelines for the conduct of socio-economic research) and the Declaration of Helsinki (http://www.share-project.org/fileadmin/pdf_documentation/Declaration_of_Helsinki.pdf, accessed on 10 January 2022) (a set of ethical principles regarding human experimentation developed for the medical community by the World Medical Association, last revised at the 64th WMA Meeting held in Fortaleza, Brazil in October 2013). As SHARE users, we were familiar with the fundamental principles of research ethics (e.g., see guidelines mentioned above) and we take them into account in an appropriate manner when conducting research using SHARE data [46].

2.1. Participants

The participants of this study were recruited voluntarily. Our sampling strategy is defined in the paper by Bergmann et al. [47]. The geographically dispersed sample consisted of rural Spanish people aged ≥ 65 years, the majority of whom were living in rural areas ($n = 3389$) and were recruited for the W6 SHARE project [48]. Women represented 53.8% of the sample ($n = 1826$), and 60.4% of the participants were limited in activities as a result of mobility difficulties or chronic diseases. In relation to tobacco use, which can aggravate chronic diseases, e.g., diabetes (see [49], among others) or cardiovascular diseases (see [50], among others), 43.81% of the sample reported smoking daily. The mean body mass was 27.41 (SE = 0.057). As for drugs related to COVID-19 risk diseases, 20.09% were taking drugs for diabetes, 13.45% for coronary diseases and 3.95% for chronic bronchitis. The regression analysis was gender-stratified.

2.2. Data Analysis

The variables were selected to represent self-rated health and measure quality of life in older aged people. Quality of life is divided in four main subscales: control, autonomy, pleasure and self-realization. The measure for quality of life used the CASP-12 scale [51–53], to which we added the self-realization variable 'How often do you feel satisfied with activities (volunteering, hobbies etc …)?' The dependent variable was the perceived health status. The dependent and independent variables used a Likert-type scale from 1 to 4 (where 1 signifies 'not at all' and 4 is 'a lot'), except in the variable 'How often do you feel satisfied with activities (volunteering, hobbies etc …)?', in which case a 1-to-10 scale was used (1 being 'not at all' and 10 being 'a lot'). As all of these variables were single Likert-type scale variables, stringent levels were considered in the first calculation (https://www.theanalysisfactor.com/can-likert-scale-data-ever-be-continuous/, accessed on 10 January 2022): Spearman's test, the Kaiser–Meyer–Olkin (KMO) test and a uniqueness criterion not stratified by gender. This analysis provides an idea of the ageing perception and self-rated health with Likert-type scale variables. Spearman's test was performed with age-related perception variables and self-rated health. The selected variables covered the following requirements: Spearman's test for two p-value probability levels, $p < 0.005$ and $p < 0.001$ (both very strong evidence for rejecting H0), and a uniqueness criterion for values ≥ 0.4 with optimal criteria for ≥ 0.6 and KMO ($0 < KMO < 1$). Second, all predictors were entered into the two linear regression models, simultaneously stratifying by gender.

3. Results

Table 1 shows the mean and standard deviation of the dependent variable 'Self-perceived health' and the 13 selected independent variables. The variables relating to ageing perception have a positive trend for both sexes in the case of 'How often do you feel that life is full of opportunities?', 'How often do you feel/think that the future looks good for you?' and 'How often do you feel/think satisfied with activities?' Some of the negative variables are 'How often do you feel/think that family responsibilities prevent you from doing what you want to do?', 'How often do you feel/think that what happens to you is out of your control?', 'How often do you feel/think left out of things?' and 'How often do you feel/think that shortage of money stops you from doing the things you want to do?', among others.

Table 1. Variables selected from the SHARE questionnaire: mean and standard deviation.

Subscale	Variables	Total (N = 3389) (Mean $\pm$ SE)	Men (N = 1563) (Mean $\pm$ SE)	Women (N = 1826) (Mean $\pm$ SE)
Self-rated health	Self-perceived health	3.31 $\pm$ 0.02	3.21 $\pm$ 0.03	3.41 $\pm$ 0.03
Control	How often do you think your age prevents you from doing the things you would like to do?	2.44 $\pm$ 0.03	2.49 $\pm$ 0.04	2.38 $\pm$ 0.04
Control	How often do you feel that what happens to you is out of your control?	3.01 $\pm$ 0.02	3.10 $\pm$ 0.03	2.92 $\pm$ 0.04
Control	How often do you feel left out of things?	3.40 $\pm$ 0.02	3.44 $\pm$ 0.03	3.37 $\pm$ 0.03

Table 1. *Cont.*

Subscale	Variables	Total (N = 3389) (Mean $\pm$ SE)	Men (N = 1563) (Mean $\pm$ SE)	Women (N = 1826) (Mean $\pm$ SE)
Autonomy	How often do you think that you can do the things that you want to do?	2.04 $\pm$ 0.02	1.99 $\pm$ 0.03	2.08 $\pm$ 0.03
Autonomy	How often do you think that family responsibilities prevent you from doing what you want to do?	3.21 $\pm$ 0.02	3.22 $\pm$ 0.03	3.20 $\pm$ 0.03
Autonomy	How often do you think that shortage of money stops you from doing the things you want to do?	2.59 $\pm$ 0.03	2.57 $\pm$ 0.04	2.61 $\pm$ 0.04
Pleasure	How often do you look forward to each day?	1.63 $\pm$ 0.02	1.60 $\pm$ 0.03	1.67 $\pm$ 0.03
Pleasure	How often do you feel that your life has meaning?	1.54 $\pm$ 0.02	1.48 $\pm$ 0.02	1.60 $\pm$ 0.03
Pleasure	How often, on balance, do you look back on your life with a sense of happiness?	1.72 $\pm$ 0.02	1.66 $\pm$ 0.03	1.78 $\pm$ 0.03
Self-realization	How often do you feel full of energy these days?	2.02 $\pm$ 0.02	1.91 $\pm$ 0.03	2.14 $\pm$ 0.03
Self-realization	How often do you feel that life is full of opportunities?	2.15 $\pm$ 0.02	2.09 $\pm$ 0.03	2.21 $\pm$ 0.03
Self-realization	How often do you feel that the future looks good for you?	2.21 $\pm$ 0.02	2.13 $\pm$ 0.03	2.28 $\pm$ 0.03
Self-realization	How often do you feel satisfied with activities (volunteering, hobbies etc …)?	8.09 $\pm$ 0.04	8.08 $\pm$ 0.05	8.10 $\pm$ 0.06

Table 2 shows the results of the SHARE questionnaire. We observed a Spearman's coefficient in all variables with a moderate range of -0.40 to 0.48. Uniqueness results between 0.41 and 0.84 were obtained. The total KMO obtained was 0.87.

Table 2. Variables selected from the SHARE questionnaire and statistical analysis: Spearman, KMO and uniqueness.

Variables How Often Do You Feel/Think … ?	Spearman	KMO	Uniqueness
How often do you think your age prevents you from doing the things you would like to do?	−0.40 ***	0.87	0.57
How often do you feel that what happens to you is out of your control?	−0.31 ***	0.83	0.47
How often do you feel left out of things?	−0.22 ***	0.84	0.57
How often do you think that you can do the things that you want to do?	0.27 ***	0.93	0.81
How often do you think that family responsibilities prevent you from doing what you want to do?	−0.01	0.70	0.79
How often do you think that shortage of money stops you from doing the things you want to do?	−0.16 ***	0.80	0.80
How often do you look forward to each day?	0.33 ***	0.84	0.41
How often do you feel that your life has meaning?	0.31 ***	0.84	0.41
How often, on balance, do you look back on your life with a sense of happiness?	0.09 ***	0.89	0.78
How often do you feel full of energy these days?	0.48 ***	0.91	0.41
How often do you feel that life is full of opportunities?	0.29 ***	0.87	0.51
How often do you feel that the future looks good for you?	0.34 ***	0.88	0.42
How often do you feel satisfied with activities (volunteering, hobbies etc …)?	−0.14 ***	0.94	0.84
Total		0.87	

Note: *** $p < 0.001$.

In the regression models of perceived health (Table 3), significant variables ($p < 0.001$) were 'How often your age prevents you from doing the things you would like to do?' (βmen, women $= -0.18$), and 'How often do you feel/think full of energy these days?' (βmen $= 0.30$, βwomen $= -0.04$). This model explained the 24.66% variance for men and 28.66% variance for women. In the case of women, the retained variables were 'How often do you feel/think that you can do the things that you want to do?' ($\beta = 0.06$). In men, there was a high coincidence for (with $p < 0.001$) 'How often do you feel/think look back on your life with a sense of happiness?' ($\beta = -0.13$), and $p < 0.01$ for the variable 'How often do you feel/think that family responsibilities prevent you from doing what you want to do?' ($\beta = 0.09$). The model obtained a slightly better fit for women (F, 13, 771 = 28.66 and R2 = 0.33) than men (F, 13, 781 = 24.66 and R2 = 0.29).

Table 3. Linear standardized regression models of perceived health status by gender using SHARE data. Standardized β coefficients.

Variables	Men (N = 1563)		Women (N = 1826)	
	B	SE	B	SE
How often do you think your age prevents you from doing the things you would like to do?	−0.18	0.03 ***	−0.18	0.03 ***
How often do you feel that what happens to you is out of your control?	−0.06	0.04	−0.06	0.04
How often do you feel left out of things??	0.03	0.04	0.03	0.04
How often do you think that you can do the things that you want to do?	0.03	0.03	0.06	0.03 *
How often do you think that family responsibilities prevent you from doing what you want to do?	0.09	0.03 *	0.04	0.03
How often do you think that shortage of money stops you from doing the things you want to do?	−0.06	0.03	−0.06	0.03
How often do you look forward to each day?	0.11	0.05	0.04	0.05
How often do you feel that your life has meaning?	0.09	0.06	0.04	0.05
How often, on balance, do you look back on your life with a sense of happiness?	−0.13	0.04 ***	−0.07	0.04
How often do you feel full of energy these days?	0.30	0.05 ***	0.33	0.04 ***
How often do you feel that life is full of opportunities?	0.08	0.05	0.04	0.04
How often do you feel that the future looks good for you?	0.02	0.04	0.00	0.05
How often do you feel satisfied with activities (volunteering, hobbies etc …)?	−0.02	0.02	−0,01	0.02
Constant	2.81	0.29 ***	3.07	0.27 ***
	F (13, 781)	24.66	F (13, 771)	28.66
	Adj R-squared	0.29	Adj R-squared	0.33
	Root MSE	0.81	Root MSE	0.78

Note: * $p < 0.01$; *** $p < 0.001$; Prob > F = 0.00 for both models. SE = Standard Error.

4. Discussion

This study responded to the question of how older Spanish people perceived their health, and compared their perceptions based on their sex. The main results are centred on the responses to statements such as 'How often do you feel/think full of energy these days?' and 'How often your age prevents you from doing the things you would like to do?' for both genders. In the case of women, it is remarkable the response to the statement 'How often do you feel/think that you can do the things that you want to do?'. In men, the statements 'How often do you feel/think look back on your life with a sense of happiness?' and 'How often do you feel/think that family responsibilities prevent you from doing what you want to do?' garnered more positive responses.

First, the results of this analysis showed how health is related to a negative assessment of activities; that is, health perception is linked more to agreement with negative ideas such as 'How often your age prevents you from doing the things you would like to do?' than to agreement with the positive idea that 'How often do you feel/think full of energy these days?'. In other words, people who say that age stops them from doing things perceive themselves as having poorer health, while people who consider that age gives them energy see themselves as enjoying better health.

Second, ageing is associated with increased musculoskeletal- or mobility-related diseases or problems, which is consistent with previous studies [54–56]. This study shows that 60.4% of the participants were limited in activities as a result of mobility difficulties or chronic diseases.

Third, the results displayed a lower association of perceived health with statements such as 'How often do you feel that your life has meaning?' and 'How often do you feel satisfied with activities (volunteering, hobbies etc …)?', particularly in women. In addition, because women generally have more family and care commitments than men when they are younger, they may find that having more time for other activities as they become older makes them feel more enthusiastic and positive; unfortunately, they may also find that they do not have as much vitality as they would like at this life stage, because of the previously mentioned physical changes they experience.

All of these data also justify the need for person-centred studies, i.e., ergonomic and usability studies for products designed to increase mobility with the goal of fostering greater independence. Moreover, the use of technology in health, relational and psychosocial support devices for older people highlights the need for further research on the applicability and use of systems that improve their effectiveness and ergonomics. This is much more relevant for specific socio-demographic profiles, such as women living in rural areas or in less accessible areas with fewer services, but who also have positive experiences at the relational level [18,56].

Another aspect to consider is that ageing being connected to fewer family responsibilities is perceived more positively in relation to health by men than by women, which is undoubtedly linked to the ongoing care provided by women, who continue to look after family members and dependents even as they themselves age [18]. In addition, it is worth remembering the increase in dependent ageing in rural areas also increases the need of care and social resources, due to the increase in recent years of dependent persons who are over 65 years old. It is estimated that, in the coming years, this problem will most likely grow, giving rise to new group profiles of older people who are at risk of being dependent persons, especially in the context of the current pandemic and as a direct result of having suffered from the virus [57,58]. In short, there are differences in terms of how life is perceived with greater family responsibilities and commitments when one is a woman, but women are also slightly more optimistic and have a more positive sense of purpose, or 'ikigai'.

Middle-aged and older people express greater concern about the general economic situation in Spain than they do about their own personal financial situation and future, as shown by their responses to the prompts 'Degree of concern about the situation' and 'The degree of optimism'. That is to say, there is a certain vitality and optimism that can be observed in older people which, together with the analysed perception of health, is related to their need to feel useful and valued in society [15]; this is especially observed in older women. Unfortunately, the current socio-political and economic context of the pandemic may trigger a negative perception of health and the future in groups that have experienced a high rate of vulnerability/mortality in connection with the virus; this will affect these senior citizens' quality of life, as they may see their future as more uncertain.

With regard to the limitations of this study, the first limitation is linked to the Likert scale variables used: all of the ageing perception variables are Likert data. The second limitation refers to the lack of more recent surveys or questionnaires due to the pandemic; such studies would be useful in updating the data reported in this study. Furthermore, qualitative studies might help to analyse the direct discourses of older people or of the carers, either in the homes or in different residential environments. A third limitation of this study is related to the current pandemic: COVID-19 has slowed down this successful process and is regarded as the most important public health disease of the last 100 years for the world's population; it is the main cause of hospitalization and death, particularly for older people, and has reduced life expectancy at all latitudes [59]. In particular, the high mortality rate due to illness and complications caused by the SARs-CoV-2 virus in the Spanish population is bringing about a negative change in future prospects, especially for men [60]. Furthermore, complications are even greater for those who suffer from heart disease, diabetes and/or chronic bronchitis [61–63]. In addition, people affected (directly or indirectly, for example, women and residential carers) by the new disease of COVID-19 may be more vulnerable to suffering certain motor pathologies as a consequence of the disease, for example, due to thrombosis in infected persons [64–66], necrosis [67] due to complications, such as avascular necrosis [68], along with other issues of a sensory nature. Unfortunately, however, in the current socio-political and economic context of the pandemic, a negative perception of health is emerging in groups with a high COVID-19 mortality/vulnerability rate. Obviously, this new disease will affect not only their health, quality of life and the future (perceived and real) of adults and older people, but also that of their carers, who are still mostly women, whether they are family members or professional caregivers [18,69–72]. Moreover, the effects could have a negative impact on the continuity of care in dependency situations, which adds to the inequalities experienced by those living in rural areas. For example, rural disparities in accessibility impede mobility [73] and access to healthcare services, and they hinder continuity of care [74]. These barriers could have negative repercussions on the healthcare coverage of patients with diabetes and hypertension [75]. In this regard, there are some positive experiences with chronic disease management in rural areas where nurses intervene to provide more continuous care [76]. This type of program could be implemented in both social and healthcare resources in rural areas for patients with chronic diseases.

As far as the continuity of care goes, some of the great social and health solutions that have been implemented involve using technological resources, such as telemedicine [77,78] and emerging technologies focused on improving the autonomy [72] of older people. These emerging technologies include design concepts, such as accessibility, usability, safety, ergonomics, people preferences and also gender perspectives [72,78]. In addition to providing greater continuity of care, these technologies can be used to reduce the problems of loneliness and isolation (which are often experienced by older people) [19].

Within the context of pandemic, it is still unclear if rural spaces are healthier environments, or if there may be less chance of being infected by the virus in these areas. The social perception of health is influenced by the belief that urban dwellers are at a higher risk of COVID-19 infection than people living in rural areas [79]. However, there are other health issues in rural areas. The limited access and difficulty reaching rural healthcare and support, poor Internet connectivity and the still-existing digital divide do carry other risks that affect those living outside urban areas: the inequities in healthcare for older people [80], the loneliness, isolation and other problems caused by social distancing [81].

Clearly, more research regarding the current perception of health in the context of the pandemic and different socio-spatial scenarios (including rural/urban areas) could be developed, particularly considering the positive impact of emerging technologies that will affect the future of older people and their caregivers.

Finally, this study is only centred on the gender perspective. Other analysis could be entered into new models (e.g., socio-economic status, health status, pain status, living with others vs. alone, mental health status) which are included in SHARE questionnaire.

5. Conclusions

Broadly speaking, older people's perceptions with regard to their health and future was negative before the COVID-19 pandemic, but there were glimpses of optimism in some responses to prompts regarding their personal situation. Above all, there are gender differences and, even though women have a longer life expectancy, the continuity and pressure of family care commitments continues to be high at this stage of life (≤ 50) for the female population. This persistence of gender-based roles, among other explanations or variables, leads middle-aged and older women to perceive their health and quality of life more negatively, which also leads them to feel less vital and more dissatisfied with activities such as volunteering and hobbies. In sum, older women die later than their male peers, but this does not imply that their quality of life and perception of health is more positive or negative. Rural older women make a positive evaluation of their health and feel at their age more enthusiastic but less vital than they would like to be, while rural older men connect good health to passivity and reminiscence.

Consequently, there is a need for further research on psycho-social issues with regard to the impact of SARS-CoV-2 virus. Adopting a gender, socio-spatial, inclusive and intergenerational approach could be a key factor in better understanding health perception and targeting programmes/policies for better health and social sustainability, satisfaction and quality of life for older people in different regions.

Author Contributions: All the authors have equally contributed to the final version of this paper. All authors have read and agreed to the published version of the manuscript.

Funding: This work is part of the 'Quality of life and Ageing in Spain, Sweden and Portugal'. The QASP research project is funded by the Institute of Health Carlos III, Intramural Strategical Action in Health AESI 2018, Ref: PI18CIII/00046, PR: MR Forjaz. https://qaspresearch.wixsite.com/blog (accessed on 20 November 2021). We appreciate the support of the 'Active Aging, Quality of Life and Gender. Promoting a positive image of old age and aging combating ageism' Program—ENCAGEn cm R&D Activities Program. Ref. H2019/HUM-5698. Funded by Programs of R&D in Community of Madrid Social Sciences and Humanities, co-financed with the European Social Fund.PR: G. Fernández-Mayoralas, C Rodriguez-Blazquez, D. Zamarrón, M.S. Agulló-Tomás, M.A. Molina. https://encage-cm.es/ (accessed on 20 November 2021).

Institutional Review Board Statement: Not applicable.

Informed Consent Statement: Not applicable.

Data Availability Statement: As indicated, data used the Survey of Health, Ageing and Retirement, SHARE: http://www.share-project.org/home0.html (accessed on 15 November 2021).

Conflicts of Interest: The authors declare no conflict of interest.

References

1. The World Bank Data. Population Ages 65 and Above (% of Total Population). 2022. Available online: https://data.worldbank.org/indicator/SP.POP.65UP.TO.ZS (accessed on 21 November 2021).
2. WHO. Ageing and Health. 2021. Available online: https://www.who.int/news-room/fact-sheets/detail/ageing-and-health (accessed on 16 January 2022).
3. WHO. UN Decade of Healthy Ageing. 2021. Available online: https://www.who.int/initiatives/decade-of-healthy-ageing (accessed on 16 January 2022).
4. Gabriel, Z.; Bowling, A.N.N. Quality of life from the perspectives of older people. *Ageing Soc.* **2004**, *24*, 675–691. [CrossRef]
5. Bowling, A.; Fleissig, A.; Gabriel, Z.; Banister, D.; Dykes, J.; Dowding, L.M.; Sutton, S.; Evans, O. Let's ask them: A national survey of definitions of quality of life and its enhancement among people aged 65 and over. *Int. J. Aging Hum. Dev.* **2003**, *56*, 269–306. [CrossRef] [PubMed]
6. Connolly, S.; O'shea, E. The Perceived Benefits of Participating in Voluntary Activities Among Older People: Do They Differ by Volunteer Characteristics? *Act. Adapt. Aging* **2015**, *39*, 95–108. [CrossRef]
7. Havighurst, R.J. Successful aging. In *Processes of Aging*; Williams, R.H., Tibbits, C., Donahue, W., Eds.; Atherton Press: New York, NY, USA, 1963; Volume 1, pp. 299–320.
8. Spreitzer, E.; Snyder, E.E. Correlates of life satisfaction among the aged. *J. Gerontol.* **1974**, *29*, 454–458. [CrossRef] [PubMed]
9. Brandtstädter, J.; Baltes-Götz, B. Personal control over development and quality of life perspectives in adulthood. In *Successful Aging: Perspectives from the Behavioral Sciences*; Baltes, M.M., Baltes, P.B., Eds.; European Network on Longitudinal Studies on Individual Development; Cambridge University Press: Cambridge, UK, 1990; pp. 197–224.
10. WHO. *Report of the Second World Assembly on Ageing*; United Nations: New York, NY, USA, 2002.
11. Bowling, A. *Ageing Well: Quality of Life in Old Age*; Open University Press: Maidenhead, UK, 2005.
12. Walker, A. *Understanding Quality of Life in Old Age*; Open University Press: Maidenhead, UK, 2005.
13. Cattan, M.; Hogg, E.; Hardill, I. Improving quality of life in ageing populations: What can volunteering do? *Maturitas* **2011**, *70*, 328–332. [CrossRef] [PubMed]
14. Fernández-Mayoralas, G.; Rojo-Pérez, F.; Martínez-Martín, P.; Prieto-Flores, M.E.; Rodríguez-Blázquez, C.; Martín-García, S.; Rojo-Abuín, J.M.; Forjaz, M.J. Active ageing and quality of life: Factors associated with participation in leisure activities among institutionalized older adults, with and without dementia. *Aging Ment. Health* **2015**, *19*, 1031–1041. [CrossRef]
15. Zorrilla-Muñoz, V.; Blanco-Ruiz, M.; Criado-Quesada, B.; Fernandez-Sanchez, M.; Merchan-Molina, R.; Agulló-Tomás, M.S.A.; Género y envejecimiento desde el prisma de las organizaciones que trabajan con mayores. Rev. *Prism. Soc. 2018, 500–510.* Available online: https://revistaprismasocial.es/article/view/2468 (accessed on 16 December 2021).
16. Agulló-Tomás, M.S.; Zorrilla-Muñoz, V.; Gómez-García, M.V. Aproximación socio-espacial al envejecimiento ya los programas para cuidadoras/es de mayores. *Rev. INFAD Psicol. Int. J. Dev. Educ. Psychol.* **2019**, *2*, 211–228. [CrossRef]
17. Chien, W.-T.; Chiu, Y.; Lam, L.-W.; Ip, W.-Y. Effects of a needs-based education programme for family carers with a relative in an intensive care unit: A quasi-experimental study. *Int. J. Nurs. Stud.* **2006**, *43*, 39–50. [CrossRef]
18. Agulló-Tomás, M.S.; Zorrila-Muñoz, V.; Gomez-Garcia, M.V.; Género y evaluación de programas de apoyo para cuidadoras/es de mayores. Prism. *Soc. Rev. Investig. Soc. 2018, 391–415.* Available online: https://revistaprismasocial.es/article/view/2469 (accessed on 16 December 2021).
19. Blusi, M.; Asplund, K.; Jong, M. Older family carers in rural areas: Experiences from using caregiver support services based on Information and Communication Technology (ICT). *Eur. J. Ageing* **2013**, *10*, 191–199. [CrossRef]

20. Walker, A. *The New Generational Contract: Intergenerational Relations, Old Age and Welfare*; ULC Press: London, UK, 1996.
21. von Steinbüchel, N.; Lischetzke, T.; Gurny, M.; Eid, M. Assessing quality of life in older people: Psychometric properties of the WHOQOL-BREF. *Eur. J. Ageing* **2006**, *3*, 116–122. [CrossRef]
22. Lucas-Carrasco, R.; Laidlaw, K.; Power, M.J. Suitability of the WHOQOL-BREF and WHOQOL-OLD for Spanish older adults. *Aging Ment. Health* **2011**, *15*, 595–604. [CrossRef]
23. Prieto-Flores, M.-E.; Moreno-Jiménez, A.; Fernandez-Mayoralas, G.; Rojo-Perez, F.; Forjaz, M.J. The relative contribution of health status and quality of life domains in subjective health in old age. *Soc. Indic. Res.* **2012**, *106*, 27–39. [CrossRef]
24. Forjaz, M.J.; Rodriguez-Blazquez, C.; Ayala, A.; Rodriguez-Rodriguez, V.; de Pedro-Cuesta, J.; Garcia-Gutierrez, S.; Prados-Torres, A. Chronic conditions, disability, and quality of life in older adults with multimorbidity in Spain. *Eur. J. Intern. Med.* **2015**, *26*, 176–181. [CrossRef] [PubMed]
25. Santos Alexandre, R. Being and landscape: An ontological inquiry into a Japanese rural community. *Asia Pac. J. Anthropol.* **2019**, *20*, 232–246. [CrossRef]
26. Foreman, K.J.; Marquez, N.; Dolgert, A.; Fukutaki, K.; Fullman, N.; McGaughey, M.; Pletcher, M.A.; Smith, A.E.; Tang, K.; Yuan, C.-W. Forecasting life expectancy, years of life lost, and all-cause and cause-specific mortality for 250 causes of death: Reference and alternative scenarios for 2016–40 for 195 countries and territories. *Lancet* **2018**, *392*, 2052–2090. [CrossRef]
27. Wenger, G.C. Myths and realities of ageing in rural Britain. *Ageing Soc.* **2001**, *21*, 117–130. [CrossRef]
28. Banjare, P.; Dwivedi, R.; Pradhan, J. Factors associated with the life satisfaction amongst the rural elderly in Odisha, India. *Health Qual. Life Outcomes* **2015**, *13*, 201. [CrossRef] [PubMed]
29. Kim, H.J. Health promotion behaviors of rural elderly women living alone and their life satisfaction. *J. Korean Acad. Community Health Nurs.* **2016**, *27*, 254–261. [CrossRef]
30. Giang, L.T.; Nguyen, T.T.; Nguyen, N.T. Social support and self-rated health among older men and women in Vietnam. *J. Popul. Ageing* **2020**, *13*, 427–442. [CrossRef]
31. Dassah, E.; Aldersey, H.; McColl, M.A.; Davison, C. Factors affecting access to primary health care services for persons with disabilities in rural areas: A "best-fit" framework synthesis. *Glob. Health Res. Policy* **2018**, *3*, 1–13. [CrossRef]
32. Pino, L.; González-Vélez, A.E.; Prieto-Flores, M.E.; Ayala, A.; Fernandez-Mayoralas, G.; Rojo-Perez, F.; Martinez-Martin, P.; Forjaz, M.J. Self-perceived health and quality of life by activity status in community-dwelling older adults. *Geriatr. Gerontol. Int.* **2014**, *14*, 464–473. [CrossRef] [PubMed]
33. Ingrand, I.; Paccalin, M.; Liuu, E.; Gil, R.; Ingrand, P. Positive perception of aging is a key predictor of quality-of-life in aging people. *PLoS ONE* **2018**, *13*, e0204044. [CrossRef] [PubMed]
34. Sexton, E.; King-Kallimanis, B.L.; Morgan, K.; McGee, H. Development of the Brief Ageing Perceptions Questionnaire (B-APQ): A confirmatory factor analysis approach to item reduction. *BMC Geriatr.* **2014**, *14*, 44. [CrossRef] [PubMed]
35. Yaghoobzadeh, A.; Sharif Nia, H.; Hosseinigolafshani, Z.; Mohammadi, F.; Oveisi, S.; Torkmandi, H. Associated factors of ageing perception among elderly in Qazvin, 2015. *J. Gerontol.* **2017**, *1*, 1–10. [CrossRef]
36. Santini, Z.I.; Koyanagi, A.; Tyrovolas, S.; Haro, J.M.; Koushede, V. The association of social support networks and loneliness with negative perceptions of ageing: Evidence from the Irish Longitudinal Study on Ageing (TILDA). *Ageing Soc.* **2019**, *39*, 1070–1090. [CrossRef]
37. Mohammadpour, A.; Sadeghmoghadam, L.; Shareinia, H.; Jahani, S.; Amiri, F. Investigating the role of perception of aging and associated factors in death anxiety among the elderly. *Clin. Interv. Aging* **2018**, *13*, 405. [CrossRef] [PubMed]
38. Warmoth, K.; Tarrant, M.; Abraham, C.; Lang, I.A. Relationship between perceptions of ageing and frailty in English older adults. *Psychol. Health Med.* **2018**, *23*, 465–474. [CrossRef]
39. Freeman, A.T.; Santini, Z.I.; Tyrovolas, S.; Rummel-Kluge, C.; Haro, J.M.; Koyanagi, A. Negative perceptions of ageing predict the onset and persistence of depression and anxiety: Findings from a prospective analysis of the Irish Longitudinal Study on Ageing (TILDA). *J. Affect. Disord.* **2016**, *199*, 132–138. [CrossRef]
40. Warmoth, K.; Tarrant, M.; Abraham, C.; Lang, I.A. Older adults' perceptions of ageing and their health and functioning: A systematic review of observational studies. *Psychol. Health Med.* **2016**, *21*, 531–550. [CrossRef] [PubMed]
41. Hickey, A.; O'Hanlon, A.; McGee, H. Quality of life in community-dwelling older people in Ireland: Association with ageing perceptions, physical health and psychological well-being. *Ir. J. Psychol.* **2010**, *31*, 135–150. Available online: http://salmandj.uswr.ac.ir/article-1-1011-en.html (accessed on 17 December 2021). [CrossRef]
42. Sadegh Moghadam, L.; Foroughan, M.; Mohammadi, F.; Ahmadi, F.; Farhadi, A.; Nazari, S.; Sadeghi, N. Aging perception in older adults. *Iran. J. Ageing* **2016**, *10*, 202–209. Available online: http://salmandj.uswr.ac.ir/article-1-1011-en.html (accessed on 17 December 2021).
43. Slotman, A.; Cramm, J.M.; Nieboer, A.P. Validation of the Dutch Aging Perceptions Questionnaire and development of a short version. *Health Qual. Life Outcomes* **2015**, *13*, 54. [CrossRef] [PubMed]
44. Jaafar, M.H.; Villiers-Tuthill, A.; Sim, S.H.; Lim, M.A.; Morgan, K. Validation of the brief ageing perceptions questionnaire (B-APQ) in Malaysia. *Aging Ment. Health* **2020**, *24*, 620–626. Available online: http://www.share-project.org/uploads/tx_sharepublications/201804_SHARE-WAVE-6_MFRB.pdf (accessed on 15 November 2021). [CrossRef] [PubMed]
45. Malter, F.; Börsch-Supan, A. *SHARE Wave 6: Panel Innovations and Collecting Dried Blood Spots*; MEA, Max Planck Institute for Social Law and Social Policy: Munich, Germany, 2017; Available online: http://www.share-project.org/uploads/tx_sharepublications/201804_SHARE-WAVE-6_MFRB.pdf (accessed on 15 November 2021).

46. SHARE. SHARE Conditions of Use. 2022. Available online: http://www.share-project.org/data-access/share-conditions-of-use.html (accessed on 10 January 2022).

47. Bergmann, M.; Kneip, T.; De Luca, G.; Scherpenzeel, A. *Survey Participation in the Survey of Health, Ageing and Retirement in Europe (SHARE), Wave 1-6*; Working Paper Series 41-2019; Munich Center for the Economics of Aging: Munich, Germany, 2019; Available online: http://www.share-project.org/uploads/tx_sharepublications/WP_Series_41_2019_Bergmann_et_al.pdf (accessed on 15 December 2021).

48. Börsch-Supan, A. *Survey of Health, Ageing and Retirement in Europe (SHARE) Wave 6; Release Version: 7.0.0*; SHARE-ERIC, Munich Center for the Economics of Aging (MEA): Munich, Germany, 2019; Available online: http://www.share-project.org/data-documentation/waves-overview/wave-6.html (accessed on 15 November 2021).

49. López Zubizarreta, M.; Hernández Mezquita, M.Á.; Miralles García, J.M.; Barrueco Ferrero, M. Tobacco and diabetes: Clinical relevance and approach to smoking cessation in diabetic smokers. *Endocrinol. Diabetes Nutr.* **2017**, *64*, 221–231. [CrossRef] [PubMed]

50. Gritz, E.R.; Vidrine, D.J.; Fingeret, M.C. Smoking cessation: A critical component of medical management in chronic disease populations. *Am. J. Prev. Med.* **2007**, *33*, S414–S422. [CrossRef] [PubMed]

51. Wiggins, R.D.; Netuveli, G.; Hyde, M.; Higgs, P.; Blane, D. The evaluation of a self-enumerated scale of quality of life (CASP-19) in the context of research on ageing: A combination of exploratory and confirmatory approaches. *Soc. Indic. Res.* **2008**, *89*, 61–77. [CrossRef] [PubMed]

52. Rodríguez-Blázquez, C.; Ribeiro, O.; Ayala, A.; Teixeira, L.; Araújo, L.; Forjaz, M.J. Psychometric properties of the CASP-12 scale in Portugal: An analysis using SHARE data. *Int. J. Environ. Res. Public Health* **2020**, *17*, 6610. [CrossRef] [PubMed]

53. Von dem Knesebeck, O.; Hyde, M.; Higgs, P.; Kupfer, A.; Siegriest, J. Quality of life and wellbeing. In *Health, Ageing and Retirement in Europe: First Results from the Survey of Health, Ageing and Retirement in Europe*; Borsch-Supan, A., Nrugiavini, A., Jurges, H., Mackenbach, J., Siegriest, J., Weber, G., Eds.; MEA: Mannheim, Germany, 2005; pp. 199–203.

54. Cimas, M.; Ayala, A.; Sanz, B.; Agulló-Tomás, M.S.; Escobar, A.; Forjaz, M.J. Chronic musculoskeletal pain in European older adults: Cross-national and gender differences. *Eur. J. Pain* **2018**, *22*, 333–345. [CrossRef] [PubMed]

55. Metz, D.H. Mobility of older people and their quality of life. *Transp. Policy* **2000**, *7*, 149–152. [CrossRef]

56. Zorrilla-Muñoz, V.; Agulló-Tomás, M.S.; Garcia-Sedano, T. Análisis socio-ergonómico en la agricultura. Evaluación del sector oleico desde una perspectiva de género y envejecimiento. *ITEA Inf. Técnica Económica Agrar. Rev. Asoc. Interprofesional Para El Desarro. Agrar. (AIDA)* **2019**, *115*, 83–104. [CrossRef]

57. Mitrani, R.D.; Dabas, N.; Goldberger, J.J. COVID-19 cardiac injury: Implications for long-term surveillance and outcomes in survivors. *Heart Rhythm* **2020**, *17*, 1984–1990. [CrossRef] [PubMed]

58. Paliwal, V.K.; Garg, R.K.; Gupta, A.; Tejan, N. Neuromuscular presentations in patients with COVID-19. *Neurol. Sci.* **2020**, *41*, 3039–3056. [CrossRef] [PubMed]

59. Marois, G.; Muttarak, R.; Scherbov, S. Assessing the potential impact of COVID-19 on life expectancy. *PLoS ONE* **2020**, *15*, e0238678. [CrossRef] [PubMed]

60. Ruiz-Cantero, M.T. Las estadísticas sanitarias y la invisibilidad por sexo y de género durante la epidemia de COVID-19. *Gac. Sanit.* **2021**, *35*, 95–98. [CrossRef] [PubMed]

61. Chudasama, Y.V.; Gillies, C.L.; Zaccardi, F.; Coles, B.; Davies, M.J.; Seidu, S.; Khunti, K. Impact of COVID-19 on routine care for chronic diseases: A global survey of views from healthcare professionals. *Diabetes Metab. Syndr. Clin. Res. Rev.* **2020**, *14*, 965–967. [CrossRef] [PubMed]

62. Jordan, R.E.; Adab, P.; Cheng, K. COVID-19: Risk factors for severe disease and death. *BMJ* **2020**, *368*, m1198. [CrossRef]

63. Schett, G.; Sticherling, M.; Neurath, M.F. COVID-19: Risk for cytokine targeting in chronic inflammatory diseases? *Nat. Rev. Immunol.* **2020**, *20*, 271–272. [CrossRef] [PubMed]

64. Connors, J.M.; Levy, J.H. COVID-19 and its implications for thrombosis and anticoagulation. *Blood* **2020**, *135*, 2033–2040. [CrossRef] [PubMed]

65. Hashemi, A.; Madhavan, M.V.; Bikdeli, B. Pharmacotherapy for Prevention and Management of Thrombosis in COVID-19. *Semin. Thromb. Hemost.* **2020**, *46*, 789–795. [CrossRef] [PubMed]

66. McFadyen, J.D.; Stevens, H.; Peter, K. The Emerging Threat of (Micro)Thrombosis in COVID-19 and Its Therapeutic Implications. *Circ. Res.* **2020**, *127*, 571–587. [CrossRef] [PubMed]

67. Feldmann, M.; Maini, R.N.; Woody, J.N.; Holgate, S.T.; Winter, G.; Rowland, M.; Richards, D.; Hussell, T. Trials of anti-tumour necrosis factor therapy for COVID-19 are urgently needed. *Lancet* **2020**, *395*, 1407–1409. [CrossRef] [PubMed]

68. Agarwala, S.R.; Vijayvargiya, M.; Pandey, P. Avascular necrosis as a part of 'long COVID-19'. *BMJ Case Rep. CP* **2021**, *14*, e242101. [CrossRef] [PubMed]

69. Campbell, S.M.; Roland, M.O.; Buetow, S.A. Defining quality of care. *Soc. Sci. Med.* **2000**, *51*, 1611–1625. [CrossRef] [PubMed]

70. Di Rosa, M.; Kofahl, C.; McKee, K.; Bień, B.; Lamura, G.; Prouskas, C.; Döhner, H.; Mnich, E. A typology of caregiving situations and service use in family carers of older people in six European countries. *GeroPsych* **2011**, *24*, 5–18. [CrossRef]

71. Agulló-Tomás, M.S.; Zorrilla-Muñoz, V. Technologies and Images of Older Women. In *Human Aspects of IT for the Aged Population. Technology and Society. HCII 2020. Lecture Notes in Computer Science 12209*; Gao, Q., Zhou, J., Eds.; Springer: Cham, Switzerland, 2020; pp. 163–175. [CrossRef]

72. Zorrilla-Muñoz, V.; Agulló-Tomás, M.S.; Forjaz, M.J.; Fernandez, E.; Rodriguez-Blazquez, C.; Ayala, A.; Fernandez-Mayoralas, G. Technology, Gender and COVID-19. Analysis of Perceived Health in Adults and Older People. In *Human Aspects of IT for the Aged Population. Supporting Everyday Life Activities. HCII 2021. Lecture Notes in Computer Science*; Gao, Q., Zhou, J., Eds.; Springer: Cham, Switzerland, 2021; pp. 363–379. [CrossRef]
73. Camarero, L.; Oliva, J. Thinking in rural gap: Mobility and social inequalities. *Palgrave Commun.* **2019**, *5*, 95. [CrossRef]
74. Apostolopoulos, N.; Ratten, V.; Stavroyiannis, S.; Makris, I.; Apostolopoulos, S.; Liargovas, P. Rural health enterprises in the EU context: A systematic literature review and research agenda. *J. Enterprising Communities People Places Glob. Econ.* **2020**, *14*, 4, 563–582. [CrossRef]
75. Carrillo-Balam, G.; Cantoral, A.; Rodríguez-Carmona, Y.; Christensen, D.L. Health-care coverage and access to health care in the context of type 2 diabetes and hypertension in rural Mexico: A systematic literature review. *Public Health* **2020**, *181*, 8–15. [CrossRef] [PubMed]
76. Davisson, E.A.; Swanson, E.A. Patient and Nurse Experiences in a Rural Chronic Disease Management Program. *Prof. Case Manag.* **2018**, *23*, 10–18. [CrossRef] [PubMed]
77. Miele, G.; Straccia, G.; Moccia, M.; Leocani, L.; Tedeschi, G.; Bonavita, S.; Lavorgna, L.; Digital Technologies Web; Social Media Study Group of the Italian Society of Neurology. Telemedicine in Parkinson's disease: How to ensure patient needs and continuity of care at the time of COVID-19 pandemic. *Telemed. e-Health* **2020**, *26*, 1533–1536. [CrossRef] [PubMed]
78. Criado-Quesada, B.; Zorrilla-Muñoz, V.; Agulló-Tomás, M.S. El uso de tecnologías de asistencia sanitaria digital por parte de la población mayor desde una perspectiva de género e intrageneracional. *Teknokultura* **2021**, *18*, 103–113. [CrossRef]
79. Chisty, M.A.; Islam, M.A.; Munia, A.T.; Rahman, M.M.; Rahman, N.N.; Mohima, M. Risk perception and information-seeking behavior during emergency: An exploratory study on COVID-19 pandemic in Bangladesh. *Int. J. Disaster Risk Reduct.* **2021**, *65*, 102580. [CrossRef] [PubMed]
80. Henning-Smith, C. The unique impact of COVID-19 on older adults in rural areas. *J. Aging Soc. Policy* **2020**, *32*, 396–402. [CrossRef] [PubMed]
81. Fuller, H.R.; Huseth-Zosel, A. Older adults' loneliness in early COVID-19 social distancing: Implications of rurality. *J. Gerontol. Ser. B Psychol. Sci. Soc. Sci.* **2021**, gbab053. [CrossRef] [PubMed]

Article

The Revised Memory and Behavior Problems Checklist for Nursing Homes: Validation among Different Spanish Territories

Cristina Velasco [1,*], Javier López [1], Gema Pérez-Rojo [1], Cristina Noriega [1] and José Ángel Martínez-Huertas [2]

[1] Department of Psychology, School of Medicine, Universidad San Pablo-CEU, CEU Universities, 28925 Madrid, Spain; jlopezm@ceu.es (J.L.); gema.perezrojo@ceu.es (G.P.-R.); cristina.noriegagarcia@ceu.es (C.N.)

[2] Faculty of Psychology, Universidad Autónoma de Madrid, Calle Iván Pavlov, 6, Ciudad Universitaria de Cantoblanco, 28049 Madrid, Spain; josea.martinez@uam.es

* Correspondence: cristina.velascovega@ceu.es

Citation: Velasco, C.; López, J.; Pérez-Rojo, G.; Noriega, C.; Martínez-Huertas, J.Á. The Revised Memory and Behavior Problems Checklist for Nursing Homes: Validation among Different Spanish Territories. *Land* **2022**, *11*, 251. https://doi.org/10.3390/land11020251

Academic Editors: Vanessa Zorrilla-Muñoz, Maria Silveria Agulló-Tomás, Eduardo Fernandez, Blanca Criado Quesada, Sonia De Lucas Santos and Jesus Cuadrado Rojo

Received: 7 January 2022
Accepted: 4 February 2022
Published: 8 February 2022

Publisher's Note: MDPI stays neutral with regard to jurisdictional claims in published maps and institutional affiliations.

Abstract: Memory and behavioral difficulties among older people living in nursing homes can cause burden and other consequences in professional caregivers. There is a lack of instruments that evaluate these behaviors and their influence in formal caregivers. The aim of this study is to develop and psychometrically test—the Revised Memory and Behavior Problems Checklist for Nursing Homes (RMBPC-NH). A cross-sectional study was carried out. The sample was made up of 312 formal caregivers working in nursing homes from different territories in Spain, 87.5% were women and 12.5% were men. The average age of participants was 39 years (SD = 12.2). The sample was recruited from January 2019 to March 2020. Participants completed a self-administered questionnaire, which included sociodemographic information, and assessed quality of technical equipment, level of training, experience of working with older people, job satisfaction, professional quality of life, burnout, and conception of negative stereotypes held towards aging. The four-factor structure of the RMBPC-NH showed a good fit, namely in relation to memory, functional, and emotional factors, and other problems. It has shown adequate psychometric properties, internal consistency, and validity (correlations with professional quality of life, job satisfaction, burnout, and negative stereotypes). The RMBPC-NH is a useful instrument to evaluate the frequency of older people's memory and behavior problems and professional caregivers' burden. The practical application in nursing homes is discussed.

Keywords: behavior problems; memory problems; long-term care; nursing homes; professional quality of life; professional caregivers; burnout; job satisfaction; territory and instrument development

1. Introduction

Behavior and memory problems (BMP) in institutionalized older people are frequent. These problems may be associated with processes of cognitive decline and dementia diagnoses, and these symptoms normally increase over time [1]. Studies focused on long-term care residents indicate that prevalence of dementia is 58%, and the prevalence of behavioral and psychological symptoms of dementia is 78% among residents with dementia diagnoses [2]. Not only dementia is associated with BMP, for example, major depressive disorder and other psychiatric diagnoses, such as anxiety, could be associated with BMP in long-term care residents [1,3]. In Spain, around 62% of older people in institutional settings are diagnosed with dementia. Due to increased life expectancy among the general population, as people living longer into old age, these problems will become more prevalent [3,4]. Traditionally, the presence of BMP problems is a predictor of burden among informal caregivers [5]. The presence of BMP in institutionalized older people may

---

also affect professionals, specifically in terms of their quality of life and job satisfaction and stress [6,7].

1.1. Job Satisfaction, Professional Quality of Life, Stress, and Stereotypes towards Old Age among Professional Caregivers in Nursing Homes

Professional caregivers' commitment and job satisfaction have been associated with different factors. A well-organized and pleasant work environment and low levels of staff turnover have been linked to high job satisfaction [8]. Job satisfaction has also been strongly associated with good job environment, teamwork, a safe climate, and adequate staffing resources [9]. When comparing job satisfaction between different types of staff, nurses are more satisfied with their job than nursing assistants. Nursing assistants show less satisfaction because of work monotony, position-related strain, and under recognition of work efforts [10].

Professional' quality of life results from the balance between work demands and psychological, social, and employment resources. Recently, a study has highlighted the relationship between care burden and quality of working life in professional caregivers of older people [11].

In this sense, suffering significant stress could affect the quality of care provided by formal and informal caregivers. People who suffer from dementia could have memory and behavior problems and functional problems. Specifically, Sun et al. [7] point that formal caregivers of people with dementia find some environmental factors to be sources of stress, for example, other staff or residents' family members. However, they identified personal factors as feelings of not being able to provide good care, the lack of resources, or not find the opportunity to do best for the people they care for, causing strain. In general, formal caregivers experience job satisfaction, but they also experience significant stress that can affect the quality of care they provide [12].

Furthermore, negative stereotypes held towards aging may influence the relationship established with older people and facilitate overprotection and paternalist attitudes, unnecessary interventions, and may reduce the autonomy of the older adults [13]. These attitudes could relate to BMP, as well as create more dependence in residents, which may produce lower job satisfaction and stress. Higher levels of training reduced the presence of negative stereotypes held towards aging. Being a woman aged between 45 and 59 years has been associated with higher levels of negative stereotypes held towards aging. This has been the typical profile for professional caregivers in nursing homes in the past, woman around middle age [14].

Interestingly, there is a significant amount of research analyzing the relationship between the BMP of institutionalized older people and professional caregivers' burden, quality of professional life, and job satisfaction. Moreover, Islam et al. [15] have pointed out the importance of training professional caregivers, because higher levels of training have been associated with higher levels of well-being and lower levels of burden.

1.2. The Revised Memory and Behavior Problems Nursing Homes

The first questionnaire to evaluate MBP was "The Revised Memory and Behavior Problems" (RMBP; [16]) questionnaire, that was used to assess informal caregivers of dementia patients. In this research line, Ray et al. [17] created the Nursing Home Behavior Problems Scale (NHBPS). It was designed to be completed by professional caregivers (nurses and nursing assistants) with 29 items concerning serious behavior problems. The NHBPS [17] has 6 subscales: uncooperative or aggressive behavior, irrational or restless behavior, sleep problems, annoying behavior, inappropriate behavior, and dangerous behavior. This inventory was very difficult to implement among nursing home staff. Later, Wagner et al. [18] used the Memory and Behavior Problems Checklist (MBPC-NH) which was an adaptation of the MBPC-NH (Orr-Rainey and Terri, unpublish manuscript).

The Revised Memory and Behavior Problems Checklist for Nursing Homes (RMBPC-NH) was modified by Allen et al. [19]. The RMBPC-NH was designed to be answered by professional caregivers. It included a 6-point Likert scale, with response options from 0 (not at all/not applicable) to 5 (several times a day). Adequate psychometric properties were obtained [19], and it has been related with users' variables (cognitive ability, activities of daily life) and professional caregivers' variables (depression and job stress). However, the RMBPC-NH has not been validated in the Spanish population. Additionally, this topic is very important in this moment, because the consequences of the COVID-19 pandemic may have a negative impact on cognitive and functional capabilities of the older population, especially among dementia patients, provoking the presence of more memory and behavioral symptoms [20].

The purpose of this study was to validate the RMBPC-NH [19] in a Spanish sample of professional caregivers in nursing homes from different territories. We examined the factor structure, the psychometric properties, and their relationship with other measures, such as professionals' quality of life, job satisfaction, burden, and negative stereotypes held towards aging.

2. Materials and Methods

2.1. Participants

From January 2019 to March 2020, 312 professional caregivers working in nursing homes in Spain participated in the study. A proportion of 87.5% of participants were women and the average age was 39 years (SD = 12.2). Additional demographic information is presented in Table 1.

Table 1. Sample characteristics.

	N	%
Gender		data
Male	39	12.5
Female	273	87.5
Job position		
Nursing assistants	202	58.3
Graduates (doctors, nurses, psychologists, physiotherapists, etc.)	104	33
Others (cooks, cleaners and/or administration)	15	5.4
Territory		
Community of Madrid	174	55.8
Catalonia	66	21.2
Extremadura	38	12.2
Valencian Community	25	8
Andalusia	9	2.9
Care professional experience		
Less than a year	100	32
Less than 2 years	32	10.3
2–5 years	29	12.5
More than 5 years	141	45.2
Level of professional training		
Much training	86	27.5
Enough training	159	50.8
Some training	55	17.9
No training	12	3.8
Quality in technical equipment		
Very appropriate	95	30.4
Enough appropriate	126	40.3
Some appropriate	84	26.8
No appropriate	8	2.6

Note: N = 312.

2.2. Instruments

- *The Revised Memory and Behavior Problems Checklist for Nursing Homes* (RMBPC-NH) [19]: This 42-item instrument was used to measure the frequency of resident BMP (cognitive, emotional, functional, and other problems) and its relationship with staff's burden and other residents' well-being.
- *Professional Quality of Life Questionnaire* (QPL-35) [21]: This 35-item instrument was used to measure the balance between work demands and personal resources. Items were scaled from "none" (values 1 and 2), to "some" (values 3, 4, and 5), to "quite a lot" (values 6, 7, and 8), to "a lot" (values 9 and 10). In this sample, a high reliability was found for the general scale ($\alpha = 0.91$) and its subscales: management support ($\alpha = 0.92$), workload ($\alpha = 0.89$), and intrinsic motivation ($\alpha = 0.82$).
- *Job Satisfaction Questionnaire S10/12* [22]: This 12-item questionnaire was used to measure job satisfaction. Items were scored from 1 (very unsatisfied) to 7 (very satisfied). In this sample, a high reliability was found for the general scale ($\alpha = 0.92$) and its subscales: supervisory practices ($\alpha = 0.93$,), working environment ($\alpha = 0.71$), and benefits received ($\alpha = 0.81$).
- *Maslach Burnout Inventory* (MBI) [23]: This 22-item questionnaire requests that professionals indicate the frequency with which they experienced some statements of "job-related" feelings. Items were scored from 0 (never experienced such a feeling) to 6 (experience such feelings every day). In this sample, a high reliability was found for the general scale ($\alpha = 0.85$) and its subscales: emotional exhaustion ($\alpha = 0.86$), depersonalization ($\alpha = 0.49$), and personal accomplishment ($\alpha = 0.78$).
- *Negative Stereotypes towards Aging Questionnaire* (CENVE) [24]: This 15-item measure was used to evaluate negative stereotypes held towards aging. Items were scored from 1 (strongly disagree) to 4 (strongly agree). In this sample, a high reliability was found for the general scale ($\alpha = 0.90$) and its subscales: health ($\alpha = 0.83$), character–personality ($\alpha = 0.77$), and motivational–social ($\alpha = 0.66$).

2.3. Procedure

A cross-sectional study was carried out in different nursing homes in Spain. We contacted managers of nursing homes and gave them verbal and written information about the research, and they invited workers to take part in the study. The participants were volunteers, and one person from the research team explained to the nursing home staff how to complete the questionaries. The time taken to complete the self-administered questionnaire was around 30 minutes. The validation process of the Spanish version of the RMBPC-NH [19] followed the defined guidelines for adapting tests [25,26] (Hambleton, 2005; ITC, 2016). First, two authors of this study translated the original English scale into Spanish. Then, a bilingual independent translator performed the back translation. Discrepancies emerging between the original and the Spanish version were discussed, and the research team of CEU adjusted the translation.

2.4. Data Analysis

In order to study the psychometric properties of the Spanish version of the RMBPC-NH, we conducted different confirmatory factor analyses (CFAs) using R's lavaan package [27]. We examined and compared the two different factor structures proposed in the literature: the unidimensional model and the four-factor model. Given that items presented high kurtosis and skewness, the weighted least mean square and variance adjusted (WLSMV) estimator was used because it is a robust estimator for ordered data and does not assume normal distributions (e.g., Muthén and Muthén [28]). The reliabilities of the scale and its subscales were analyzed through Cronbach's α 2, using R's CTT [29] and psych [30] packages, respectively. Different descriptive analyses were conducted to describe the state of the professional caregivers. The statistical differences between territories (autonomous communities—AACC) in total of frequency of behavior and memory problems and care-related burdens were analyzed with univariate ANOVAs. Different mixed-effects models with random intercepts for participants were conducted to analyze potential differences in the burden scale of RMBPC-NH and subscales, considering the different levels of the participants, using R's lme4 package [31]. We analyzed the construct validity of RMBPC-NH scale scores, exploring its relationships with different variables using Pearson's correlation coefficients and mixed-effects models with random intercepts for participants. All the statistical analyses were performed in R software version 3.6.1 [32]. The statistical significance was corrected using the Holm–Bonferroni correction when multiple comparisons were conducted in mixed-effects models [33].

2.5. Ethical Considerations

The study was approved by the University Ethics Committee. Informed consent was obtained from all respondents, and confidentiality was explicitly guaranteed. Participants were volunteers, and they were asked to fill out a self-administered questionnaire, which included sociodemographic information, their staff category, level of training, and caring experience.

3. Results

3.1. Factor Structure

Two CFAs were fitted to test the unidimensional and the four-factor models for the Spanish validation of the RMBPC-NH scale by frequency and care-related burden scores. Model fit results can be found in Table 2. In the case of frequency scores, the unidimensional model showed a moderate fit to the data, whilst the four-factor structure, proposed by Wagner et al. [18], obtained a good fit. A robust χ^2 difference for the nested model comparison [34] showed that the four-factor structure obtained significantly better performance than the unidimensional one ($\Delta\chi^2 = 68.66$, $\Delta df = 6$, $p < 0.001$). In the case of care-related burden, the same pattern of results was found: the unidimensional model showed a good fit to the data, but the four-factor structure [18] obtained a better fit. The same robust χ^2 difference for the nested model comparison was conducted, showing that the four-factor structure obtained significantly better performance than the unidimensional one ($\Delta\chi^2 = 17.27$, $\Delta df = 6$, $p < 0.01$). We only present care-related burden results because social–emotional burden results did not present relevant differences. Moreover, we explored item descriptive analysis (mean and standard deviation) and factor loadings for the four-factor structures. All factor loadings were adequate for RMBPC-NH subscales.

Table 2. Results of confirmatory factor analysis for unidimensional and four-factor structures in frequency and care-related burden scores of the RMBPC-NH scale.

RMBPC-NH Score	Factor Structure	Model Fit					
		$\chi^2(df)$	p	*CFI*	*TLI*	*RMSEA [90% IC]*	*SRMR*
Frequency	Unidimensional	$\chi^2(819) = 1964.31$	<0.001	0.93	0.93	0.07 [0.06, 0.08]	0.114
	Four-factors	$\chi^2(813) = 1687.73$	<0.001	0.95	0.95	0.06 [0.05, 0.06]	0.099
Care-related burden	Unidimensional	$\chi^2(819) = 2012.13$	<0.001	0.98	0.98	0.07 [0.07, 0.08]	0.053
	Four-factors	$\chi^2(813) = 1968.75$	<0.001	0.98	0.98	0.07 [0.06, 0.07]	0.050

Note: $N = 312$.

3.2. Reliability and Descriptive Analysis

Table 3 presents descriptive analysis of frequency, care-related burden, and social–emotional burden for each subscale of the adaptation of the RMBPC-NH for Spanish professional caregivers working in nursing homes. As can be observed, the reliability of the general scale and its subscales was high (Cronbach's α ranging from 0.80 to 0.99). Means and medians showed a positive skewness for item responses in both frequency and burden scores. Accordingly, the descriptive results of the different subscales of care-related and social–emotional burdens were very similar, which explains their lack of differences in the following statistical analyses.

Table 3. Descriptive analysis and reliability of RMBPC-NH scale and its subscales.

| Variable | | | Descriptive Analysis | | | | | Reliability |
Score	Problem	Items	*M*	*SD*	*Mdn*	Range	Empirical Range	Cronbach's α
Frequency	Total	42	52.46	39.17	41	0–210	0–185	0.96
	Memory	12	18.48	16.42	14	0–60	0–60	0.94
	Functional	11	11.60	9.87	9	0–55	0–52	0.84
	Emotional	12	17.49	12.04	15	0–60	0–56	0.85
	Other	7	4.90	5.79	2	0–35	0–34	0.80
Care-related burden	Total	42	32.96	40.56	17	0–168	0–167	0.99
	Memory	12	9.75	11.82	5	0–48	0–48	0.96
	Functional	11	8.62	10.76	5	0–44	0–44	0.95
	Emotional	12	10.01	12.21	5	0–48	0–48	0.96
	Other	7	4.59	7.06	1	0–28	0–28	0.95
Social–emotional burden	Total	42	30.67	38.79	15	0–168	0–168	0.99
	Memory	12	8.94	11.27	5	0–48	0–48	0.97
	Functional	11	7.39	10.51	2	0–44	0–44	0.96
	Emotional	12	9.85	11.45	6	0–48	0–48	0.96
	Other	7	4.49	6.87	1	0–28	0–28	0.95

3.3. Validity Evidence

As it was previously shown, all the problems presented a similar distribution of care-related and social–emotional burden experienced by professional caregivers. In this way, mixed-effects models with random intercepts for participants revealed no statistically significant differences between problems for care-related burden staff ($F(3,1122) = 1.46$, $p = 0.23$) nor for social–emotional burden staff ($F(3,1120) = 1.72$, $p = 0.16$).

The relationship between the frequency of problems and the care-related burden and its subscales was analyzed. Pearson correlation coefficients revealed a medium relationship between the reported frequency of problems and how this affected the care-related burden of professional caregivers for all problems ($r = 0.40$, $p < 0.01$), memory problems ($r = 0.32$, $p < 0.01$), functional problems ($r = 0.51$, $p < 0.01$), emotional problems ($r = 0.40$, $p < 0.01$), and other problems ($r = 0.49$, $p < 0.01$). There were important individual differences related to subjectively judging how those problems increased their care-related burden. No substantive differences were found for social–emotional burden scales, compared with care-related burdens.

We analyzed the validity of the Spanish version of the RMBPC-NH scale and its subscales for professional caregivers in Spanish nursing homes, exploring its relationship with quality of professional life, job satisfaction, negative stereotypes held towards aging, and burnout variables, using Pearson correlation coefficients. Results can be found in Table 4. All the correlation coefficients were medium, but they were in accordance with theoretical predictions (these results will be explained in more detail in Discussion Section). It is worth mentioning here that, in general, correlation coefficients with these variables were higher for care-related burden than for problem frequency. Again, Table 4 only reports care-related burden results, because social–emotional burden results did not present substantive differences with care-related burden.

Table 5 presents the results of different mixed-effects models with random intercepts for participants to explore the differences in RMBPC-NH scale scores in different categorial and ordinal covariates (see Table 1). Specifically, a mixed-effects model was conducted for each score and covariate, correcting the statistical significance of fixed effects with the Holm–Bonferroni correction [33]. Results for the sex variable show that women tend to report more care-related burden than men, but no differences were found in the frequency of problems. Results for care experience with older people (in months) indicate a negative relationship between time and the frequency of reported memory problems, and a positive relationship between time and care-related burden (that is, the more time the person has been working in nursing homes, the more burden he/she felt). Results for technical equipment show that, although it is only related to the frequency of reported functional problems, the lack of quality in technical equipment has a great impact on professional caregivers' care-related burden. Results for the level of training show that people with a higher level of training report lower frequency of general, functional, emotional, and other problems, and they also have less staff burden related to functional and emotional problems.

Table 4. Pearson correlation coefficients between RMBPC-NH factor scores and different work-related burnout, quality of professional life, and job satisfaction.

Variables	Frequency					Care-related Burden				
	All	**Memory**	**Functional**	**Emotional**	**Other**	**All**	**Memory**	**Functional**	**Emotional**	**Other**
Professional quality of life (PQL)	−0.15 *	−0.06	−0.21 **	−0.16 **	−0.14 *	−0.22 **	−0.20 **	−0.20 **	−0.24 **	−0.20 **
Workload (PQL-WL)	−0.06	0.03	−0.121 *	−0.09	−0.11	−0.18 **	−0.14 *	−0.16 **	−0.22 **	−0.19 **
Intrinsic motivation (PQL-IM)	−0.08	−0.09	−0.08	−0.07	−0.03	−0.06	−0.08	−0.04	−0.03	−0.02
Managerial support (PQL-MS)	−0.19 **	−0.10	−0.25 **	−0.18 **	−0.16 **	−0.22 **	−0.19 **	−0.22 **	−0.23 **	−0.19 **
Job satisfaction (total)	−0.19 **	−0.10	−0.27 **	−0.17 **	−0.17 **	−0.24 **	−0.22 **	−0.24 **	−0.26 **	−0.22 **
Supervisory practices	−0.12 *	−0.04	−0.19 **	−0.11	−0.10	−0.14 *	−0.11	−0.13 *	−0.16 **	−0.15 *
Working environment	−0.16 **	−0.11	−0.20 **	−0.13 *	−0.14 *	−0.22 **	−0.21 **	−0.23 **	−0.24 **	−0.17 **
Benefits received	−0.22 **	−0.11	−0.31 **	−0.20 **	−0.21 **	−0.25 **	−0.24 **	−0.24 **	−0.24 **	−0.23 **
Negative stereotypes held towards ageing	0.14 *	0.11	0.17 **	0.09	0.16 **	0.14 *	0.17 **	0.21 **	0.09	0.11
Health stereotypes	0.14 *	0.12 *	0.17 **	0.08	0.13 *	0.11	0.15 *	0.17 **	0.04	0.07
Motivational–social stereotypes	0.05	0.03	0.10	0.00	0.07	0.15 *	0.17 **	0.21 **	0.11	0.12 *
Character–personality stereotypes	0.19 **	0.15 **	0.19 **	0.16 **	0.21 **	0.13 *	0.13 *	0.21 **	0.09	0.11
Burnout (MBI)	0.08	0.03	0.09	0.14 *	0.02	0.10	0.08	0.10	0.12 *	0.08
Burnout emotional exhaustion (EE)	0.00	−0.06	0.05	0.04	0.01	0.08	0.03	0.08	0.11 *	0.07
Burnout depersonalization (DP)	0.12 *	0.05	0.16 **	0.15 **	0.13 *	0.21 **	0.20 **	0.21 **	0.21 **	0.19 **
Burnout personal Accomplishment (PA)	0.02	0.05	−0.04	0.06	−0.07	−0.07	−0.05	−0.06	−0.07	−0.08

Note: ** = $p < 0.01$. * = $p < 0.05$.

Table 5. Estimations (and standard errors) of mixed-effects models for sex, care experience, technical equipment, and level of training of mixed-effects models.

	RMBPC-NH	Sex (Reference: Male)	Care Experience	Technical Equipment	Level of Training
Frequency	All	0.21 (0.18)	−0.03 (0.05)	0.08 (0.07)	−0.11 * (0.05)
	Memory	0.19 (0.25)	−0.17 * (0.07)	−0.07 (0.10)	−0.04 (0.07)
	Functional	0.29 (0.18)	0.03 (0.05)	0.16 * (0.07)	−0.15 ** (0.05)
	Emotional	0.25 (0.19)	0.04 (0.05)	0.07 (0.08)	−0.11 * (0.05)
	Other	0.07 (0.17)	0.01 (0.04)	0.17 * (0.07)	−0.11 * (0.05)
Care-related burden	All	0.45 * (0.19)	0.14 ** (0.05)	0.33 ** (0.08)	−0.09 (0.05)
	Memory	0.38 * (0.19)	0.09 ^{t} (0.05)	0.28 ** (0.08)	−0.07 (0.06)
	Functional	0.44 * (0.19)	0.11 * (0.05)	0.31 ** (0.08)	−0.11 * (0.06)
	Emotional	0.55 * (0.19)	0.21 ** (0.05)	0.39 ** (0.08)	−0.10 ^{t} (0.06)
	Other	0.42 * (0.21)	0.15 ** (0.05)	0.38 ** (0.08)	−0.09 (0.06)

Note: $N = 312$. A mixed-effects model was conducted for each RMBPC-NH scale score and covariate. The statistical significance of fixed effects was corrected with the Holm–Bonferroni correction [29]. ** = $p < 0.01$. * = $p < 0.05$. ^{t} = $p < 0.10$.

Additionally, we analyzed the differences between AACC levels using univariate ANOVAS. No differences between AACC levels were found for frequency total ($F(4,262) = 1.056, p = 0.379, \eta^2 = 0.016$), frequency cognitive ($F(4,277) = 1.379, p = 0.241, \eta^2 = 0.020$), frequency emotional ($F(4,276) = 1.558, p = 0.186, \eta^2 = 0.022$), nor frequency other ($F(4,279) = 1.355, p = 0.250, \eta^2 = 0.019$). Only frequency functional presented statistically significant differences between AACC levels with a small effect size ($F(4,270) = 2.465, p = 0.045, \eta^2 = 0.035$). On the contrary, we found significant differences between AACC levels for care-related burden ($F(4,244) = 11.263, p < 0.001, \eta^2 = 0.156$), care-related burden cognitive ($F(4,263) = 8.150$,

$p < 0.001, \eta^2 = 0.110$), care-related burden functional ($F(4,257) = 11.560, p < 0.001, \eta^2 = 0.152$), care-related burden emotional ($F(4,260) = 11.697, p < 0.001, \eta^2 = 0.153$), and care-related burden other ($F(4,266) = 10.559, p < 0.001, \eta^2 = 0.137$). All these differences had relevant effect sizes. Pairwise comparisons were observed between Catalonia and Community of Madrid ($p < 0.001$) and Catalonia and Valencian Community ($p < 0.001$) for total, cognitive, and other care-related burden. Catalonia showed the same differences with Community of Madrid and Valencian Community, but also with functional care-related burden ($p = 0.024$) and emotional care-related burden ($p = 0.018$) with Extremadura.

4. Discussion

There is a lack of instruments for measuring behavioral and memory problems and their impact in formal caregivers. For this reason, we set out to validate and analyze the psychometric characteristics of RMBPC-NH. Our results provide satisfactory evidence for the reliability and validity of the RMBPC-NH, consistent with the standards for psychological testing [35]. Additionally, for this reason, the RMBPC-NH may be considered an adequate scale to evaluate the frequency of BMP in older people living in nursing homes and how it affects professional caregivers. Questionnaires adapted to different sociocultural contexts are necessary. The four-factor structure of the RMBPC-NH proposed by Wagner et al. [18] (namely: memory, functional, emotional, and other problems) has shown excellent psychometric properties in formal caregivers working in nursing homes in Spain. Item descriptive analysis (mean and standard deviation) and factor loadings for this four-factor structure were appropriate. Internal consistency reliability was high, not only in the global scale but also in all subscales. Similar results were found in the original scale [19].

The Spanish version of the RMBPC-NH is a useful tool to evaluate the prevalence of common BMP in nursing homes, as well as the impact on professionals' burden. The two questions associated with every problem in evaluating staff burden are especially important for focusing not only on the presence/absence of the problem, but also on showing which problems affect professional caregivers or their work environment.

Regarding the validity analyses, results showed a negative relationship between total frequency problems, functional problems, emotional and others, and professional quality of life. According to Sun et al. [7], behavioral and psychological symptoms impact directly over caregiver quality of life. Moreover, previous research published has pointed out the relationship between professional quality of life and stress and burden in formal caregivers [9,21]. Furthermore, this relationship was stronger in care-related burden experienced by professional caregivers in all the subscales. Kalanlar and Kuru [11] point out how care-related burden directly influences the daily lives of professional caregivers working in nursing homes. Moreover, feelings of emotional demand and poor quality of team supervision predict burnout in professional caregivers. Burnout has been associated with elderly abuse in nursing homes [36]. The validity of the Spanish version of the RMBPC-NH has been established using the mediating role of burden to promote good care in institutionalized older adults [37].

Higher job satisfaction in professional caregivers has been linked to a lower frequency of BMP. These findings are in line with the study by Allen et al. [19]. Burden and stress in both formal and informal caregivers of people with dementia have been described traditionally [5]. However, the present study highlights the importance of job satisfaction associated with the prevalence of BMP. A strong correlation between depersonalization and frequency of BMP and care-related burden has been found. Gallego-Alberto et al. [38] have highlighted depersonalization as a significant predictor of anxiety in professional caregivers.

Another relevant finding in the current study is that professional caregivers who experience more care-related burden have lower levels of professional quality of life and job satisfaction, and higher levels of negative stereotypes held against aging. It seems that objective frequency of memory and behavioral problems is not related so much to quality of life and job satisfaction, but is related to the professional interpretation of these situations.

The frequency of memory problems does not have a relationship with any of the other variables studied, excluding health and character–personality negative stereotypes. These results may be explained by the fact that memory problems do not affect job performance as much as other types of problems do. In this line, behavior problems, aggressiveness, irritation, and disinhibition are those that require more supervision time by the caregivers [39,40].

Female professional caregivers tend to present with more care-related burden than men. However, they do not report more frequency of problems in older people. Similarly, people with more experience working with older adults report lower frequency of problems, but higher levels of burden. This may be because the problems that residents present are repetitive, and they may feel that they do not have enough strategies to handle them, which, in turn, is related to burden [7,40]. In this line, the lack of quality in technical equipment is significantly related to staff burden. This result was found also in previous studies. Adequate technical equipment is required to promote good care practices in older people nursing homes [41,42].

Finally, professionals with a higher level of training report lower frequency of problems and they also have lower care-related burden. This result also demonstrates the importance of training as an essential component in professional caregivers of nursing homes [6,7,40]. Carrying out adequate assessments and interventions is quite important when approaching BMP among residents living in nursing homes [43].

There were significant differences found among levels for care-related burden between AACCs. These differences may be due to the differences in health care in the different AACCs in Spain. This result warrants further analysis of factors related to care-related burden, because health decisions are implemented specifically in each AACC [44].

The study presents some methodological limitations which may limit the external validity of the instrument. We will consider increasing the sample of participants to include professional caregivers from different territories and different cultures, to analyze the results and to promote the external validation of the Spanish questionnaire.

Interventions that aim to increase job satisfaction, stress, and professional quality of life, as well as reduce negative stereotypes held towards aging, may have direct benefits for professional caregivers and indirect benefits for users [8,10,37]. Focusing on these variables would allow professionals to be more involved in their work and promote the good care of older people. For this, training professional caregivers is the main tool to promote good practices in the care of older people [7,45].

5. Conclusions

The Spanish version of the RMBPC-NH shows excellent psychometric properties and provides information about the care-related and social burdens of professional caregivers, associated with residents' memory and behavior problems. The evaluation of the frequency of memory and behavior problems can be an innovative proposal in nursing homes because of their relationship with factors related to occupational mental health. Professionals with higher levels of quality of life and job satisfaction will provide better care for older adults. In this line of thought, concrete intervention models could be generated to promote good practices for both professionals and older people in nursing homes. Moreover, nursing homes have been heavily affected by COVID-19, which has provoked an increase in behavioral and psychological symptoms in older people with dementia; consequently, caregivers have experienced higher levels of burnout and job dissatisfaction, and lower quality of life, which could have an impact on quality of care.

Author Contributions: Conceptualization, C.V., J.L., and G.P.-R.; methodology, C.V., J.L., G.P.-R., C.N., and J.Á.M.-H.; software, J.Á.M.-H.; validation, C.V., J.L., G.P.-R., and C.N.; formal analysis, C.V. and J.Á.M.-H.; investigation, C.V., J.L., G.P.-R., and C.N.; resources, G.P.-R. and J.L.; data curation, C.V., G.P-R. and C.N.; writing—original draft preparation, C.V., J.L. and G.P.-R.; writing—review and editing, C.V., J.L, G.P.-R., C.N., and J.Á.M.-H.; visualization, C.V.; supervision, C.V., J.L., and G.P.-R.;

project administration, G.P.-R. and J.L.; funding acquisition, G.P.-R. and J.L. All authors have read and agreed to the published version of the manuscript.

Funding: This research was funded by Universidad San Pablo, CEU, CEU Universities (CEU-Santander, grant number MCOV20V3).

Acknowledgments: The authors thank all the nursing homes and participants in the study.

Conflicts of Interest: The authors declare no conflict of interest.

References

1. Lanctôt, K.L.; Amatniek, J.; Ancoli-Israel, S.; Arnold, S.E.; Ballard, C.; Cohen-Mansfield, J.; Ismail, Z.; Lyketsos, C.; Miller, D.S.; Musiek, E.; et al. Neuropsychiatric signs and symptoms of Alzheimer's disease: New treatment paradigms. *Alzheimer's Dement. Transl. Res. Clin. Interv.* **2017**, *3*, 440–449. [CrossRef] [PubMed]
2. Shiells, K.; Pivodic, L.; Holmerová, I.; Van den Block, L. Self-reported needs and experiences of people with dementia living in nursing homes: A scoping review. *Aging Ment. Health* **2019**, *24*, 1553–1568. [CrossRef] [PubMed]
3. Seitz, D.; Purandare, N.; Conn, D. Prevalence of psychiatric disorders among older adults in long-term care homes: A systematic review. *Int. Psychogeriatr.* **2010**, *22*, 1025–1039. [CrossRef]
4. Martínez-Lage, P.; Martín-Carrasco, M.; Arrieta, E.; Rodrigo, J.; Formiga, F. Map of Alzheimer's disease and other dementias in Spain. MapEA Project. *Rev. Esp. Geriatr. Gerontol.* **2018**, *53*, 26–37. [CrossRef] [PubMed]
5. Zarit, S.H.; Reever, K.E.; Bach-Peterson, J. Relatives of the impaired elderly: Correlates of feelings of burden. *Gerontologist* **1980**, *20*, 649–655. [CrossRef] [PubMed]
6. Kunkle, R.; Chaperon, C.; Berger, A.M. Formal Caregiver Burden in Nursing Homes: An Integrative Review. *West. J. Nurs. Res.* **2020**, *43*, 877–893. [CrossRef] [PubMed]
7. Sun, M.; Mainland, B.J.; Ornstein, T.J.; Sin, G.L.; Herrmann, N. Correlates of nursing care burden among institutionalized patients with dementia. *Int. Psychogeriatr.* **2018**, *30*, 1549–1555. [CrossRef] [PubMed]
8. Kurniawaty, K.; Ramly, M.; Ramlawati, R. The effect of work environment, stress, and job satisfaction on employee turnover intention. *Manag. Sci. Lett.* **2019**, *9*, 877–886. [CrossRef]
9. Schwendimann, R.; Dhaini, S.; Ausserhofer, D.; Engberg, S.; Zúñiga, F. Factors associated with high job satisfaction among care workers in Swiss nursing homes—A cross sectional survey study. *BMC Nurs.* **2016**, *6*, 15–37. [CrossRef] [PubMed]
10. Estévez-Guerra, G.J.; Núñez-González, E.; Fariña-López, E.; Marrero-Medina, C.D.; Hernández-Marrero, P. Determinants of job satisfaction in long-term care facilities. *Rev. Esp. Geriatr. Gerontol.* **2007**, *42*, 285–292. [CrossRef]
11. Kalanlar, B.; Kuru Alici, N. The effect of care burden on formal caregiver's quality of work life: A mixed-methods study. *Scand. J. Caring Sci.* **2020**, *34*, 1001–1009. [CrossRef]
12. Edberg, A.-K.; Anderson, K.; Orrung Wallin, A.; Bird, M. The Development of the strain in dementia care scale (SDCS). *Int. Psychogeriatr.* **2015**, *27*, 2017–2030. [CrossRef]
13. Fernández-Ballesteros, R.; Sánchez-Izquierdo, M.; Olmos, R.; Huici, C.; Casado, J.M.R.; Jentoft, A.C. Paternalism vs. autonomy: Are they alternative types of formal care? *Front. Psychol.* **2019**, *10*, 1–4. [CrossRef]
14. Menéndez Álvarez-Dardet, S.; Cuevas-Toro, A.M.; Pérez-Padilla, J.; Lorence Lara, B. Assessment of negative stereotypes about old age in young people and adults. *Rev. Esp. Geriatr. Gerontol.* **2016**, *51*, 323–328. [CrossRef]
15. Islam, M.S.; Baker, C.; Huxley, P.; Russell, I.T.; Dennis, M.S. The nature, characteristics and associations of care home staff stress and wellbeing: A national survey. *BMC Nurs.* **2017**, *16*, 22. [CrossRef]
16. Teri, L.; Truax, P.; Logsdon, R.; Uomoto, J.; Zarit, S.; Vitaliano, P.P. Assessment of behavioral problems in dementia: The revised memory and behavior problems checklist. *Psychol. Aging* **1992**, *7*, 622–631. [CrossRef]
17. Ray, W.A.; Taylor, J.A.; Lichtenstein, M.J.; Meador, K.G. The Nursing Home Behavior Problem Scale. *J. Gerontol.* **1992**, *47*, M9–M16. [CrossRef]
18. Wagner, A.W.; Teri, L.; Orr-Rainey, N. Behavior Problems among Dementia Residents in Special Care Units: Changes over Time. *J. Am. Geriatr. Soc.* **1995**, *43*, 784–787. [CrossRef]
19. Allen, R.S.; Burgio, L.D.; Roth, D.L.; Ragsdale, R.; Gerstle, J.; Bourgeois, M.S.; Dijkstra, K.; Teri, L. The Revised Memory and Behavior Problems Checklist—Nursing Home: Instrument development and measurement of burden among certified nursing assistants. *Psychol. Aging* **2003**, *18*, 886–895. [CrossRef]
20. Ryoo, N.; Pyun, J.M.; Baek, M.J.; Suh, J.; Kang, M.J.; Wang, M.J.; Youn, Y.C.; Yang, D.W.; Kim, S.Y.; Park, Y.H.; et al. Coping with Dementia in the Middle of the COVID-19 Pandemic. *J. Korean Med. Sci.* **2020**, *35*, e383. [CrossRef]
21. Martín, J.; Gómez, T.; Martínez, C.; del Cura, M.I.; Cabezas, M.C.; García, S. Measurement of the evaluative capacity of the CVP-35 questionnaire for perceiving the quality of professional life. *Aten. Primaria* **2008**, *40*, 327–334.
22. Meliá, J.; Peiro, J. The S10/12 Job Satisfaction Questionnaire: Factorial structure, reliability and validity. *Rev. Psicol. Del Trab. Y Las Organ.* **1989**, *4*, 179–187.
23. Maslach, C.; Jackson, S.E.; Leiter, M.P. *Maslach Burnout Inventory Manual*, 3rd ed.; Consulting Psychologists Press: Palo Alto, CA, USA, 1996.

24. Mena, M.J.B.; Palacios, C.S.; Trianes, M.V. Questionnaire for negative old age stereotypes. *Rev. Multidiscip. Gerontol.* **2005**, *15*, 212–220.

25. Hambleton, R.K. Issues, designs and technical guidelines for adapting tests into multiple languages and cultures. In *Adapting Educational and Psychological Tests for Cross-Cultural Assessment*; Hambleton, R.K., Merenda, P.F., Spielberger, S.D., Eds.; Lawrence Erlbaum Associates: Mahwan, NJ, USA, 2005; pp. 3–38.

26. ITC (International Test Commission). *The ITC Guidelines for Translating and Adapting Tests*, 2nd ed.; ITC: Kolkata, India, 2016. Available online: www.InTestCom.org (accessed on 16 November 2021).

27. Rosseel, Y. lavaan: An R Package for Structural Equation Modeling. *J. Stat. Softw.* **2012**, *71*, 472–498. [CrossRef]

28. Muthén, L.K.; Muthén, B.O. *Mplus User's Guide*, 7th ed.; Muthén & Muthén: Los Angeles, CA, USA, 2015.

29. Willse, J.T. CTT: Classical Test Theory Functions, R Package Version 2.3.3. 2018. Available online: https://CRAN.R-project.org/package=CTT (accessed on 16 November 2021).

30. Revelle, W. *Psych: Procedures for Personality and Psychological Research, R package Version 1.8.12*; Northwestern University: Evanston, IL, USA, 2018. Available online: https://CRAN.R-project.org/package=psych (accessed on 16 November 2021).

31. Bates, D.; Mächler, M.; Bolker, B.; Walker, S. Fitting Linear Mixed-Effects Models Using lme4. *J. Stat. Softw.* **2015**, *67*, 1–48. [CrossRef]

32. R Development Core Team. *R: A Language and Environment for Statistical Computing*; R Foundation for Statistical Computing: Vienna, Austria, 2019. Available online: http://www.R-project.org/ (accessed on 16 November 2021).

33. Holm, S. A Simple Sequentially Rejective Multiple Test Procedure. *Scand. J. Stat.* **1979**, *6*, 65–70.

34. Satorra, A. *Scaled and Adjusted Restricted Tests in Multi-Sample Analysis of Moment Structures BT–Innovations in Multivariate Statistical Analysis: A Festschrift for Heinz Neudecker*; Heijmans, R.D.H., Pollock, D.S.G., Satorra, A., Eds.; Springer US: Boston, MA, USA, 2000; pp. 233–247, ISBN 978-1-4615-4603-0.

35. American Psychological Association, American Educational Research Association & National Council on Measurement in Education. *Standards for Educational and Psychological Testing*; American Psychological Association: Washington, DC, USA, 2014.

36. Andela, M.; Truchot, D.; Huguenotte, V. Work Environment and Elderly Abuse in Nursing Homes: The Mediating Role of Burnout. *J. Interpers. Violence* **2018**, *36*, 5709–5729. [CrossRef] [PubMed]

37. López, J.; Pérez-Rojo, G.; Noriega, C.; Velasco, C. Personal and Work-Related Factors Associated with Good Care for Institutionalized Older Adults. *Int. J. Environ. Res. Public Health* **2021**, *18*, 820. [CrossRef] [PubMed]

38. Gallego-Alberto, L.; Losada, A.; Vara, C.; Olazarán, J.; Muñiz, R.; Pillemer, K. Psychosocial Predictors of Anxiety in Nursing Home Staff. *Clin. Gerontol.* **2018**, *41*, 282–292. [CrossRef]

39. Okura, T.; Langa, K.M. Caregiver Burden and Neuropsychiatric Symptoms in Older Adults with Cognitive Impairment: The Aging, Demographics, and Memory Study (ADAMS). *Alzheimer Dis. Assoc. Disord.* **2011**, *25*, 116–121. [CrossRef]

40. Monthaisong, D. Nurses' Experiences Providing Care for People with Dementia: An Integrative Literature Review. *J. Nurs. Care* **2018**, *7*, 1–5. [CrossRef]

41. Kayser-Jones, J.; Schell, E.S.; Porter, C.; Barbaccia, J.C.; Shaw, H. Factors Contributing to Dehydration in Nursing Homes: Inadequate Staffing and Lack of Professional Supervision. *J. Am. Geriatr. Soc.* **1999**, *47*, 1187–1194. [CrossRef] [PubMed]

42. Rada, A.G. Covid-19: The precarious position of Spain's nursing homes. *BMJ* **2020**, *369*, m1554. [CrossRef] [PubMed]

43. Lee, S.J.; Kim, M.S.; Jung, Y.J.; Chang, S.O. The Effectiveness of Function-Focused Care Interventions in Nursing Homes: A Systematic Review. *J. Nurs. Res.* **2019**, *27*, 1–13. [CrossRef] [PubMed]

44. Medeiros Figueiredo, A.; Daponte-Codina, A.; Moreira Marculino Figueiredo, D.C.; Toledo Vianna, R.P.; Costa de Lima, K.; Gil-García, E. Factors associated with the incidence and mortality from COVID-19 in the autonomous communities of Spain. *Gac. Sanit.* **2021**, *35*, 445–452. [CrossRef] [PubMed]

45. Pérez-Rojo, G.; Lopez, J.; Noriega, C.; Velasco, C.; Carretero, I.; Horrillo, C. From elder abuse to good treatment towards older people. A new assessment paradigm. *Rev. Vict.* **2017**, *6*, 57–80. [CrossRef]

Article

Anti-Ageism Social Actions: Lights and Shadows

Macarena Sánchez-Izquierdo [1,*], Rodrigo García-Sánchez [2] and Rocío Fernández-Ballesteros [3]

[1]   UNINPSI Clinical Psychology Center, and Department of Psychology, Comillas Pontifical University,
     C. Universidad Comillas, 3-5, 28049 Madrid, Spain
[2]   Department of Psychology, Comillas Pontifical University, C. Universidad Comillas, 3-5, 28049 Madrid, Spain;
     201802899@alu.comillas.edu
[3]   Department of Psychobiology and Health, Autonomous University of Madrid,
     Ciudad Universitaria de Cantoblanco, 28049 Madrid, Spain; r.fballesteros@uam.es
*   Correspondence: msizquierdo@comillas.edu; Tel.: +34-(91)-7343950

Abstract: Ageism refers to the stereotypes, prejudices and discrimination towards others or oneself due to age, and it is the most prevalent type of social disadvantage, even more so than those due to gender and race, with negative effects worldwide. Ageism is an evidently real social problem that needs to be addressed and fought. Our study has two main objectives: firstly, to study to what extent programmes, projects or actions have been developed to combat ageism around the world; and secondly, to what extent they have been evaluated via the positive outcomes registered by the participants. Two different information sources were used: (a) a literature review of ageism programmes or interventions; and (b) an online questionnaire sent to international and national institutions surveying their policies or programmes against ageism. Our results show a relatively high number of actions combating ageism but a lack of a systematic evaluation of the outcomes of those actions. In conclusion, first, it is necessary to develop programmes and actions combating ageism, and the evaluation of these programmes is urgently needed in order to identify strategies that truly and effectively tackle ageism. There is a need to urge institutions to perform external evaluations of their anti-ageism social policies and to encourage scientists to conduct randomized and controlled studies.

Keywords: images; ageism; aging stereotypes; social policies; programmes; differences by countries

Citation: Sánchez-Izquierdo, M.;
García-Sánchez, R.;
Fernández-Ballesteros, R.
Anti-Ageism Social Actions: Lights
and Shadows. *Land* **2022**, *11*, 195.
https://doi.org/10.3390/
land11020195

Academic Editors:
Vanessa Zorrilla-Muñoz,
Maria Silveria Agulló-Tomás,
Eduardo Fernandez,
Blanca Criado Quesada,
Sonia De Lucas Santos
and Jesus Cuadrado Rojo

Received: 15 December 2021
Accepted: 25 January 2022
Published: 26 January 2022

Publisher's Note: MDPI stays neutral
with regard to jurisdictional claims in
published maps and institutional affiliations.

1. Introduction

Ageism is a concept defined by Butler [1] through three psychosocial dimensions, attributed and/or associated with others or oneself, based on age: stereotypes, prejudice and discrimination. These three psychological constructs referring to older adults imply how they, because of their chronological age, are perceived, what emotions they provoke, and what type of behaviours they usually experience from others.

A recent study based on the European Social Survey has found that ageism is the most prevalent type of discrimination, as it reported by almost 35% of all participants over the age of 18 [2]. Sexism and racism represent relatively stable categories that do not vary across the life course of an individual. Age, on the other hand, changes with time, and people are likely to change age group affiliation with the passage of time. Hence, in contrast to the other two "isms" (sexism and racism), everyone is susceptible to experiencing ageism if they live long enough [3,4]. Moreover, ageism might exacerbate the negative impact of other forms of discrimination, including sexism and racism, influencing the quality of older adults' life [5], with devastating effects on older women [6].

As confirmed by recent history [7], a pandemic is an extremely dangerous situation, not only because human life is at risk, but because other negative socio-economic events concerning the future are involved, and also because a pandemic is a historical event in which extreme shared emotions come to light, ageism among them. This was warned against, predicted, and denounced very early on, at the start of 2020, by an international group of social scientists [8] asking us to engage in *Avoiding Ageism and Fostering Intergenerational Solidarity*. They denounced that "with the pandemic there has been a parallel outbreak of ageism" (p. 1). This statement is based on the analysis of the public discourse in social media and public announcements made by government representatives, emphasising the extreme frailty of older adults and their cost to society and overgeneralising some of their traits of the entire older population, which is, precisely, highly diverse. Older adults are a high-risk group; thus, the health and safety behaviour modifications implemented during the pandemic have been more restrictive for them than for people of other ages, influencing, at the same time, how society views older adults (i.e., ageist stereotypes) and agitating intergenerational tensions [8–10], although chronological age is not an objective formulation for a policy implementation since we can find a diversity of health statuses within older age groups [11], even there are older people that are relatively less vulnerable than many younger people and vice versa [12].

Nevertheless, the repercussion from this overgeneralisation could have a boomerang effect in society, thus following Levy's embodiment theory [13], which posits that "the extension of older adults' negative stereotypes can be internalised by people of all ages and when these views become self-relevant, influencing older persons' beliefs about their own aging, they can detrimentally impact health" (p. 1), as a diversity of studies have supported (among others: [14–16]. Along the same line, a recent study assessing negative cultural views about older adults in 29 European countries has highlighted that they are negatively associated with active ageing both at the individual and population level. This is also supported by the negative cultural view of older adults being considered a health threat in themselves, at the population and individual level [17]. Finally, it must be emphasised that the paper by Ayalon et al. [8] does not end with negative predictions based on psychosocial theories, but with recommendations for combating ageism and increasing intergenerational solidarity. The key question here is whether predictions and recommendations coming from social science have had a positive influence on what is happening across the world.

After reviewing several studies from different sources, ageism seems to be growing as the COVID-19 pandemic continues [18–21]. At the same time, a strong call from international organisations (such as WHO [22]) demanding public and private social agents for anti-ageism movements, policies and actions has been spread around the world. Nevertheless, there is no information regarding the effect of these calls and to what extent new anti-ageism policies emerge, so much more research must be conducted because, as expressed by several institutions and scientists, "policies and laws are among the most important strategies to include in any effort to combat ageism" ([22] p. 104).

This paper aims to obtain sufficient information to evaluate the programmes and their effectiveness, including (1) to what extent policies, programmes or actions have been developed to combat ageism around the world; and (2) to what extent positive effects in the participants have been registered.

2. Materials and Methods

Two different sources were used: (a) a literature review of ageism programmes or interventions; and (b) an online questionnaire sent to international and national institutions surveying their policies or programmes against ageism.

2.1. Source of Data 1: Literature Review of Ageism Programmes
2.1.1. Procedure

The literature search strategy and selection criteria were the following. We conducted a literature review of ageism programmes with the following databases: PscyInfo and EBSCO. The search strategy was guided by a specific question: which programmes or actions have been developed to combat ageism around the world and have had positive effects? The search strategy combined key terms related to "ageism", "age discrimination", "age prejudice", "age stereotype", or "social exclusion" and "interventions" or "programme" with terms related to "elder" or "older adults" that have been used from January 2018 onwards. Eligible studies met the following inclusion criteria: (1) they evaluated an intervention designed to reduce ageism, (2) they examined at least one ageism outcome in relation to older adults and (3) they were published after 2018, when the meta-analysis *Interventions to Reduce Ageism Against Older Adults* was carried out [23].

2.1.2. Method

Database searches were conducted in July 2021. Following an initial phase of removing duplicates and completely irrelevant records not meeting the inclusion criteria, one reviewer screened records for potentially eligible titles and abstracts and subsequently reviewed full texts to determine their inclusion in the literature review.

2.2. Source of Data 2: An Online Questionnaire for International and National Institutions

2.2.1. Procedure

Civil society organisations of/for older adults who might have developed programmes or policies against ageism were searched via Google, and the following principal international associations were contacted: International Federation on Aging, Helpage International European Federation of Older Persons, AGE Platform Europe, Association Age Well Foundation and American Federation for Aging Research, and their members were also contacted. To collect the sample, in August 2020 and July 2021 we contacted the 369 identified institutions, including 69 in the European Union, 3 in the UK (United Kingdom), 21 in the USA (United States of America), 3 in Canada, 196 in Latin America, 35 in Africa, 11 in Australia and 31 in Asia ($n = 369$). A letter of invitation was sent to all these institutions, describing the purpose of our research project, along with a letter of introduction to inform and request their collaboration in the study:

> *"We are addressing you, and your Association, on behalf of the Research project XXXXXXX. This project has been developed under the auspices of the Government of the Madrid Region by the Spanish National Research Council, with the collaboration of the Autonomous University of Madrid. As you know, the COVID-19 pandemic has increased negative views about older adults related to discrimination and ageism. One of the objectives of this project, therefore, is to find strategies, policies and/or programmes, at national and international levels, with which to fight against ageism and age discrimination.*
>
> *As member of the Association, could you please answer the 6-ITEM GOOGLE FORMS SURVEY regarding this subject? It will only take you a few minutes to complete: https://forms.office.com/r/icNeccNtxz (last accessed date 28 June 2021)*
>
> *If you have any questions, and to receive information about XXXXX (http://encage-csic.es/, accessed on 1 December 2021), please contact us; and if you have any further information you could offer, please write to us at the address below".*

2.2.2. Method

The online questionnaire contained the following eight questions: 1. Name of the reporting entity. 2. Do you have any strategy, policy or programme with the purpose of fighting against ageism or age discrimination? 3. If "yes": could you provide its name? 4. What is/are its general objective/s? 5. Could you please briefly describe it (actions, materials, people involved, etc.)? 6. Do you have any public information about it (URL, flyer, etc.)? 7. Could you give us the name and contact details of the contact person? 8. In case of multiple strategies, policies or programmes, could you please repeat these six questions/answers as many times as needed?

An intensive examination of their webpages was performed, seeking evidence regarding outcome evaluation (effectiveness, efficacy and efficiency) of the programmes facilitated (different types of activities carried out, number of people involved in each project and results obtained or objective reach).

The information collected was stored following ethical requirements. The study followed the Declaration of Helsinki's principles [24] at all times. The research was carried out abiding by the ethical principles related to personal data, consent, confidentiality and their use. All participants gave their consent in writing after being informed of the voluntary nature of their participation and their total freedom to withdraw from the study. All participants received and signed the same informed consent form to participate in this study.

3. Results

In this section we firstly present the results from the literature review of ageism programmes. Secondly, we review the results from the online questionnaire for international and national institutions.

3.1. Results from the Literature Review of Ageism Programmes

The database identified 1232 total articles, and we identified 21 records for a full-text review following removal of duplicates and irrelevant records, and only 13 fulfilled our inclusion criteria.

Ageism intergenerational interventions demonstrated a significant effect on negative stereotypes about ageing [25–29], attitudes [25,30–35], positive behaviour towards older adults [30], knowledge [33], comfort [35], aging anxiety [27,28], death anxiety [34] and well-being in older adults [26], but no significant effects on working with older adults [30] or in affective attitudes toward older adults [28] were found. Although the Intergenerational Artistic Installation of Madrigal et al. [36] found an improvement in younger adults' attitudes toward older adults, but not in younger adults' attitudes toward aging. An empathy-building intervention by Bailey et al. [37] did not find self-reported ageism and aging anxiety after the intervention.

Following the World Health Organisation's outstanding strategies [22], programmes were classified into three groups: policy and law; intergenerational programmes; and educational programmes. We have not found any policy or law interventions. Ageism interventions were mostly intergenerational interventions [25,27,35,36], educational programmes [30–32,37] and interventions that included both educational and intergenerational contact components [28,29,33]. We also found cognitive behavioural therapy [34].

The majority of studies (n = 9) used a quasi-experimental design, and only four studies were randomised controlled trials (see Table 1).

Table 1. Studies included.

			No. of Participants		
Study	**Intervention**	**Design**	**Control Group**	**Intervention Group**	**Ageism Outcome**
[26]	Intergenerational	Randomised controlled trials	25 older institutionalised adults and 24 young students	21 older institutionalised adults and 24 young students	Negative stereotypes about ageing and emotional well-being in older adults
[27]	Intergenerational	Post-test only research design		61 undergraduate Students and at least 19 older adults	Stereotypes about ageing and ageing anxiety
[35]	Intergenerational programme	Quasi-experimental design. pre- and post-test control group (CG) design	n = 151: 73 older and 78 young participants	n = 161: 77 older and 84 young participants	Attitude and sense of comfort
[25]	Intergenerational contact	Quasi-experimental study design		302 participants aged 18–29	Stereotypes about ageing and ageist attitudes

Table 1. *Cont.*

Study	Intervention	Design	No. of Participants		Ageism Outcome
			Control Group	**Intervention Group**	
[36]	Intergenerational	Quasi-experimental study design		34 undergraduate students	Attitudes toward aging and attitudes toward older adults
[31]	Educational intervention	Pre-test–post-test trial model and quasi-experimental study design		38 care staff members in one nursing home	Attitudes toward aging
[30]	Educational intervention	Randomised control trial	104 undergraduate Students	83 undergraduate Students	Ageism and knowledge of aging and behaviour toward older adults and willingness to work with older adults
[32]	Educational programme	Pre-test–post-test trial model and quasi-experimental study design		134 medical students	Attitudes Toward Aging
[37]	Educational: transformative learning intervention	Quasi-experimental study		197 students (18–48 years)	Stereotypes about ageing
[33]	Education about aging, extended intergenerational contact, combined condition	Randomised control trial	Study 1: 88 undergraduates Study 2: 132	Study 1: 266 undergraduates: 86 education, 86, Ext. contact 94 combined Study 2: 505 community participants ages 18–59: 122 education, 125 ext. contact, 128 combined	Attitude and stereotypes and aging anxiety and anxiety about interacting with older Adults and aging knowledge
[28]	Interventions that included both educational and intergenerational contact components	Quasi-experimental study design		14 undergraduate students	Stereotypes about ageing and aging anxiety and psychological concerns about aging and affective attitudes toward older adults
[34]	Cognitive behavioural therapy	Randomised controlled trial	55 nurses	55 nurses	Death anxiety and ageism
[29]	Interventions that included both educational and intergenerational contact components	Quasi-experimental study design		23 undergraduate students	Stereotypes about ageing

3.2. Results from the Online Questionnaire for International and National Institutions

Of the 369 institutions, 21 agreed to participate in our research (Table 2), signing an informed consent form and providing contact information and filling out the online questionnaire (see link to the online questionnaire in the procedure section), but three were not carrying out programmes focused on older adults. Finally, 18 institutions carrying out some programmes focused on older adults were included (Figure 1). The data provided by participants were carefully managed, preserving anonymity and confidentiality.

Again, following the World Health Organisation's outstanding strategies [22], programmes were classified into three groups: (1) policy and law; (2) intergenerational programmes; and (3) educational programmes. Additionally, we added two more blocks: a fourth block regarding social participation, and a fifth regarding anti-discrimination programmes during the COVID-19 pandemic.

Table 2. Institutions that completed the online questionnaire and their programmes.

	Country	Institution	Programmes
	USA (New York)	**Long Term Care Community Coalition**	**Policy and law**
European Union	European Union	HelpAge International	• Policy and law: defence of human rights. • Policy and law: health and active ageing programmes. • Intergenerational programmes. Social participation • Educational interventions
	Greece	50plus Hellas	• Policy and law: defence of human rights. • Policy and law: health and active ageing programmes. • Intergenerational programmes • Anti-discrimination programmes during the COVID-19 pandemic
	Spain	Euskofederpen: Territorial Federation of Provincial Associations of Pensioners and Retirees of Álava, Guipúzcoa and Vizcaya	• Policy and law: health and active ageing programmes. • Intergenerational programmes. Social participation
	Italy	National Association of Social Centres, Committees of the Elderly, and Gardens (ANCeSCAO—APS)	• Intergenerational programmes • Anti-discrimination programmes during the COVID-19 pandemic
	Portugal	APRe! Associação de Aposentados, Pensionistas e Reformado	• Policy and law: defence of human rights
	Serbia and Austria	Serbian Red Cross	• Policy and law: defence of human rights • Intergenerational programmes • Educational interventions • Anti-discrimination programmes during the COVID-19 pandemic
	Ukraine	"Turbota pro-Litnih v Ukraine" (Age Concern Ukraine)	• Policy and law: defence of human rights • Policy and law: health and active ageing programmes. • Intergenerational programmes. Social participation • Anti-discrimination programmes during the COVID-19 pandemic
	Russia	Eduard Kariukhin	• Policy and law: defence of human rights • Policy and law: health and active ageing programmes.
	Russia	Irina Khalay Association	• Intergenerational programmes • Educational interventions
Asia	Bangladesh	Nobo Jatra Foundation	• Policy and law: defence of human rights
	Nepal	National Senior Citizen Federation (NASCIF)	• Intergenerational programmes

Table 2. *Cont.*

	Country	Institution	Programmes
	USA (New York)	**Long Term Care Community Coalition**	**Policy and law**
Africa	Togolese Republic	Association Nos Années de Vie (ANAVIE)	• Policy and law: defence of human rights • Intergenerational programmes. Social participation
	Liberia	Coalition of Caregivers and Defenders of the Elderly in Liberia (COCAEL).	• Policy and law
	Uganda	Uganda Reach the Aged Association (URAA)	• Policy and law
	Kenia	Kenyan Ministry of Labour, Social Security and Services	• Intergenerational programmes.

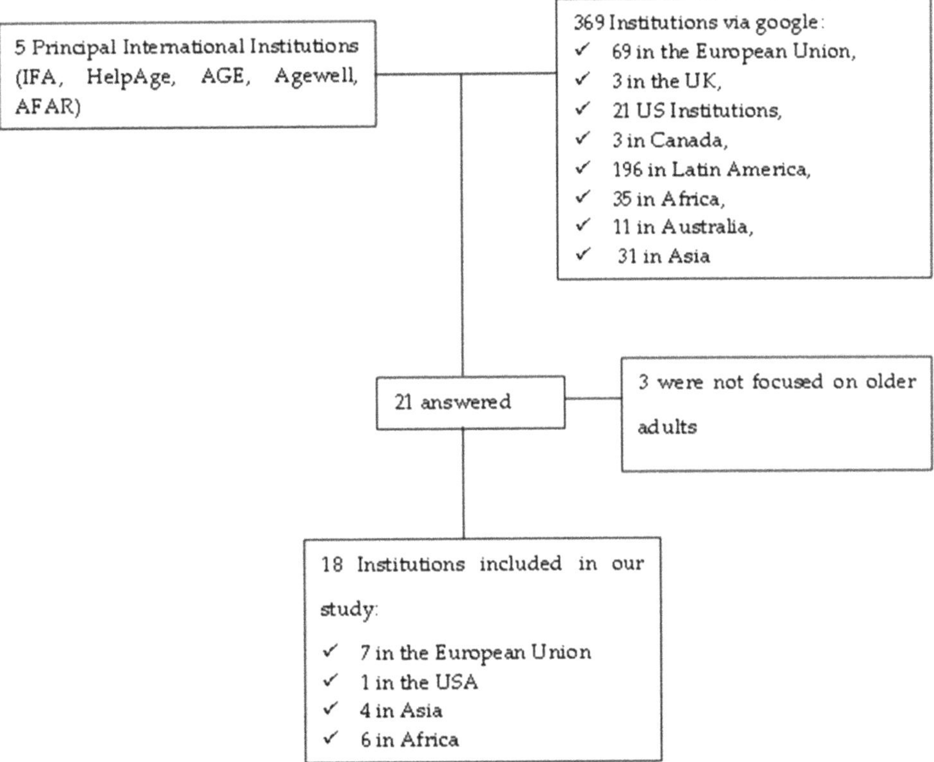

Figure 1. PRISMA from the online questionnaire by international and national institutions.

Here, we describe the main features (active principles) of each programme category, providing examples of them, as well as, in the case that evaluation data are presented (directly through the questionnaire or in their annual reports), providing information about the programme's effects.

1. Policy and law

Here, we found two programme blocks: (1) programmes focused on the defence of human rights and the defence of rights programmes; and (2) programmes focused on health and active ageing programmes.

Programmes focused on the defence of human rights and the defence of rights programmes, logically, contain an essential principle: the protection of the rights of older people.

In the European Union, the Associação de Aposentado, Pensionistas e Reformados of Portugal (Association of Retired Persons and Pensioners—APRe!) defence of rights programme stands out, with its programme "An APRe! on the move" and efforts to support all demands for better living conditions for retirees and pensioners, to pressure the political power regarding pensions and warning actions (to pressure the political power, dialogue with central and local authorities to define policies) to improve continuous and long-term palliative care. The Serbian Red Cross, with its programme "Empowerment of older women: prevention of violence by challenging social norms in Serbia and Austria (EmPreV)", carried out 35 workshops to sensitise health and social staff and volunteers about this problem, including 35 information sessions with older women.

The 50plus Hellas institution in Greece, with its programme "The Europe we want is for all ages (2019)" and the Ukrainian Charity's program "Turbota pro-Litnih v Ukraine" (Age Concern Ukraine), focused on revealing and preventing abuse towards older people with the *Elder Abuse Prevention* programme in the Ukraine, a programme of 300 volunteers who brought these mistreatments to light. In addition, the Strengthening the Voices of Older People in Ukraine programme helped older people better understand their rights and learn how to defend them. To this end, they questioned 1500 people who suffered from age discrimination and low self-esteem.

Two programmes carried out by HelpAge International stand out: (1) the programme "Making older women count", where they monitored the achievement of gender equality in the Sustainable Development Goals (SDG) segregated by sex and age, and other reasons of discrimination; and (2) the programme on ageing and SDGs: six steps for the inclusion of older people to promote the rights and inclusion of older people;programme 1500 members were directly participating in the campaign, but there were 350,000 more participants engaging with the programme via social networks, and 7.5 million readers were registered for ageism articles.

In addition, the Russian institution Eduard Kariukhin was tasked with informing society and the state about ageism and its negative consequences on the quality of life of older people, seeking a legislative, executive and civil response.

Likewise, in Asia, Bangladesh wanted to focus on document and resource management and strengthening nursing home establishments, obviously not forgetting to safeguard of the rights of this very vulnerable group in this continent. The Nobo Jatra Foundation of Bangladesh also collaborated with these types of programmes, implementing the rights of older people via advocacy, education and training, research, consulting and direct support services. In addition, the foundation organised a meeting to share and observe the International Day of the Elderly (about 50 older adults, men and women attended).

In Africa, the Association Nos Années de Vie (ANAVIE) of the Togolese Republic stands out with its programme that developed policies to improve the quality of life of the elderly in Liberia. However, the most striking of these were two top-down proposals. One of them advocated for the inclusion of older people in the Affirmative Action Law Project to improve services and rights and to finance national programmes. They provided services to the elderly during the Ebola crisis (2014) by creating the Coalition of Caregivers and Defenders of the Elderly in Liberia (COCAEL). Uganda Reach the Aged Association (URAA), with the National Endowment for Democracy (NED) and VOICE programmes, understood and challenged age discrimination. Likewise, along with the work of the paralegal advisers and support for access of the rights of older refugees in Adjumani, they supported more senior people and their homes with access to justice and vindication of rights; thanks to this, they sensitised at least 400 older people to cases involving the violation of rights. Likewise, the advisers resolved 51 issues of gender violence.

In short, nine international institutions have fought to defend the rights of older people. For this, programmes have been carried out to empower this social group and make their rights, which are not currently being respected, visible. In addition, the HelpAge

International association focused on the rights of older people, in general, and older women, in particular. This social group is discriminated against by both age and gender.

The second programme block is focused on health and active ageing programmes. In USA, the New York Long Term Care Community Coalition is dedicated to improving nursing home care to prevent neglect and abuse of the elderly in residential care settings.

In the European Union, the HelpAge International stands out, which, via the Health Outcomes Tool and Better Health programmes, has managed to serve 2 million older people; provided community help to 170,000 older people; and improved facilities to serve 30,000 people with disabilities and 20,000 refugees. Older people benefited from all the improvements in health, rehabilitation and psychological counselling service.

In addition, its Stay Active festivals programme, in which 1140 older people (75% women) participated, has created festivals and workshops in several cities in Greece to empower older people to engage in a more active and healthier lifestyle. In addition, 50plus Hellas in Greece created Android applications to obtain valid information on issues related to seniors with the Stay Active programme.

In Russia, the active ageing programmes stand out. These aim to raise awareness concerning the need to improve the quality of life of older people by increasing their knowledge of civil law. In addition, they celebrate events such as the International Day of the Elderly. The Active Generation (62 events and 200 participants) and the Golden Age School (30 events and 150 participants) programmes received donations from individuals.

Additionally, in the European Union, the Ukrainian Charity's progamme "Turbota pro-Litnih v Ukraine" (Age Concern Ukraine) wanted to promote an active and healthy lifestyle among older people thanks to the programme "Support for the Madrid International Plan of Action on Aging Relating to Ukraine" (UNFPA). Finally, in Spain, Euskofederpen: Territorial Federation of Provincial Associations of Pensioners and Retirees of Álava, Guipúzcoa and Vizcaya increased the voluntary activity of older people in their social centres, thus creating a network of social intervention programmes at a regional level aimed at active and healthy ageing, and cognitive stimulation programmes for adults with or without cognitive impairment through the Burulogy programme.

Various programmes have been developed within the African continent. For example, the Association Nos Années de Vie (ANAVIE) (Togolese Republic), with its "Aging in good health" programme, was able to bring together 35 sick, older people to carry out regular monitoring at their homes, as well as emergency tasks such as screening for contagious diseases.

Some institutions carried out more specific programmes aimed at universal access to health. Along these lines, we have the Community Initiative for Peace and Development: food from South Sudan with the provision of non-food items (NFI) materials to the elderly, or the Uganda Reach the Aged Association (URAA) with the social protection of older people and Saidia Wazee Karagwe (SAWAKA) (Tanzania) facilitating access to equitable and affordable health services.

Within this group, we found associations that have sought to promote policies eradicating any form of discrimination against older people. For example, Kenya's Ministry of Labour, Social Security and Services incorporates laws and policies, including geriatric research and studies in education and training curricula, to expand the health care support system for older people, and seeks to guarantee access to a healthy diet for the elderly.

In summary, programmes have been collected from eleven international institutions whose objectives were to promote active and healthy ageing of the elderly and avoid discrimination in access to health programmes due to age. The most notable association has been HelpAge International, which has served more than two million older adults thanks to their international programmes. In addition, they managed to provide community help, improve health facilities and promote a more active and healthy life in older people.

2. Intergenerational programmes

In continuation, we highlight the most outstanding initiatives regarding intergenerational programmes.

Within the European Union, Greece and Italy stand out, including 50plus Hellas from Greece, who, via the DIGITOL and "the value of time" programmes, work with the elderly to develop digital skills and encourage recognising what older people bring to the young. In Italy, the only entity that offers some data is the National Association of Social Centres, Committees of the Elderly, and Gardens (ANCeSCAO—APS). This is an association with activities that, through the Time to Care programme (still running), promotes generational exchange using telephone contact, assistance activities and telecare service. The number of participants involved has varied according to the area of Italy. Still, the most critical data were available for Matera, where 1500 young people under 35 years of age participated in this intergenerational programme. The municipality of Castenaso carried out a task clocking up 250 h.

In Russia, the Irina Khalay Association develops the Active Generation and School of the Golden Age programmes, which are dedicated to sharing the life experience of older people with younger people. On the other hand, we have the blog published by the Serbian Red Cross on the importance of intergenerational solidarity.

In Asia, the National Senior Citizen Federation (NASCIF) brings together various associations throughout Nepal, developing multiple programmes. Within the intergenerational area, it seeks to promote dialogue and discourse to reduce the generation gap.

In summary, programmes have focused on fostering mutual respect between generations and promoting the participation of older people in society, valuing their knowledge and skills. In addition, the Italian institution ANCeSCAO—APS attracted many participants in the Italian municipalities of Matera and Castenaso. It advocated promoting generational exchange using electronic devices, among other ideas.

3. Educational interventions

The educational interventions we refer to (all in the European Union) focus on self-development. Personal development programmes aim to develop the skills and self-development of older people as an essential principle.

The HelpAge International institution has carried out three programmes: (1) the Older Citizen Monitoring (OCM) programme, for which they developed a training manual to strengthen the knowledge and skills of the elderly and in which 200 older refugees participated. (2) With the Rohingya Response Project, they were able to help 2000 older refugees with skills training, as well as providing them with loans to help them start fishing net businesses, setting up small shops, etc. (3) With the Community Safe Spaces programme, they created a meeting space in which older people could learn new skills and carry out social activities, serving 700 older people altogether.

In Russia, the Irina Khalay institution held forums and round tables and master classes under the sewing programme for older women to increase this group's legal literacy and quality of life.

Finally, the Serbian Red Cross has contributed to digital inclusion via older peoples' access to information and communication technologies. There were ten older adult participants in each of the 29 Serbian municipalities.

In summary, three institutions for self-development stand out, whose objectives were to develop the skills and regular progress of the elderly. The programmes are from a Russian institution, a Serbian institution and HelpAge International, which carried out three high-participation programmes to train older people in new skills.

4. Social participation programmes

Some institutions have focused on social participation. Logically, social participation programmes focus on an essential principle: promoting participation in society by older people, and, therefore, intergenerational contact.

In the European Union, the entity HelpAge has carried out two programmes: (1) The first, called Maintaining Our Dignity, with the participation of 300 older people, was tasked with presenting HelpAge and its network members in the 11th session of the OEWG (open-ended working group) on ageing and your rights to justice and work; (2) the second programme, called Improving urban environments for older people, was an investigation of urban problems, in which 1300 people participated.

Additionally, in Europe, the Ukrainian Charity association "Turbota pro-Litnih v Ukraine" (Age Concern Ukraine), on the other hand, developed a collaboration with local authorities for older people to help them decide on local priorities. This was achieved thanks to the programme "Improving Government Accountability Through Older Citizens' Monitoring in Ukraine".

Additionally, in Spain, Euskofederpenha has actively worked on transmission between institutions and social centres, claiming the right of older people to be active and improving conditions in social centres, as well as a digital transformation so that they remain connected.

In Africa, the Association Nos Années de Vie (ANAVIE) of the Togolese Republic stands out. This Association has created clubs for older people (Tabligbo & Clubs d'Agomé-Kpodzi et de Kpalimé-Tsivé). It has encouraged the participation of older people in social and cultural events within the community. It has also carried out psychotherapeutic programmes, psychological consultations and capacity-building workshops. For example, older adults can learn to use digital banking channels such as mobile money accounts. It has also been concerned about the fight for their rights and the fight against abuse, ensuring a minimum income for older people. It should be noted that this entity has published various documents and reports such as *Why is it time for a convention on the rights of the elderly* and the *Report of the conference on strengthening the rights of the world's elderly*. In addition, in Kenya, the Kenyan Ministry of Labour, Social Security and Services has developed the adult education programme In Liberia, the Centre for Community Advancement and Family Empowerment (CECAFE), which provides social services to older adults in order to develop partnerships with communities and government and non-governmental organisations to ameliorate older peoples' social problems and challenges.

In summary, five institutions at the international level have opted for the participation of older people in society through different programmes. The most prominent is HelpAge International, where 300 more senior people and 1300 people of all ages participated. All the institutions involved have fought, through their different programmes, to make older people visible and encourage their participation in society.

5.   Anti-discrimination programmes during the COVID-19 pandemic

There have also been more specific programmes, such as those against discrimination during the COVID-19 pandemic, such as one carried out by HelpAge International, where they developed a series of guidelines for the elderly, families, caregivers and nursing homes to support them, as well as making the age discrimination faced by many older people throughout the pandemic visible. During the pandemic, these programmes against discrimination were joined by the National Association of Social Centres, Aged and Garden Committees (ANCeSCAO—APS) in Italy, the Uganda Reach the Aged Association (URAA), Serbian Red Cross and 50plus Hellas in Greece.

In addition, programmes were carried out for the provision of material and food that supplied sanitary and health products and hygiene kits and supported the elderly confined at home in daily tasks, highlighting the work of Europe, specifically the Ukrainian Charity "Turbota pro-Litnih v Ukraine" (Age Concern Ukraine), which was tasked with delivering food and hygiene items to more than 1000 older people in various non-government-controlled areas in Donbass. The donors were Germany, with the EVZ Grant from the German Federal Fund with 50,000 euros and France, with the French Embassy contributing 5000 euros.

In summary, during the COVID-19 pandemic, anti-ageism programmes have been carried out to make the existing discrimination visible, in addition to providing health and nutritional aid. Eight institutions worldwide participated.

4. Discussion

Research on ageism is important given that ageism may have a negative effect on both individuals and society, and it has increased since the COVID-19 pandemic [9,10,38] Moreover, the WHO has identified reducing ageism as a key target for improving human health [39]. Although we can find numerous interventions and programmes to combat ageism, the overall effectiveness of these programmes is still unknown [40], especially considering the variation across different cultures and among older adults themselves [23]. This paper has examined policies and programmes developed to combat ageism through two sources of information: a literature review, searching for updated academic studies and an online questionnaire, addressed to worldwide institutions to obtain information about programmes carried out in different countries (and cultures) and their results.

The literature review has shown, as Burnes et al. [23] have already highlighted, that ageism interventions are mostly intergenerational interventions, educational programmes and interventions that include both educational and intergenerational contact components. Burnes et al. [23] reviewed 63 studies and showed that these three interventions had an effect on attitudes towards ageing (including stereotypes towards ageing and prejudice towards older people), and increasing knowledge about ageing. Additionally, they showed that the greatest change in attitudes towards aging, knowledge of aging and well-being in dealing with older adults occurred when education and intergenerational contact were combined, but these changes did not occur in terms of anxiety about their own aging, or in terms of interest in working in the field of the elderly.

Our findings suggest that these interventions reduced negative stereotypes about ageing and improved attitudes toward ageing, combating ageism and, in turn, improving the health and well-being of older people, although only four of the studies developed were randomised, with the rest using a quasi-experimental design. As Burnes et al. [23] concluded in their review, "more rigorous designs to examine the effects of interventions are strongly recommended" (p. 10). In contrast, we did find that an educational intervention might provoke more positive behaviour toward older adults, and more willingness among social work students to consider a career in geriatrics [30]. Furthermore, intergenerational service learning has emerged as an intervention that can significantly decrease ageism scores in undergraduate students [37], and in addition of the opportunity to engage in "real-life" contact experiences with older adults, the students also engaged in personal exploration of older adults' service settings. Having the opportunity to increase knowledge of ageing and familiarity with the reality of the older adults' world may encourage empathy and a more positive vision of older adults in students and younger people [41]. The promotion of empathy emerges as a construct that might provoke the willingness to work with older adults.

On the other hand, most of the studies reviewed by Burnes et al. [23] were from the United States; therefore, the authors recommended that research be undertaken in different places around the world, taking into account the possible variations across different cultures. In our study, we performed an updated literature review of ageism programmes from different databases, searching for studies around the world. We found a study comparing ageism during the COVID-19 pandemic in three English-speaking countries (Australia, the United Kingdom and the United States) finding no variation [42], but there were no more studies from different cultures. This is one of the reasons for our second data source: an online questionnaire sent to international and national institutions surveying their policies or programmes against ageism worldwide.

The online questionnaire sent to the international and national institutions that we surveyed regarding their policies or programmes against ageism was answered by a small percentage of the entities, although it was sent to five continents.

The institutions that completed the questionnaire were from all around the world, developing different programmes and actions in five blocks during the COVID-19 pandemic: (1) policy and law; (2) intergenerational programmes; (3) educational programmes; (4) social participation programmes, and (5) anti-discrimination programmes. Additionally, although for some of them we could ascertain how many people participated or the number of programmes developed, we found no analysis of how the situation changed (an experimental design with a pre- and a post-evaluation), or an analysis of the outcomes.

The discrimination that older adults experience increases in the case of women, who are subject to greater discrimination [43,44]. Therefore, gendered ageism is an increasing issue that must be attended to. In this study we found that there are international institutions that are working towards the human rights of older women, which indicates that society is going in the right direction, although these programmes have not been evaluated.

In sum, there is no global analysis of strategies that can work towards addressing ageism, and, therefore, it is necessary to analyse what strategies and policies exist. Although Dixon and Sibthorpe [45] emphasised the importance and role of policy makers and scientists in promoting health and well-being, there is still a need to evaluate the programmes and policies that have been carried out in order to identify the strategies that truly and effectively tackle ageism.

Social policy is concerned with the ways societies across the world meet human needs for security, education, work, health and well-being. Social policy has the goal of addressing how societies respond to important challenges, such as the growth of the elderly population, demographic and economic change, poverty, pensions, health and social care. Social policy considers the different roles of national governments, the family, civil society, the market and international organisations in providing services and support across the course of life of individuals.

In the present moment, living in the midst of the COVID-19 pandemic, it is even more important, since we have seen evidence of openly ageist discourses, and we have even seen hashtags such as #BoomerRemover on social media platforms [46], or other ageist phrases, such as coffin dodger and boomer doomer, expressing younger adults' hostility toward older adults, blaming older adults as the culprits of all the health and safety behaviour modifications implemented during the pandemic [42]; thus, this reflects younger peoples' views that the pandemic is an issue for older people ("old people problem") [42]. These ageist discourses and social media messages are contributing to feelings of worthlessness in older people and a sense of having no value [10], and at the same time, it has reinforced paternalistic perceptions that infer older adults are fragile and vulnerable [41,47].

In this sense, some society support (mainly long-term care and adult children) has been directed towards protect older adults with overaccommodative polices and/or behaviours (such as avoiding contact with or sequestering older adults) that may undermine older adults' autonomy, their right to make their own health-based decisions and even their social and emotional wellbeing [21,47]. Vale et al. [48] showed that both hostile and benevolent ageism predict divergent responses to the pandemic; while hostile ageism is associated with less pandemic-related health and safety precautions, benevolent ageism is related to increased behaviour changes, but only as a result of increased pandemic-related fear. These findings are fundamental for developing programmes or policies representing older people in the context of the pandemic, as well as in deciding which messages to use, because they may have indirect consequences on how older people are viewed and thus treated [42,47,48].

Furthermore, taking into account the determinants that seem to contribute to other and self-directed forms of ageism, interpersonal contact with older adults emerges as fundamental, specifically the quality of the contact over frequency and in regards to the importance of how older individuals are presented [49]. Therefore, several aspects need to be addressed: firstly, stimulating intergenerational contact in a positive context and secondly, promoting the presentation of more positive images of older adults. These are themes that must be taken into consideration by social policies.

In our study, we determined that it is necessary to carry out more rigorous studies on the effect of anti-ageism programmes. While it is true that studies carried out by scientists are more "objective", only a small number of them are randomised controlled trials, which are necessary to obtain proven data on the benefits of these programmes. On the other hand, although we can only discuss the institutions that answered our questionnaire (constituting a small percentage of the existing entities), they did not offer the results of the programmes carried out, and in the few cases that they did, if at all, it was only via the annual reports of the entities. Nonetheless, we can affirm that there are entities on four continents that carry out policy and laws, intergenerational and social participation and educational and specific programmes during the COVID pandemic: Asia, Africa, America and Europe, although we only obtained a small number of answers to our online questionnaire.

As Scriven stated ([50], p. 1) "Evaluation is the process of determining the merit, worth and value of things, and evaluations are the products of that process" and is considered a basic methodology of social sciences. Furthermore, social policies refer to those sets of ideas, plans, projects or programmes developed for responding to social needs, aimed at reducing inequalities in access to services between citizens, independently of their socio-economic status, race, ethnicity, migration status, gender, sexual orientation, disability or age.

Weiss [51] is one of the pioneers in linking social policies and programme evaluation through the concept of policy analysis, having created the journal *Educational Evaluation and Policy Analysis* in 1979, which addresses many of the crucial and complex issues that concern evaluators and public managers. Weiss explains how the political context in which social programmes work and evaluations are implemented could help programme managers and evaluators improve. More specifically, she examines how evaluation research can help improve public policy making and how programme evaluation studies can be utilised. She also describes theory-based evaluation and why it matters, as well as the implications of the political nature of public programmes. In Spain, Agulló-Tomás et al. [52] have analysed 439 programmes for older people caregivers, which were mostly carried out by non-profit institutions and public Spanish entities; the authors highlighted several identified weaknesses: the lack of adaptation to different contexts (social and cultural and political and institutional), a lack of specific targets as outcome indicators and the need for more outcome evaluations and better quality of the evaluation itself.

The methodology resources do not appear to have been extensively or systematically utilised in the study of social policies. As pointed out previously by Officer & de la Fuente-Núñez [40], programmes and policies need to be evidence-based to understand the nature of the problem, who is affected and how, and which actions have been successful. Furthermore, they should be supported via long-term funding to ensure sustained actions to combat ageism. Therefore, programmes would benefit from the optimal use of evidence, which could reliably inform practice and policy.

The COVID-19 pandemic has had an overwhelming impact on older persons. One lesson to learn is that addressing ageism is critical for creating a more equal world in which the dignity and rights of every human being are respected and protected. As several authors have highlighted, the pandemic provides an optimal opportunity to work on ageism in natural situations, not in artificial research scenarios [8,21,42,48].

In sum, the state of the art of anti-ageism social policies could be summarised as some lights and plenty of shadows. We are going to discuss the light and shadows at an academic, social and political level.

In academic studies, we found the following lights: (1) Interpersonal contact with older adults is crucially important, specifically the quality of the contact over frequency and also the importance of how older individuals are presented. (2) The greatest changes occurred when education and intergenerational contact are combined, which may reduce negative stereotypes about ageing and improve attitudes toward ageing, combating ageism and, in turn, improve the health and well-being of older people. Moreover, educational intervention might provoke more positive behaviour toward older adults and more willingness in students to consider a career in geriatrics.

On the other hand, we found the following shadows: (1) There is a lack of studies across different cultures. There is a need for more worldwide research, so scientists can analyse the possible variation across different cultures, and therefore, design programmes and policies personalized to each culture, if necessary. Older adults' images may vary between cultures; therefore, it is necessary to take this into consideration in the design of these programmes, in terms of their effectiveness. For example, a programme designed for Asian people, in an occidental country, may not be as effective as in an oriental country. (2) Although there are numerous studies focused on combating ageism, very few of them use randomised and controlled studies.

Focusing on social and political programmes, there are a number of programmes combating ageism around the world, although there is a need to know what is really working and what is not. Between the lights we found: There are numerous programmes and actions carried out in different countries in five blocks: (1) policy and law; (2) intergenerational programmes; (3) educational programmes; (4) social participation programmes, and (5) anti-discrimination programmes during the COVID-19 pandemic

Between the shadows we found: There is a lack of more rigorous studies on the effect of anti-ageism programmes. There is no global analysis of the strategies that can work towards addressing ageism and, therefore, it is necessary to analyse the strategies and policies that are being carried out. We found numerous programmes worldwide, but there is no evidence of the overall efficacy, effectiveness and efficiency of such programmes.

4.1. Limitations

The data obtained are not fully representative, since only a small percentage of the institutions which work for older adults answered our questionnaire. We must take into account that this study was carried out during the COVID-19 pandemic, and the information obtained is important for reflecting on two important points. First, institutions should be urged to perform external evaluations of their anti-ageism social policies. Furthermore, institutions should also self-evaluate their programmes regarding policy and comments that might underlay ageist attitudes and promote age discrimination: for example, is an age limit proposed to restrict access to intensive care? Should older people isolate themselves instead of requiring widely implemented social distancing measures? [12,47]. Secondly, it is necessary to urge scientists to use randomised and controlled studies.

4.2. Future Research

Future academic studies should focus on several important points. Firstly, ageism programmes around the world should be evaluated, taking into account the possible variation across different cultures. Moreover, it is important to urge socio-political scientists to standardize outcome evaluation of all types of socio-political actions and programmes trying to implement empirical methods and controlled studies.

Secondly, future studies need to evaluate the programmes and policies that have been carried out in order to identify the strategies that truly and effectively tackle ageism; there are many programmes carried by different institutions and international government, but they have not been evaluated.

Regarding social policies, there are two themes that need to be taken into consideration: firstly, stimulating intergenerational contact in a positive context and secondly, promoting the presentation of more positive images of older adults. Programmes and policies need to be evidence-based to understand the nature of the problem: who is affected and how, and which actions have been successful.

5. Conclusions

Research about the evaluation of programmes combating ageism emerges as a fundamental task worldwide. Although we can find numerous actions, interventions and programmes combating ageism worldwide, the overall results of these programmes are still unknown. It is important to standardize outcome evaluations of all types of socio-political actions and programmes, as well as to evaluate programmes and policies that have been carried out in order to identify those strategies that truly and effectively impact ageism.

Author Contributions: R.F.-B. and M.S.-I. designed the systematic review and online questionnaire. M.S.-I. extracted the information from the studies from the systematic review. R.G.-S. and M.S.-I. extracted the information from the online questionnaire. R.F.-B., M.S.-I. and R.G.-S. contributed to the interpretation of the results. R.F.-B., M.S.-I. and R.G.-S. wrote the manuscript. All authors have read and agreed to the published version of the manuscript.

Funding: This work is part of the R&D Activities Program ENCAGEn-CM: "Active Ageing, Quality of Life and Gender. Promoting a positive image of old age and ageing, and combating ageism". Funded by the Community of Madrid, Programs of R&D on Social Sciences and Humanities, and co-financed with the European Social Fund (Ref. H2019/HUM-5698) (https://encage-cm.es/, accessed on 1 December 2021).

Informed Consent Statement: Informed consent was obtained from all subjects involved in the study.

Conflicts of Interest: The authors declare no conflict of interest.

Statement of Ethical Approval: This type of study does not require approval by an ethics committee. It is not a direct intervention, but a self-report. All participants were informed of the research and the voluntary nature of participation and were informed about the possibility of dropping out of the survey at any time. They provided informed consent prior to accessing the questionnaire. Additionally, all information is encoded or anonymised.

References

1. Butler, R.N. Age-Ism: Another Form of Bigotry. *Gerontologist* **1969**, *9*, 243–246. [CrossRef]
2. Ayalon, L. Feelings towards Older vs. Younger Adults: Results from the European Social Survey. *Educ. Gerontol.* **2013**, *39*, 888–901. [CrossRef]
3. Palmore, E. The Ageism Survey: First Findings. *Gerontologist* **2001**, *41*, 572–575. [CrossRef] [PubMed]
4. Palmore, E.B. Ageism Comes of Age. *Gerontologist* **2003**, *43*, 418–420. [CrossRef]
5. Kim, J.-H.; Song, A.; Chung, S.; Kwak, K.B.; Lee, Y. The Comparative Macro-Level Agism Index: An International Comparison. *J. Aging Soc. Policy* **2021**, *33*, 571–584. [CrossRef]
6. Global AgeWatch Index 2015: Insight Report, Summary and Methodology | Reports | Global AgeWatch Index 2015. Available online: https://www.helpage.org/global-agewatch/reports/global-agewatch-index-2015-insight-report-summary-and-methodology/ (accessed on 6 December 2021).
7. Fineberg, H.V. Pandemic Preparedness and Response—Lessons from the H1N1 Influenza of 2009. *N. Engl. J. Med.* **2014**, *370*, 1335–1342. [CrossRef] [PubMed]
8. Ayalon, L.; Chasteen, A.; Diehl, M.; Levy, B.R.; Neupert, S.D.; Rothermund, K.; Tesch-Römer, C.; Wahl, H.-W. Aging in Times of the COVID-19 Pandemic: Avoiding Ageism and Fostering Intergenerational Solidarity. *J. Gerontol. Ser. B* **2020**, *76*, e49–e52. [CrossRef]
9. Ayalon, L. There Is Nothing New under the Sun: Ageism and Intergenerational Tension in the Age of the COVID-19 Outbreak. *Int. Psychogeriatr.* **2020**, *32*, 1221–1224. [CrossRef]
10. Brooke, J.; Jackson, D. Older People and COVID-19 Isolation, Risk and Ageism. *J. Clin. Nurs.* **2020**, *29*, 2044–2046. [CrossRef] [PubMed]
11. Oliver, D. David Oliver: What the Pandemic Measures Reveal about Ageism. *BMJ* **2020**, *369*, m1545. [CrossRef]
12. Fletcher, J.R. Chronological Quarantine and Ageism: COVID-19 and Gerontology's Relationship with Age Categorisation. *Ageing Soc.* **2021**, *41*, 479–492. [CrossRef]
13. Levy, B. Stereotype Embodiment: A Psychosocial Approach to Aging. *Curr. Dir. Psychol. Sci.* **2009**, *18*, 332–336. [CrossRef] [PubMed]
14. Levy, B.R.; Slade, M.D.; Kunkel, S.R.; Kasl, S.V. Longevity Increased by Positive Self-Perceptions of Aging. *J. Personal. Soc. Psychol.* **2002**, *83*, 261–270. [CrossRef]
15. Chasteen, A.L.; Bhattacharyya, S.; Horhota, M.; Tam, R.; Hasher, L. How Feelings of Stereotype Threat Influence Older Adults' Memory Performance. *Exp. Aging Res.* **2005**, *31*, 235–260. [CrossRef] [PubMed]
16. Siebert, J.S.; Braun, T.; Wahl, H.W. Change in Attitudes toward Aging: Cognitive Complaints Matter More Than Objective Performance. *Psychol. Aging* **2020**, *35*, 357–368. [CrossRef] [PubMed]
17. Fernández-Ballesteros, R.; Olmos, R.; Pérez-Ortiz, L.; Sánchez-Izquierdo, M. Cultural Aging Stereotypes in European Countries: Are They a Risk to Active Aging? *PLoS ONE* **2020**, *15*, e0232340. [CrossRef] [PubMed]
18. Bergman, Y.S.; Bodner, E.; Cohen-Fridel, S. Cross-Cultural Ageism: Ageism and Attitudes toward Aging among Jews and Arabs in Israel. *Int. Psychogeriatr.* **2013**, *25*, 6–15. [CrossRef]
19. Kornadt, A.E.; Albert, I.; Hoffmann, M.; Murdock, E.; Nell, J. Perceived Ageism during the Covid-19-Crisis Is Longitudinally Related to Subjective Perceptions of Aging. *Front. Public Health* **2021**, *9*, 679711. [CrossRef]
20. Walker, A. Why the UK Needs a Social Policy on Ageing. *J. Soc. Policy* **2018**, *47*, 253–273. [CrossRef]
21. Morrow-Howell, N.; Galucia, N.; Swinford, E. Recovering from the COVID-19 Pandemic: A Focus on Older Adults. *J. Aging Soc. Policy* **2020**, *32*, 526–535. [CrossRef]

22. World Health Organization. *Global Report on Ageism*; World Health Organization, Ed.; World Health Organization: Geneva, Switzerland, 2021.

23. Burnes, D.; Sheppard, C.; Henderson, C.R.; Wassel, M.; Cope, R.; Barber, C.; Pillemer, K. Interventions to Reduce Ageism against Older Adults: A Systematic Review and Meta-Analysis. *Am. J. Public Health* **2019**, *109*, E1–E9. [CrossRef] [PubMed]

24. WMA—The World Medical Association. Declaration of Helsink: Ethical Principles for Medical Research Involving Human Subjects. *JAMA* **2013**, *310*, 2191–2194. [CrossRef] [PubMed]

25. Cadieux, J.; Chasteen, A.L.; Packer, D.J. Intergenerational Contact Predicts Attitudes toward Older Adults through Inclusion of the Outgroup in the Self. *J. Gerontol. Ser. B* **2018**, *74*, 575–584. [CrossRef] [PubMed]

26. Carcavilla, N.; Meilán, J.J.G.; Llorente, T.E. The Impact of International Videoconferencing among Older Adults and Secondary Students. *Gerontol. Geriatr. Educ.* **2019**, *41*, 352–366. [CrossRef] [PubMed]

27. Cesnales, N.I.; Dauenhauer, J.A.; Heffernan, K. Everything Gets Better with Age: Traditional College-Aged Student Perspectives on Older Adult Auditors in Multigenerational Classrooms. *J. Intergener. Relatsh.* **2020**, 1–12. [CrossRef]

28. Lytle, A.; Nowacek, N.; Levy, S.R.; Lytle, A. Instapals: Reducing Ageism by Facilitating Intergenerational Contact and Providing Aging Education Instapals: Reducing Ageism by Facilitating Intergenerational Contact and Providing Aging Education. *Gerontol. Geriatr. Educ.* **2020**, *41*, 308–319. [CrossRef]

29. Yoelin, A.B. Intergenerational Service Learning within an Aging Course and Its Impact on Undergraduate Students' Attitudes about Aging. *J. Intergener. Relatsh.* **2021**, 1–16. [CrossRef]

30. Even-Zohar, A.; Werner, S. The Effect of Educational Interventions on Willingness to Work with Older Adults: A Comparison of Students of Social Work and Health Professions. *J. Gerontol. Soc. Work.* **2020**, *63*, 114–132. [CrossRef]

31. Gök Uğur, H.; Hendekci, A. Effect of Planned Training Provided to Care Staff in Nursing Homes on Their Attitudes toward the Elderly. *Turk Geriatr. Derg.* **2019**, *22*, 376–383. [CrossRef]

32. Jeste, D.V.; Avanzino, J.; Depp, C.A.; Gawronska, M.; Tu, X.; Sewell, D.D.; Huege, S.F. Effect of Short-Term Research Training Programs on Medical Students' Attitudes toward Aging. *Gerontol. Geriatr. Educ.* **2018**, *39*, 214–222. [CrossRef]

33. Lytle, A.; Levy, S.R. Reducing Ageism: Education about Aging and Extended Contact with Older Adults. *Gerontologist* **2019**, *59*, 580–589. [CrossRef] [PubMed]

34. Rababa, M.; Alhawatmeh, H.; Al, N.; Manal, A. Testing the Effectiveness of Cognitive Behavioral Therapy in Relieving Nurses' Ageism toward Older Adults: A Randomized Controlled Trial. *Cogn. Ther. Res.* **2020**, *45*, 355–366. [CrossRef] [PubMed]

35. Sun, Q.; Lou, V.W.; Dai, A.; To, C.; Wong, S.Y. The Effectiveness of the Young–Old Link and Growth Intergenerational Program in Reducing Age Stereotypes. *Res. Soc. Work. Pract.* **2019**, *29*, 519–528. [CrossRef]

36. Madrigal, C.; Fick, D.; Mogle, J.; Hill, N.L.; Bratlee-, E.; Belser, A.; Madrigal, C.; Fick, D.; Mogle, J.; Hill, N.L.; et al. Disrupting Younger Adults' Age-Based Stereotypes: The Impact of an Intergenerational Artistic Installation. *J. Intergener. Relatsh.* **2020**, *18*, 399–416. [CrossRef]

37. Bailey, S.W.; Sudha, S.; Bailey, S.W. Too Little, Too Late?: Can an Integrated Empathy-Building Intervention Shift Gero-Attitudes for Undergraduates in an Online Course? *Gerontol. Geriatr. Educ.* **2021**, 1–17. [CrossRef] [PubMed]

38. Fernández-Ballesteros, R.; Sánchez-Izquierdo, M. Health, Psycho-Social Factors, and Ageism in Older Adults in Spain during the COVID-19 Pandemic. *Healthcare* **2021**, *9*, 256. [CrossRef]

39. World Health Organization. *World Report on Ageing and Health*; WHO: Geneva, Switzerland, 2015.

40. Officer, A.; de la Fuente-Núñez, V. A Global Campaign to Combat Ageism. *Bull. World Health Organ.* **2018**, *96*, 295–296. [CrossRef]

41. Rello, C.F.; López- Bravo, M.D.; Muñoz- Plata, R.M. Estereotipos sobre la edad y el envejecimiento en estudiantes y profesionales de Ciencias de la Salud. *Rev. Prism. Soc.* **2018**, *21*, 108–122. Available online: https://revistaprismasocial.es/article/view/2425 (accessed on 1 December 2021)

42. Lichtenstein, B. From "Coffin Dodger" to "Boomer Remover": Outbreaks of Ageism in Three Countries with Divergent Approaches to Coronavirus Control. *J. Gerontol. Ser. B* **2021**, *76*, e206–e212. [CrossRef]

43. Button, P. Population Aging, Age Discrimination, and Age Discrimination Protections at the 50th Anniversary of the Age Discrimination in Employment Act. In *Current and Emerging Trends in Aging and Work*; Czaja, S., Sharit, J., James, J., Eds.; Springer: Cham, Switzerland, 2020; pp. 163–188; ISBN 978-3-030-24134-6. [CrossRef]

44. Cecil, V.; Pendry, L.F.; Salvatore, J.; Mycroft, H.; Kurz, T. Gendered Ageism and Gray Hair: Must Older Women Choose between Feeling Authentic and Looking Competent? *J. Women Aging* **2021**, *4*, 1–16. [CrossRef]

45. Dixon, J.; Sibthorpe, B. How Can a Government Research and Development Initiative Contribute to Reducing Health Inequalities? *N. S. W. Public Health Bull.* **2001**, *12*, 189–191. [CrossRef] [PubMed]

46. Sparks, H. Morbid 'Boomer Remover' Coronavirus Meme Only Makes Millennials Seem More Awful. New York Post. 19 mar. 2020. 2020. Available online: https://nypost.com/2020/03/19/morbid-boomer-remover-coronavirus-meme-only-makesmillennials-seem-more-awful/ (accessed on 13 November 2020).

47. Rahman, A.; Jahan, Y. Defining a 'Risk Group' and Ageism in the Era of COVID-19. *J. Loss Trauma* **2020**, *25*, 631–634. [CrossRef]

48. Vale, M.T.; Stanley, J.T.; Houston, M.L.; Villalba, A.A.; Turner, J.R. Ageism and Behavior Change during a Health Pandemic: A Preregistered Study. *Front. Psychol.* **2020**, *11*, 3156. [CrossRef] [PubMed]

49. Marques, S.; Mariano, J.; Mendonça, J.; De Tavernier, W.; Hess, M.; Naegele, L.; Peixciro, F.; Martins, D. Determinants of Ageism against Older Adults: A Systematic Review. *Int. J. Environ. Res. Public Health* **2020**, *17*, 2560. [CrossRef] [PubMed]

50. Scriven, M. *Evaluation Thesaurus*; SAGE Publication: London, UK, 1991.

51. Weiss, C.H. Where Politics and Evaluation Research Meet. *Eval. Pract.* **1973**, *14*, 93–106. [CrossRef]
52. Agulló-Tomás, M.S.; Zorrilla-Muñoz, V.; Gómez, M.V. Género y evaluación de programas de apoyo para cuidadoras/es de mayores. *Rev. Prism. Soc.* **2018**, *21*, 391–415. Available online: https://revistaprismasocial.es/article/view/2469 (accessed on 1 December 2021)

Article

Territorial and Gender Differences in the Home Care of Family Members with Dementia

Sagrario Anaut-Bravo and María Cristina Lopes-Dos-Santos *

Department of Sociology and Social Work, Public University of Navarra, 31006 Pamplona, Spain; sanaut@unavarra.es
* Correspondence: cristina.lopes@unavarra.es

Abstract: The increasing prevalence of dementia is threatening the capacity of health and social service systems to provide long-term care support at the territorial level. In both rural and urban areas, specific family members (gendered care) are responsible for the daily care of their relatives. The aim of this work is to explore gender and territorial implications in the provision of in-home care by family members. To this end, family caregivers in Navarre, Spain, were administered the Psychosocial Adjustment to Illness Scale (PAIS-SR) and a semi-structured interview. The results show the good psychosocial adjustment of caregivers of relatives with dementia but the negative impacts of caregiving in the domestic, relational, and psychological domains. Moreover, the feminization of psychological distress was found to predominate in rural areas since mainly women are responsible for instrumental and care tasks, while men seek other complementary forms of support. Place of residence (rural vs. urban) was found to exert a strong effect on the respondents' conception, life experience, and provision of care. Consequently, territorial and gender differences in coping with and adjusting to care require the design of contextualized actions adapted to caregivers' needs.

Keywords: family caregiver; dementia; psychosocial adjustment; rural–urban spaces; gender; aging; Navarre

Citation: Anaut-Bravo, S.; Lopes-Dos-Santos, M.C. Territorial and Gender Differences in the Home Care of Family Members with Dementia. *Land* **2022**, *11*, 113. https://doi.org/10.3390/land11010113

Academic Editors: Vanessa Zorrilla-Muñoz, Eduardo Fernandez, Blanca Criado Quesada, Sonia De Lucas Santos, Jesus Cuadrado Rojo and Maria Silveria Agulló-Tomás

Received: 2 December 2021
Accepted: 29 December 2021
Published: 10 January 2022

Publisher's Note: MDPI stays neutral with regard to jurisdictional claims in published maps and institutional affiliations.

1. Introduction

Current evidence on family caregivers' adjustment to dementia, among other diseases [1–3], has shown that among them, place of residence and sex/gender are increasingly important variables to consider in this process.

In Spain, studies addressing psychological adjustment in the context of in-home care must be framed in an extensive tradition of both regional and urban studies linked to historical and political events that continue to resonate today [4–6]. Indeed, the country's autonomous government model, established in the Constitution of 1978, has produced significant social, economic, fiscal and cultural inequalities that warrant in-depth studies at the regional and municipal level, as well as comparative analyses of territorial differences in daily activities, such as caregiving for a relative. Since Spain joined the European Union, comparative analyses have been performed on both neighboring and non-EU states, including a growing number of publications that examine welfare systems and the welfare state model itself [7–9].

In contrast to the extensive literature on economic issues, territorial policies, or regional imbalances in Spain [6], regional studies on quality of life and welfare in the country remain scarce. This approach is recent and crosses disciplines such as anthropology, sociology, and the health sciences. As it occurs in other countries, there are few comparative studies on the care of the elderly and people with neurodegenerative diseases in rural and urban spaces [10–13]. Most studies in this line have focused on urban localities of different sizes [14] or on rural regions [15–17], while others have examined the diverse meanings of rurality and the heterogeneity of rural areas [18,19]. The social, economic, and political changes driven by globalization have led to the emergence of new scenarios, such as the so-called new rurality in the 1990s, which is dissociated from agriculture, poverty, and notions of backwardness or passivity. Indeed, the effects of these different processes of change question the validity of a definition of rural and urban reduced solely to locality size, population density, and the dominant economic sector [4,13,20–22].

According to Dickins et al. [23] and Quesada-García and Valero-Flores [24], around 70% of elderly people live at home. Specifically, Lopes [25] calculated that 62% of people diagnosed with dementia in Navarre, Spain, remain living in their own homes or in those of relatives. This reality justifies the need for territorial studies that address urban–rural differences in care, despite the conceptual limitations mentioned above [4].

Regarding studies from a gender approach, there is a vast body of literature on the feminization of care for family members of different ages [26]. Indeed, gender-based research in this field has intensified due to what is called the "care crisis" [13,27]. This crisis has been exacerbated by an aging population that requires long-term community care with a certain level of specialization [28,29] and impacts both family solidarity and sustainability [30–32]. Moreover, the progressive social reorganization of work and caregiving has not prevented the vulnerability of caregivers [33] or the influence of personal and situational variables on the process of coping and adjusting to care [3].

Analyses from a gender approach have also brought to light the increasing role of men in caregiving (Granados and Jiménez [34], Del Río Lozano [35], Aguilar, Soronellas-Masdeu and Alonso-Rey [36], Rodríguez, Samper, Marín et al [14], Mosquera et al [37], Martín-Vidaña [38] and Zygouri [39]). In general, these studies evaluate the impact of caregiving on the quality of life of the caregivers, differentiating between men and women. The approach is diverse since the increasing prevalence of neurodegenerative diseases has placed the provision of long-term care and those who provide in-home care in a prominent spot in the social and health policy agendas of most countries [40].

As recognized in the case of China [41], there are very few regional studies [42–44] that examine caregivers from rural and urban areas in relation to sex and gender differences. Even fewer studies have examined the adjustment of those who care for family members with dementia, jointly considering both the rural–urban environment and gender. This is undoubtedly the main contribution of this article, although it will not always be possible to extrapolate our results to other regions or countries due to socio-cultural differences.

A previous study carried out in Navarre, Spain, compared adjustment to disease among in-home caregivers of relatives with Parkinson's disease and dementia [45]. The authors concluded that the specific illness of the cared person, place of residence, employment status, and income, in that order, were the most influential variables. In a subsequent analysis focusing only on dementia and the effect of place of residence on the psychosocial adjustment and quality of life of caregivers of relatives with dementia [46], differences were found between residing in urban and rural areas. These differences were due to specific territorial characteristics affecting the availability of human, economic, and technical resources, as well as cultural factors related to understanding and coping with life circumstances. Based on the results, the authors hypothesized that such differences might be explained by the person's ability to adapt, cope with, and accept the meaning of caregiving tasks, and that these skills could be the result of assumed gender roles.

Following the above, this article aims to examine the interaction between place of residence and gender roles in psychosocial adaptation processes among people who care for cohabiting relatives with dementia. To this end, we analyze Navarre, a region of Spain that stands out for its distinct historical background; wide geographic, economic, and sociocultural diversity; and its own tax system and a differentiated welfare model. Given the particular characteristics of Navarre, several region-specific [47–49] and comparative cross-regional analyses have been carried out on the region, such as the recent studies of Anaut-Bravo [50,51], Pérez and Martínez [52], and García and Caballeira [53], on health and social services systems.

2. Materials and Methods

The region of Navarre (Figure 1) has been chosen for the analysis due to its climatic and geographic diversity that conditions the population distribution as well as the predominant economic activities in the territory. Navarre has a total population of 661,023 (as of January 2021), which is mainly concentrated around its capital of Pamplona and the surrounding areas (just over 50%). It is also one of the five Spanish regions with the best social protection system development rates [54]. In addition, the accessibility to the information sources needed for this study makes Navarre particularly suitable for our aims.

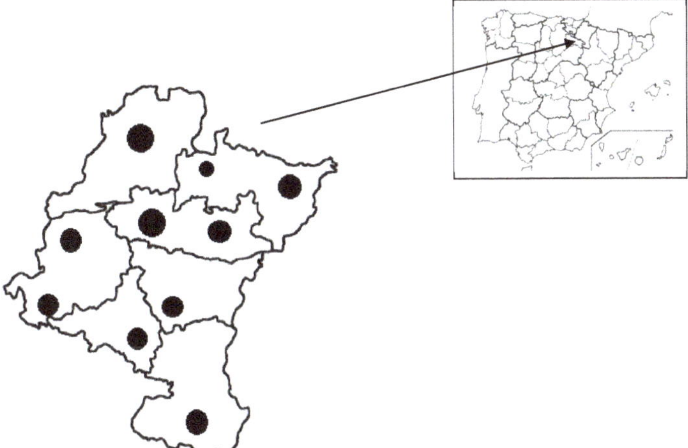

Figure 1. Territorial distribution of research participants.

The study sample was obtained (2015–2017) by non-probabilistic convenience sampling [55,56] since the research required the voluntary collaboration of primary health care and social service practitioners, as well as caregivers. A total of 61 practitioners involved in the Program for Dependency Care and Promotion of Autonomy of Basic Social Services and 55 practitioners from the Health Centers of Navarre agreed to cooperate in the research to identify and recruit family caregivers. All participants were provided the following documentation: basic information on the research content, an informed consent form, the two questionnaires that were administered for the study, the script of the interview, and the endorsement of the Ethics Committee of the Public University of Navarre (cod. PT-025-15). All of the practitioners were asked to confirm the diagnosis of dementia in the cared persons according to the International Statistical Classification of Diseases and Related Health Problems, 10th revision (ICD-10) [57].

A total of 135 telephone contacts were obtained, to which the following exclusion criteria were applied: being a relative without a direct care relationship, an occasional family caregiver, and limitations in verbal comprehension and expression. The inclusion criteria were being the main caregiver of the person with dementia and a cohabiting family member, good cognitive status to answer the questionnaire and take part in the individual interview, and voluntary participation and agreeing to sign the informed consent. The participants were from rural and urban areas and recruited at similar levels to their representativeness within the overall population of Navarre. Male and female caregivers were also recruited for the study (Table 1). The final sample was comprised of 75 caregivers of relatives with dementia residing in 29 localities of Navarre (Figure 1). Territorial representativeness was guaranteed in the distribution of participants.

In the sample obtained by the municipalities and population, participants from urban areas were clearly over-represented (Table 1), as were females. This was due to the voluntary nature of participation and some unforeseen circumstances, such as the hospitalization or illness of the family caregiver who was going to participate.

Table 1. Representation of municipalities and population of Navarre. * Rural: population, <10,000; ** urban: population, >10,000 [17]. Source: own elaboration based on data collected from the Statistics Institute of Navarre (NASTAT).

Type of Locality	Number of Municipalities	Municipalities Percentage (%)	Population	Population Percentage (%)	Participants Percentage (%)
Rural *	261	95.96	281427	43.75	18.2
Urban **	11	4.04	361.807	56.25	81.8

Each of the 75 family caregivers were administered two questionnaires to collect quantitative information. The first one gathered the following sociodemographic characteristics of the participants: age, marital status, years caring for the family member affected by dementia, employment status, kinship, and level of education. The second questionnaire was the Psychosocial Adjustment to Illness Scale (PAIS-SR) developed by Derogatis [58] and validated by Bullinger et al. [59], which has been previously administered in Spain by Portillo et al. [60]. The PAIS-SR contains 46 items grouped into seven sections (Figure 2) to collect self-reported information on the evolution of dementia, changes in care as a result of the illness, alternatives to the mentioned changes, acceptance of and adjustment to dementia, factors of influence, and support networks and life satisfaction [61]. It should be noted that the results of Section IV of the PAIS-SR scale (sexual relations) were not analyzed due to the lack of responses (85%). However, the lack of responses was analyzed to ensure that the scale maintained its internal coherence according to Cronbach's analysis [62]. The data from the questionnaires were exploited using the SPSS Statistics program version 23 and R software R together with the integrated FactoMineR package [63] and Coheris SPAD.

To examine the personal assessment of care and process of adjustment to the disease in depth, a semi-structured interview script was prepared for this research following the seven sections of the PAIS-SR self-reported questionnaire (Figure 2). Sixty of the 75 caregivers of family members with dementia responded by reaching theoretical saturation and meeting the criterion of five interviews per analyzed variable (6 sociodemographic variables and 1 corresponding to each of the scales) of Peduzzi et al. [64].

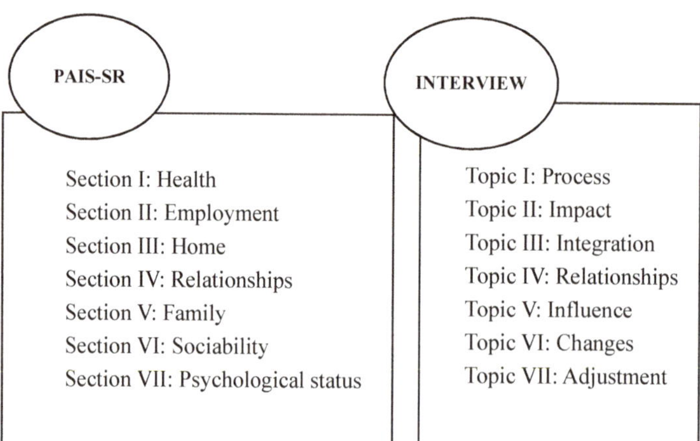

Figure 2. Equivalence of the contents of the PAIS-SR and the semi-structured interview. Source: own elaboration based on Portillo et al. [60].

The interviews were anonymized using the following codes: I (interview), F (family), M or F (male or female), R or U (rural or urban), and interview number. The data were analyzed using AQUAD 7 software. The transcripts (.docx) were transformed into .txt format files to extract the story structure and frequency of terms. The process proposed by Miles et al. was used to analyze the data [65].

3. Results

3.1. Family Caregiver Profiles

The general profile of the family caregivers is characterized by a historical trend: a greater number of females, married (64.7% men and 74% women), unemployed (70% men and 55% women), average age over 50 years, and basic level of education (51.6% women). However, differences by place of residence were detected.

In the rural areas, married caregivers predominated (Table 2), especially among women who were homemakers or engaged in full-time employment (same proportion). Additionally, male caregivers from both rural and urban area were, on average, older than female caregivers (up to 12 points of difference). It is interesting to note that the male caregivers were children of the cared person, not spouses. Regarding education, the men had only basic or vocational education, while the women had completed all educational levels, particularly basic education, followed at a certain distance by higher education.

Table 2. Sociodemographic variables of family caregivers; number by location. Source: own elaboration.

	Urban Localities	Rural Localities	Total
Sex			
Male	22	9	31
Female	63	41	104
Marital Status			
Married men	15	6	21
Married women	43	31	74
Single men	5	3	8
Single women	11	9	20
Men other	2	0	2
Women other	9	1	10
Employment Situation			
Men full-time	7	4	11
Women full-time	16	9	25
Men part-time	0	0	0
Women part-time	13	7	20
Retired men	13	3	16
Retired women	13	7	20
Male homemakers	0	0	0
Female homemakers	16	14	30
Men other	2	2	4
Women other	5	4	9
Education			
Men basic education	7	4	11
Women basic education	29	25	54
Men vocational education	4	5	9
Women vocational education	7	4	11
Men secondary education	4	0	4
Women secondary education	13	4	17
Men higher education	7	0	7
Women higher education	14	8	22
Kinship			
Sons	13	7	20
Daughters	48	24	72
Male spouses	9	2	11
Female spouses	11	13	24
Men other	0	0	0
Women other	4	4	8

Regarding participants from urban areas, most were married, especially the men. The majority of female and male participants had a basic level of education and an equal percentage of men had completed basic and higher education (23.5%). In terms of employment, most of the men were retired, while the women were engaged in full-time employment, followed at a short distance by those who were retired.

Regarding the socioeconomic differences between rural and urban areas, 47% of the male caregivers were retired and 88% lived in urban areas, while 15.6% of the female caregivers were retired and one in five lived in a rural location. Moreover, 45% of the women were employed, while 30% described themselves as "homemakers". No differences were found between urban and rural homemakers in terms of their representativeness but differences were detected in working time, with a higher percentage of full-time (56%) and part-time (70%) employed women in urban areas.

The higher employment rates of women (84.4%) point to the so-called "double shifts" and their higher qualifications. Of the female participants, 26% had a vocational or secondary education, while 22% had completed higher education, 64% of whom resided in urban areas.

3.2. Adjustment to the Care of Family Members with Dementia

3.2.1. Care Experiences

Three categories related to the participant's care experiences were identified in the discourse analysis of the interviews: coping with coexistence, available resources, and harmonious coexistence (Figure 3). The three categories reflect the caregiving process and experience, external sources of support and the emotional impact, and the management of care. Both the men and the women who cared for relatives with dementia stated that coping with the situation is difficult ("hard") and has mainly negative impacts on their relational life in the form of obligations, poor health self-care, stress, lack of freedom, interdependence, and loneliness, among others. Negativity is also associated with unstable "life changes" and the "uncertainty" caused by the evolution of dementia when coping with coexistence situations.

However, there are some differences. For women in both rural and urban areas, continuous and exclusive dedication to care leads to greater personal and social isolation. By contrast, men give higher priority to their friendships and personal space.

IFFR11: "You don't feel like going out like you used to. When they come looking for you, you don't feel like it".

IFFU12: "Nobody ever comes, no one, no one".

IFMR43: "I need to go out with my friends; be with them".

IFMU61: "Keep my space, my moments, my places, my responsibilities and tasks".

In the case of women, there is an additional nuance regarding these gender differences: a certain traditional woman-to-woman solidarity seems to endure in rural areas. Such solidarity does not appear to be based on kinship relationships but on shared experiences and physical proximity. In contrast, feelings of loneliness and stress are more common among urban women, as stated by IFFU11: "Even if you are accompanied or there are people around, you feel a bit lonely".

Something similar occurs in family relationships, particularly intragenerational families. However, clear differences were not detected between caregivers residing in rural and urban locations or between men and women. The men tended to go straight to the point and in less detail, as IFMU028 succinctly stated "Family chaos," unlike women, who provide more details.

IFFR44: "Along the way, you get angry with your family because they don't want to deal with the problem. You also see yourself without a family!"

IFFR57: "I don't have a family to turn to. [It's] too much of a burden! We don't have a life together as a couple anymore".

IFFU19: "With my sister being ill, we had many differences. If there is good communication between family members, it's easier to deal with the disease".

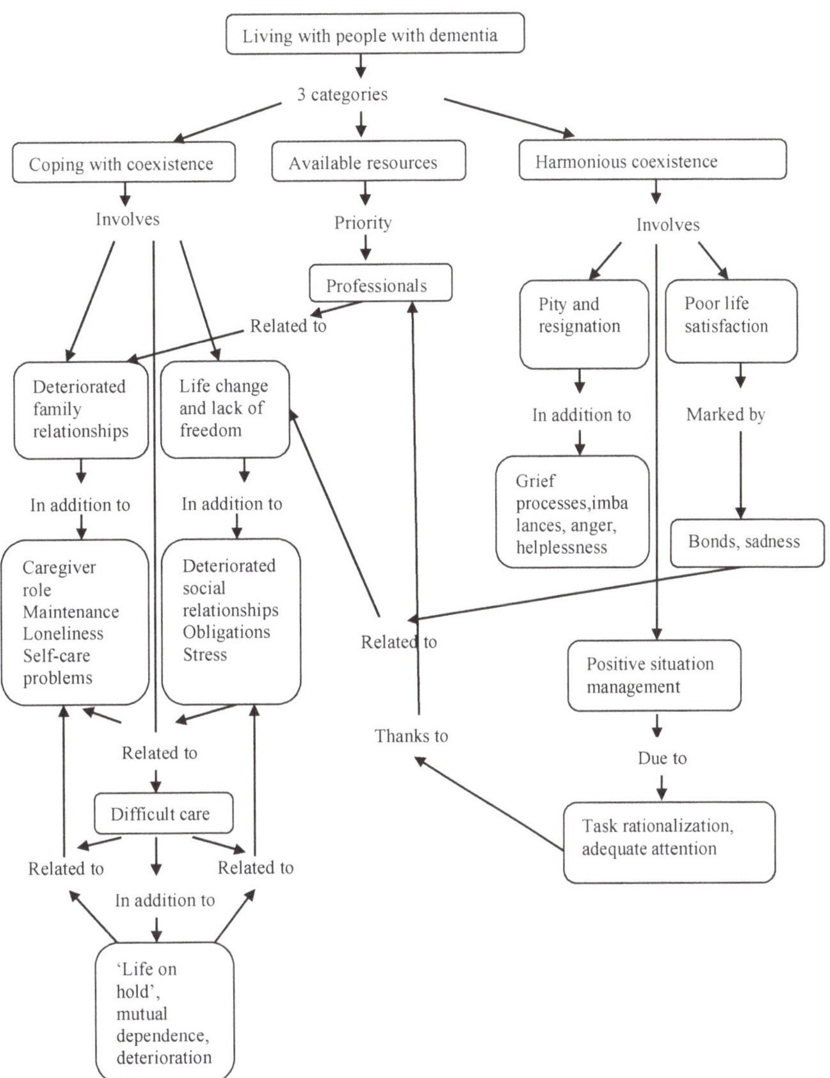

Figure 3. Conceptual map created from the discourse analysis of interviews with family caregivers. Source: own elaboration.

The discourse of family caregivers revealed the progressive deterioration of family relationships (Figure 2) and the importance of having both family support and complementary resources or services to cope with and adjust to dementia. Depending on the case, a combination of strategies may be used, such as family caregivers and support or external services (social services portfolio and/or formal caregivers). Given that there is always a family caregiver, professionals mostly provide any additional support and other family members are not involved in giving care, as shown in Table 3. This is consistent with the deteriorated family relationships mentioned by 65% of those interviewed.

Table 3. Source of care for family members with dementia; percentage distribution. Source: own elaboration based on data drawn from the interviews.

Care Provenance	Percentage (%)
Professional	50
Cohabiting caregivers	36.7
Relatives	13.3

Female caregivers from both rural and urban areas highlighted how the disease and the time they spend providing care has affected their relationship with their partners as well as produced the tremendous burden they must bear.

IFFR44: "Me, at home, my husband is very supportive, but you do get overwhelmed sometimes, the situation gets to be too much".

IFFU12: "We have very good communication with each other, but the couple suffers".

Men, however, tended to be more resolute, among other reasons, out of their desire to maintain their own space and activities, as mentioned above. Their views on the situation are similar to those of IFMU61: "I tend to seek solutions". The situation is resolved by means of external services that do both the domestic tasks performed by women prior to the illness and the specific care tasks. There are no differences by place of residence in regard to this question, nor in recognizing—as 52% of those interviewed stated—feelings that they have lost their freedom or that their life has been put on hold (Figure 2).

IFMR36: "At home things are as usual, not much has changed. What she didn't do is done by the girl who comes to clean and cook".

IFMU56: "The two hired people come, you pay them, and they help you. They are here in two shifts, and they do a lot".

Nonetheless, living in a place with available services introduces certain nuances not only in terms of the caretakers' attitudes, whose responses are conditioned by their gender, but also due to the possibility of obtaining this additional support. In this regard, women living in urban areas recognize the importance of municipal home care services (Servicio de Atención Domiciliaria, SAD) and adult daycare centers to better cope with the situation:

IFFU12: "The SAD comes two hours a day, five days a week. That help has been a lifesaver".

IFFU42: "He's in a daycare center. It's been a good thing".

Using the available public services or hiring outside help makes it easier for family caregivers to build, at least in part, their own life project, take care of tasks in accordance with their life stage, as well as ease the tension or stress and excessive work burden that caregiving involves (35% of respondents). As might be expected, the experience of those who live in rural areas is very different.

In rural environments, the scarcity of specialized social services is compounded by problems of accessibility in both their own and other localities. Mobility is often restricted by physical and geographical barriers or architectural barriers in housing, streets, and other public spaces. In some cases, the respondents also mentioned the lack of accessible public and private transport.

IFFR31: "We don't go out because we can't. We already had to take him out the window with a pulley once because he got sick".

The alternative in these cases is to hire outside help, but, as indicated by EFMR011, this only occurs when the family caregiver is no longer able to provide care: "I suffered from back pain for a long time; then we thought about hiring another person".

More than 58% of family caregivers stated that they are very or somewhat dissatisfied with their lives not only because of the impact of caregiving on their relational life but also on their emotional state (Figure 2). They find it increasingly difficult to live with their ill relative, which arouses mixed feelings of pity, resignation, helplessness, anger, rage, and anguish (47% of respondents).

IFFR11: "I felt very sorry for my mother and all I did was care for her".

IFFU12: "I get angry a lot; I get angry with myself; I get angry with her. There are days when you just want to escape from this life".

IFFU46: "It's maddening; It makes you angry, because everything revolves around the illness".

IFMR43: "Above all else, your character changes".

IFMU60: "A little helpless".

Caregivers who perceive they are coping with the situation in a coherent and adequate manner report greater life satisfaction with their tasks since they experience lower levels of stress and guilt, and demand less of themselves (Figure 2). However, 46.7% referred to how they managed the situation as "unstable" or "negative". In the interviews, no significant differences were found in this regard between men and women, or between those living in rural and urban areas. This seems to indicate, as the following comments suggest, that the key to coping lies in the skills, personality, and attitude of the family caregiver rather than in external factors.

IFFR44: "There are many days when I wake up and say: How well I am doing everything! Everything is going so well! Other times things don't work out so well".

IFFU50: "Sometimes [it's] good, sometimes [it's] crazy".

IFMR43: "I've taken everything that was bad in the house to turn it into something good, get rid of the stress and increase my energy, and do something. I'm satisfied. Some (things) are good, and others are bad, but if you have a balance of both and nothing stands out more, you swim in a calm sea, since you're okay".

IFMU60: "It's more like she's in command and you're just running behind her trying to figure out what's going on".

It is important to note that more than half of the family caregivers positively valued (or at least not in an overtly negative way) how they handled living with a family member with dementia. These caregivers did not usually suffer from general or permanent psychological distress. Opinions such as those indicated below demonstrate that the caregivers view dementia as an irrevocable fact to be accepted and a reality that will be part of their daily life for some time. They do not feel resignation but rather accept a reality. Acceptance is expressed in comments such as that of EFVR043: "I am satisfied, I wouldn't ask for more or for less".

IFFR07: "You could see it coming. With age and time these things happen".

IFFU09: "It's a question of accepting that you'll need help and that it'll get worse over time".

The three categories analyzed (Figure 2) indicate that female caregivers suffer episodes of psychosocial distress as a consequence of their greater dedication, while male caregivers employ strategies to share and delegate tasks to external help. No notable differences were detected in terms of assumed gender roles between urban and rural participants. These territorial differences were reflected in options for accessing resources to support caregiving tasks either in the form of public services or external personnel.

3.2.2. Coping with and Adjusting to Caring for a Family Member with Dementia

The results of the PAIS-SR corroborated the findings of the interviews. The Holm's test ($p < 0.001$) indicated that the three most significant sections of the PAIS-SR are the domestic environment (Section III), the social environment (Section VI) and psychological distress (Section VII). As Table 4 shows, urban caregivers have a negative perception of all three environments, while rural caregivers only perceive the domestic and social environment negatively. Regarding psychological distress (Section VII), in particular, urban caregivers manifest higher levels than rural caregivers, thus suggesting that rural caregivers adjust better to caregiving.

Table 4. Mean scores for Sections III, VI and VII of the PAIS-SR by place of residence. * The maximum score on the PAIS-SR is 24 points for Section III, 18 points for Section VI and 21 points for Section VII. Lower scores indicate better psychosocial adjustment, and higher scores indicate poorer psychosocial adjustment. Source: own elaboration.

PAIS-SR *	Urban	Rural
Section III: Domestic environment <12 good perception	12.6	12
Section VI: Social environment <9 good perception	10.2	10.6
Section VII: Psychological distress <10.5 no presence of discomfort	10.1	4.4

The mean scores presented in Table 4 were further disaggregated by both sex and place of residence to obtain more specific data regarding the impact of caregiving (Table 5). In general terms, men from rural areas have a better perception of the domestic and social environments and suffer less psychosocial distress. Compared to women, men residing in urban and rural locations adjust better to the domestic environment. Furthermore, men and women residing in urban areas show very similar scores overall, except in the social environment, which women view more positively. The results are similar for psychological distress among those living in rural areas. Therefore, those who live in rural areas generally have a more positive perception and, within this group, men cope with and adjust better to caregiving than women.

Table 5. Mean scores for Sections III, VI and VII of the PAIS-SR by sex and place of residence. Source: own elaboration.

$\bar{x}$	Domestic Environment Section III	Social Environment Section VI	Psychological Distress Section VII
Men			
Rural	11.8	9	4.55
Urban	12.12	11.12	10.15
Women			
Rural	12.47	10.81	4.76
Urban	12.41	9.36	10.32

Regarding the variances in the PAIS-SR, the most marked differences were found in all sections for caregivers residing in urban areas, indicating that the variability in caring for a family member with dementia is more pronounced in urban areas (Table 6). By sex, significant differences were observed in the domestic and social environments. In both these environments (Sections III and VI), women living in urban areas showed the greatest discrepancies with respect to the mean. The same discrepancy was found in Section VII (psychological distress), although with less intensity. Women in rural areas and men in urban areas also showed this variability in responses but to a lesser extent. Men residing in rural areas showed an almost homogeneous perception in their responses, although in psychological distress, those residing in urban areas also showed a very small deviation.

Table 6. Variance in PAIS-SR Sections III, VI and VII by place of residence and sex. Source: own elaboration.

$S^2{}_x$	Domesticenvironment Section III	Socialenvironment Section VI	Psychologicaldistress Section VII
Men			
Rural	3.33	0	0.25
Urban	7.55	6.12	0.98
Women			
Rural	9.04	5.92	2.81
Urban	13.22	10.76	10.03

Therefore, the variables territory and sex captured explicit differences in how the respondents interpret and understand the care of a family member with dementia. Gender roles and interaction with the environment (services, resources, relationships, etc.) are explanatory factors for such differences, as determined by the qualitative analysis of the discourse.

4. Discussion

The results reveal that the processes of coping and adjusting to dementia among family caregivers in Navarre differ according to place of residence (rural or urban) and sex. However, when considering all respondents, the findings indicate relatively good psychosocial adjustment (PAIS-SR), especially among caregivers living in rural areas. This result is consistent with Ehrlich et al. [1], although the authors found that satisfaction with caregiving occurs in urban and not rural localities, as has been [10] previously demonstrated.

Various studies have shown negative psychosocial adjustment in cases of long-term care and dementia [66–68]. The vast medical and nursing literature has also highlighted these negative impacts [69,70], especially on the health of family caregivers [35,71,72]. Other studies have focused on negative social and family impacts, such as stress, sleep problems, or the loss of personal independence, future expectations, and social relationships [3,73–75]. The respondents in our study referred to impacts of a social nature but attached less importance to their own health issues.

Despite the generally positive process of psychosocial adjustment among the respondents, the PAIS-SR and interviews indicated three domains in which the provision of care has a negative impact on family caregivers: the domestic, the relational, and, to a lesser extent, the psychological domains. Regarding the domestic domain, no significant differences were found between urban and rural locations. An explanation for this could be the progressive social, cultural, and economic rapprochement due, among other reasons, to greater territorial mobility [76]. This result may also be explained by the increasingly disperse family networks, weak social networks, smaller families, and the feminization of urban locations compared to the masculinization of rural ones. All these factors have led to the defamiliarization of care and limited the capacity of families to care for elderly patients who want to stay at home, as Prieto [19] and Martín and Rivera [17] have argued.

Given such changes, it is important to analyze coping strategies and adjustment in caregivers of family members with dementia from the viewpoint of psychological distress/well-being. While the PAIS-SR responses showed significant differences by place of residence and sex, in the interviews, the respondents placed greater emphasis on psychological distress, which was more pronounced depending on the caregiver's gender. Specifically, more women stated that they experienced psychosocial distress, particularly those living in urban settings. According to Losada et al. [77], the feminization of distress is due to women's greater emotional involvement and heavier care burden, which leads to a state of hypervigilance. In turn, hypervigilance affects caregivers' emotional attachment to the cared family member, as well as their life satisfaction and health status, as previous studies have shown [35,37,39,72]. The different types of bonds between the ill person and family members (i.e., spouses, daughters, and sons) detected in this study are in line with Rodríguez and Pérez [3].

An important factor in bonding is the disease itself [45]. Considering dementia involves the dynamic and progressive deterioration of cognitive function, it requires a significant effort of adjustment, flexibility, and resilience on the part of caregivers [2,3,69]. As gleaned from the interviews, caregivers' perception of being overburdened converges with the course of the disease and varying degrees of psychosocial distress, as well as the belief that professional and non-professional support for in-home care is not and will never be sufficient.

Regarding gender differences in psychosocial adjustment, our results diverge from those of Rodríguez et al. [14] as we found that certain aspects of the coping and adjustment strategies deployed by men differ from those of women. Firstly, while women assume instrumental and care tasks in their entirety, men seek complementary support more quickly. In other words, men adopt more flexible coping and adjustment strategies that reduce their psychosocial distress, as also demonstrated by Zygouri et al. [39]. Secondly, men have a more pragmatic attitude and self-manage their time and activities to not renounce their social relationships. Thirdly, care involvement by men in rural areas may be related to three features that characterize the smallest localities in our study and which coincide with other regions of Spain such as Castile and León [17]: the masculinization of the population, men's singleness, and the increased life expectancy of males. For this reason, many men living in rural areas are responsible for caring for their parents. However, when possible, they opt for external support to continue working (those who are retired continue to work their land) and enjoy a certain social life, knowing that support from family members is scarce and sporadic. The involvement of males in the care of ill family members has been detected for some decades and seems to be related to age [78]. For Spain, Abellán et al. found that there are more male than female caregivers in all types of households and forms of care from the age of 80 [79], which may explain the high average age of male caregivers in the rural areas of Navarre (64.6 years old).

The relationship between psychosocial distress and gender cannot be explained solely by weak social networks and assigned gender roles [80], nor by the fact that family support is more common in rural areas, as Manso et al. [81], Lorenzo et al. [82], and Ehrlich et al. [1] have argued. The environment where one lives and the available opportunities for biopsychosocial adjustment must also be considered [83].

Until recently, place of residence was not believed to play a significant role in caregivers' experiences. Yet, the interviews we conducted clearly indicate that the availability of services to aid in the care of ill family members, as well as family and non-family support, depend on where one lives [46]. Likewise, the physical location and geography of the place of residence affects the municipality's own accessibility (public spaces) as compared to other municipalities that offer more services [84]. In rural locations, this may partially explain the limited social and family relationships identified in the PAIS-SR, but not the lack of such support mentioned by the respondents in the interviews. In this case, both urban and rural family caregivers coincide in their assessments.

The lack of social and family support has also been reported in other studies, such as Martínez for Spain [32], Rubio et al. for Chile [75], and Wang et al. for China [41], who examined the decline in family solidarity and changes in social relationships that weakened support networks. In the interviews, the respondents highlighted problems such as family conflicts, the lack of support from other family members, and the negative impact of caregiving on relationships with other members of the nuclear family. This focus on the family rather than on social networks is related to what is called the "Mediterranean care model", that is, high family involvement supported by little formal care [2].

However, weak family, social, and community support does not seem to affect emotional and cultural ties to the geographical place in which one has lived and wants to live. There is a personal connection with the environment, that is, people build their daily lives together with others who live in the same social and territorial context to create a culture and an identity. Nonetheless, this permanence of place is only possible if access to certain professional services is available.

Around 50% of the cases studied have complementary support resources in the form of either public services (SAD or adult daycare centers) or financial aid to hire non-family caregivers through the social services portfolio of Navarre. The recognition of dependent benefits under the Dependency Law (2006) may be key to increasing such support in both urban and rural areas. However, the most important problem mentioned by both urban and rural respondents is the lack of services. This is striking given the greater number of specialized dementia care services, options for hiring non-family caregivers, and resources to support caregivers in urban areas, as Martín and Rivera have shown [17]. In addition to the scarcity or lack of services in rural municipalities, family caregivers face the problem of accessibility to neighboring municipalities where such services are available.

None of the respondents questioned the decision to remain at home as it was an essential part of their lives. Living in a familiar environment (one's home, neighborhood, or municipality) provides both caregivers and the cared person emotional and relational support, as well as a sense of personal identity. Nonetheless, more accessibility, flexibility, services, social networks, etc., are needed to adapt to the increasing presence of elderly people with neurodegenerative diseases, particularly in rural areas. These results are in line with the characteristics of rural environments pointed out by Prieto [19]: aging and over-aging; scarce and difficult access to services, infrastructures, and ICTs (Information and Knowledge Tecnologies); disperse family networks (due to emigration); the defamiliarization of care (hiring of non-family caregivers); and the masculinization of the family caregiver figure. In this sense, the territorial and social context is key to understanding each person's own coping process.

The present study has two main limitations. The first is the need to increase the size and geographical scope of the sample. In the final sample, urban dwellers and females were over-represented, which may have biased the quantitative results. The second limitation is the need to examine in greater depth the factors that most impact on care-related issues, such as gender, marital status, and the specific characteristics of households and public spaces. These and other factors play an essential role in caregivers' psychosocial adjustment to the disease and the development of collective coping strategies to ensure the sustainability of care. Despite these limitations, this article offers an adequate framework for conducting comparative studies between and within regions in the future.

5. Conclusions

Processes of psychosocial adjustment among family caregivers differ depending on their place of residence and gender. While women tend to assume the responsibility for long-term care and their socio-emotional involvement hinders social interactions and increases their emotional burden, men cope with the situation in a more pragmatic manner to maintain, to the largest possible extent, their habits and social relationships, as do women who work part or full time. Indeed, working outside the home improves the coexistence with the ill family member and reduces both the psychosocial distress and stress of providing care, despite negative feelings associated with the belief that they are not providing sufficient care.

This study has also explored the territorial dimension of psychosocial adjustment in the care of family members with dementia. The desire to remain at home speaks to notions of rootedness, identity, security, and certainty regarding private (home) and community (local) spaces. Such life choices can have impact on rural environments since they are characterized by demographic aging (the majority of inhabitants are over 65 years old) and depopulation, and many are at risk of disappearing.

However, rural environments are also defined by vital attitudes, such as the acceptance of the normal course of life and nature that helps to reduce psychosocial distress in caregivers of relatives with dementia, and greater resilience, as the interviews have shown. Moreover, not only do those who need care remain but also those who take care of them, whether they are relatives or not. In this way, the cared person and the caregiver contribute to the continuity of the local population and slow down depopulation. The results presented in this work also show that caring for a family member with dementia in rural areas is associated with lower levels of psychosocial distress than in urban locations, where caregivers often experience feelings of social isolation despite the availability of complementary care services. In fact, caregivers in rural areas—especially men—experience less psychosocial distress despite the fewer professional social and health services and the decline in family support networks, as they are compensated for by other types of resources.

Although the results point to gender differences in how in-home family care is conceived, experienced, and provided, new trends are emerging in rural localities. In this regard, neither age nor being a man appears to be a restriction for the provision of care. Indeed, the so-called "new caregivers" (men over 80 years old) demonstrate that caregiving is no longer synonymous with the loss of biopsychosocial health or poor quality of life but is more closely associated with personal and family coping strategies and adjustment to the care of family members with dementia.

Based on the above considerations and given the increasing prevalence of neurodegenerative diseases such as dementia, it will be necessary to design contextualized actions aimed at meeting caregivers' needs. These actions should promote the comprehensive care of caregivers to ensure the sustainability of family care. To achieve this, multidisciplinary, adaptive, and community services from a social co-responsibility approach are required. Additionally, we must not overlook the impact of the new rurality on the rural environment in terms of caregiver profiles and behaviors, and the demand for services, among other aspects, which are comparable to those of urban locations. In other words, we find ourselves, as A. Moreno [77] states, before new non-dichotomous urban–rural relationships that are transforming social and cultural constructs.

Author Contributions: All authors should be involved. Conceptualization, S.A.-B.; methodology, M.C.L.-D.-S. and S.A.-B.; software, M.C.L.-D.-S.; validation, M.C.L.-D.-S.; formal analysis, M.C.L.-D.-S.; investigation, M.C.L.-D.-S.; resources, M.C.L.-D.-S.; data curation, S.A.-B. and M.C.L.-D.-S.; writing—original draft preparation, M.C.L.-D.-S.; writing—review and editing, S.A.-B.; supervision, S.A.-B.; project administration, S.A.-B.; funding acquisition, S.A.-B. All authors have read and agreed to the published version of the manuscript.

Funding: This research study was funded by Fundación Bancaria la Caixa y Fundación Caja Navarra, REF P/1/19.

Informed Consent Statement: Not applicable.

Data Availability Statement: The data that support the findings of this study are available from the corresponding author upon reasonable request. NASTAT is an open access website at: http://www.navarra.es/home_es/Gobierno+de+Navarra/Organigrama/Los+departamentos/Economia+y+Hacienda/Organigrama/Estructura+Organica/Instituto+Estadistica (accessed on 14 June 2021).

Conflicts of Interest: The authors declare no conflict of interest.

References

1. Ehrilch, K.; Bstrom, A.M.; Mazaheri, M.; Heikkila, K.; Emami, A. Family caregivers' assessments of caring for a relative with dementia: A comparison of urban and rural areas. *Int. J. Older People Nurs.* **2014**, *10*, 27–37. [CrossRef] [PubMed]
2. Moreno-Cámara, S.; Palomino-Moral, P.A.; Moral-Fernández, L.; Frías-Osuna, A.; Del Pino-Casado, R. Problemas en el proceso de adaptación a los cambios en personas cuidadoras familiares de mayores con demencia. *Gac. Sanit.* **2016**, *30*, 201–207. [CrossRef] [PubMed]
3. Rodríguez, A.; Pérez, L. Estrategias de afrontamiento en cuidadoras de personas con Alzheimer. Influencia de variables personales y situacionales. *REDIS* **2019**, *7*, 153–171. [CrossRef]
4. Reig, E.; Goerlich, F.J.; Cantarino, I. *Delimitación de Áreas Rurales y Urbanas a Nivel Local. Demografía, Coberturas del Suelo y Accesibilidad*; Fundación BBVA: Bilbao, Spain, 2016. Available online: https://www.fbbva.es/wp-content/uploads/2017/05/dat/DE_2016_IVIE_delimitacion_areas_rurales.pdf (accessed on 14 June 2021).

5.   Ubago, Y.; García, I.; Iraizoz, B.; Pascual, P. Why are some spanish regions more resilient than others? *Pap. Reg. Sci.* **2019**, *98*, 2211–2231. [CrossRef]

6.   Cuadrado-Roura, J.R. European Regional Policy. What can be learned. In *Handbook of Regional Science*; Fischer, M., Nijkamp, P., Eds.; Springer: Berlin/Heidelberg, Germany, 2020; pp. 1053–1086.

7.   Amezcua-Aguilar, T.; Alberich-Nistal, T.; Sotomayor, E. Las personas mayores en el Estado de bienestar: Las políticas sociales en Alemania y España. *Cuad. Trab. Soc.* **2020**, *33*, 15–30. [CrossRef]

8.   Pardo Fernández, D. *Análisis Comparado de los Modelos de Bienestar Social Vigentes en España, Alemania, Suecia y Estados Unidos*; EAPN-Laboratorio de Alternativas: Madrid, Spain, 2018. Available online: https://www.fundacionalternativas.org/laboratorio/documentos/politica-comparada/analisis-comparado-de-los-modelos-de-bienestar-social-vigentes-en-espana-alemania-suecia-y-estados-unidos (accessed on 17 May 2021).

9.   Sousa, M.F.B.; Santos, R.; Turró-Garriga, O.; Días, R.; Dourado, M.C.N.; Conde-Salas, J.L. Factors associated with caregiver burden: Comparative study between Brazilian and Spanish caregivers of patients with Alzheimer´s disease. *Int. Psychogeriatr.* **2016**, *28*, 1363–1374. [CrossRef]

10. Bien, B.; Wojszel, B.; Sikorska-Simmons, E. Rural and urban caregivers for older adults in Poland: Perceptions of positive and negative impact of caregiving. *Int. J. Aging Hum. Dev.* **2007**, *65*, 185–202. [CrossRef]

11. Keith, J. *Burden of Care Impacting Family Caregivers of Dependent Community-Dwelling Older Adults in Rural and Urban Settings of Southern Turkey: A Mosaic of Caregiver Issues and Recommendations*; University of Dortmund: Dortmund, Germany, 2013. Available online: https://d-nb.info/1103587943/34 (accessed on 14 June 2021).

12. Madruga, M. Sintomatología psicológica en cuidadores informales en población rural y urbana. *Int. J. Dev. Educ. Psychol. (INFAD)* **2016**, *2*, 257–266. [CrossRef]

13. Agulló-Tomás, M.S.; Zorrilla-Muñoz, V.; Gómez-García, M.V. Aproximación socio-espacial al envejecimiento y a los programas para cuidadoras/es de mayores. *Int. J. Dev. Educ. Psychol. (INFAD)* **2019**, *1*, 211–228. [CrossRef]

14. Rodríguez, J.A.; Samper, T.; Marín, S.; Sigalat, E.; Moreno, A.E. Hombres cuidadores informales en la ciudad de Valencia. Una experiencia de reciprocidad. *OBETS Rev. Cienc. Soc.* **2018**, *13*, 645–670. [CrossRef]

15. Manso, M.E.; Sánchez, M.P.; Cuéllar, I. Salud y sobrecarga percibida en personas cuidadoras familiares de una zona rural. *Clínica Y Salud* **2013**, *24*, 37–45. [CrossRef]

16. Dunn, D.J.; Price, D.; Neder, S. Rural caregivers of persons with dementia. *Vis. J. Rogerian Nurs. Sci.* **2016**, *22*, 16–24. Available online: https://www.societyofrogerianscholars.org (accessed on 15 June 2021).

17. Martín, A.; Rivera, J. Feminización, cuidados y generación soporte: Cambios en las estrategias de las atenciones a personas mayores en el medio rural. *Rev. Prism. Soc.* **2018**, *21*, 219–242. Available online: https://revistaprismasocial.es/article/view/2430 (accessed on 14 June 2021).

18. Santana, P. *Introdução à Geografia da Saúde: Território, Saúde e Bem-Estar*; Universidad de Coimbra: Coimbra, Portugal, 2014. [CrossRef]

19. Prieto Villarino, M.J. Envejecimiento activo en el entorno rural. ¿Igualdad de oportunidades? *Doc. Soc.* **2017**, *185*, 149–166. Available online: https://www.caritas.es/main-files/uploads/2019/04/DocSocial185-1.pdf (accessed on 14 June 2021).

20. Wineman, A.; Alia, D.Y.; Anderson, C.L. Definitions of "rural" and "urban" and understandings of economic transformation: Evidence from Tanzania. *J. Rural Stud.* **2020**, *79*, 254–268. [CrossRef]

21. López Moreno, E. *Concepts, Definitions and Data Sources for the Study of Urbanization: The 2030 Agenda for Sustainable Development*; United Nations Expert Group Meeting on Sustainable Cities, Human Mobility and International Migration, UN/POP/EGM/2017/8: New York, NY, USA, 2017. Available online: https://www.un.org/development/desa/pd/sites/www.un.org.development.desa.pd/files/unpd_egm_201709_s2_paper-moreno-final.pdf (accessed on 16 June 2021).

22. OECD. Redefining Urban: A New Way to Measure Metropolitan Areas. 2012. Available online: https://www.oecd.org/regional/redefiningurbananewwaytomeasuremetropolitanareas.htm (accessed on 14 July 2021).

23. Dickins, M.; Goeman, D.; O'Keefe, F.; Iliffe, S.; Pond, D. Understanding the conceptualisation of risk in the context of community dementia care. *Soc. Sci. Med.* **2018**, *208*, 72–79. [CrossRef] [PubMed]

24. Quesada-García, S.; Valero-Flores, P. Proyectar espacios para habitantes con alzhéimer, una visión desde la arquitectura. *Arte Individuo Y Soc.* **2017**, *29*, 89–108. [CrossRef]

25. Lopes-Dos-Santos, M.C. La adaptación psicosocial a la enfermedad y calidad de vida de familiares cuidadores de personas con demencia en el proceso de convivencia. *Cuad. Gerontológicos* **2017**, *22*, 27–33. Available online: https://issuu.com/cuadernos_gerontologicos/docs/cuadernos_gerontolo_gicos_22 (accessed on 16 July 2021).

26. Carrasco, C.; Borderías, C.; Torns, T. *El Trabajo de Cuidado. Historia, Teoría y Políticas*; Los libros de la Catarata: Madrid, Spain, 2011.

27. Comas D'Argemir, D. La crisis de los cuidados como crisis de reproducción social. Las políticas públicas y más allá. In *Periferias, Fronteras y Diálogos. Una Lectura Antropológica de los Retos de la Sociedad Actual*; Universidad Rovira i Virgili: Barcelona, Spain, 2014; pp. 329–349. Available online: http://digital.publicacionsurv.cat/index.php/purv/catalog/book/123 (accessed on 14 June 2021).

28. Mayobre, P.; Vázquez, I. Cuidar cuesta: Un análisis del cuidado desde la perspectiva de género. *Rev. Española Investig. Sociol.* **2015**, *151*, 83–100. [CrossRef]

29. Sánchez, M. Los 'cuidados informales' de larga duración en el marco de la construcción ideológica, societal y de género de los servicios sociales de cuidados. *Cuad. Relac. Labor.* **2013**, *30*, 185–210. [CrossRef]

30. Garcés, J.; Ródenas, F.J. La teoría de la sostenibilidad social: Aplicación en el ámbito de cuidados de larga duración. *Rev. Int. Trab. Soc. Y Bienestar* **2012**, *1*, 49–59. Available online: https://revistas.um.es/azarbe/article/view/151131 (accessed on 1 September 2021).
31. Canga, A. Hacia una «familia cuidadora sostenible». *An. Del Sist. Sanit. Navar.* **2013**, *36*, 383–386. [CrossRef] [PubMed]
32. Martínez Virto, L. Crisis en la familia. Síntomas de agotamiento de la solidaridad familiar. In *VII Informe Sobre Exclusión y Desarrollo Social en España*; FOESSA; Fundación Foessa: Madrid, Spain, 2014; Documento de Trabajo 3.7. Available online: http://www.foessa2014.es/informe/uploaded/documentos_trabajo/15102014151608_2582.pdf (accessed on 15 July 2021).
33. Silva, B.; Leite, A.; Santos, R.; Diré, G. A vulnerabilidade dos cuidadores de idosos com demencia: Revisão integrativa. *Rev. Pesqui. Cuid. Fundam.* **2017**, *9*, 882–892. [CrossRef]
34. Granados, M.E.; Jiménez, A. Principios motivacionales en cuidadores informales hombres en el ámbito rural y ámbito urbano. *ENE. Rev. Enfermería* **2017**, *11*, 1. Available online: http://ene-enfermeria.org/ojs/index.php/ENE/article/view/606/cuidador_hombre (accessed on 14 June 2021).
35. Del Río Lozano, M.; García-Calvente, M.M.; Calle-Romero, J.; Machón-Sobrado, J.; Larrañaga-Padilla, M. Health-related quality of life in Spanish informal caregivers: Gender differences and support received. *Qual. Life Res.* **2017**, *26*, 3227–3238. [CrossRef]
36. Aguilar, C.; Soronellas-Masdeu, M.; Alonso-Rey, N. El cuidado desde el género y el parentesco. Maridos e hijos cuidadores de adultos dependientes. *Quad. L'institut Català D'antropologia* **2017**, *22*, 82–98. Available online: https://core.ac.uk/download/pdf/155003475.pdf (accessed on 15 July 2021).
37. Mosquera, I.; Larrañaga, I.; Del Río Lozano, M.; Calderón, C.; Machón, M.; García, M.M. Gender inequalities in the impacts of informal elderly caregiving in Gipuzkoa: CUIDAR-SE study. *Rev. Esp. Salud Pública* **2019**, *28*, 93. Available online: https://scielo.isciii.es/scielo.php?script=sci_arttext&pid=S1135-57272019000100075 (accessed on 14 June 2021).
38. Martín-Vidaña, D. Masculinidades cuidadoras: La implicación de los hombres. españoles en la provisión de los cuidados. Un estado de la cuestión. *Rev. Prism. Soc.* **2021**, *33*, 228–260. Available online: https://revistaprismasocial.es/article/view/4095 (accessed on 14 June 2021).
39. Zygouri, I.; Cowdell, F.; Ploumis, A.; Gouva, M.; Mantzoukas, S. Gendered experiences of providing informal care for older people: A systematic review and thematic synthesis. *BMC Health Serv. Res.* **2021**, *21*, 730. [CrossRef]
40. OECD-Organización para la Cooperación y el Desarrollo Económicos. *Health at Glance: OECD Indicators*; OECD: Paris, France, 2017. Available online: https://read.oecd-ilibrary.org/social-issues-migration-health/health-at-a-glance-2017_health_glance-2017-en#page1 (accessed on 15 July 2021).
41. Wang, Y.; Li, J.; Zhang, N.; Ding, L.; Feng, Y.; Tang, X.; Sun, L.; Zhou, C. Urban–rural disparities in informal care intensity of adult daughters and daughters-in-law for elderly parents from 1993–2015: Evidence from a National Study in China. *Soc. Indic. Res.* **2020**, 1–17. [CrossRef]
42. Hungdah, S. (Ed.) *Asian Countries Strategies towards the European Union in an Inter-Regionalist Context*; National Taiwan University Press: Taipei, Taiwan, 2015; ISBN 978-986-350-056-8. GPN 1010303130.
43. Tobío, C. Redes familiares, género y política social en España y Francia. *Política Y Soc.* **2008**, *45*, 87–104. Available online: http://hdl.handle.net/10016/19789 (accessed on 14 June 2021).
44. Huenchuan, S. Envejecimiento, familias y sistemas de cuidados en América Latina. In *Envejecimiento y Sistemas de Cuidados: ¿Oportunidad o Crisis?* Huenchuan, S., Roqué, M., Arias, C., Eds.; Documentos de proyectos; CEPAL–Colec: Santiago, Chile, 2006; pp. 11–28. Available online: https://repositorio.cepal.org/bitstream/handle/11362/3859/1/S2009000_es (accessed on 1 September 2021).
45. Lopes, C.; Navarta-Sánchez, M.V.; Moler, J.A.; García-Lautre, I.; Anaut-Bravo, S.; Portillo-Vega, M.C. Psychosocial adjustment of in-home caregivers of family members with dementia and Parkinson's disease: A comparative study. *Parkinson's Dis.* **2020**, *2020*, 2086834. [CrossRef]
46. Anaut-Bravo, S.; Lopes-Dos-Santos, C. El impacto del entorno residencial en la adaptación psicosocial y calidad de vida de personas cuidadoras de familiares con demencia. *OBETS Rev. Cienc. Soc.* **2020**, *15*, 43–70. [CrossRef]
47. Anaut-Bravo, S.; Laparra, M.; García, A. Desigualdades territoriales: Una realidad de largo recorrido. In *La desigualdad y la exclusión que se nos queda: II Informe CIPARAIIS Sobre el Impacto Social de la Crisis 2007–2014*; Laparra, M., de Eulate, T.G., Eds.; Bellaterra: Barcelona, Spain, 2015; pp. 169–214.
48. Martínez Virto, L.; Pérez Eransus, B. La austeridad intensifica la exclusión social e incrementa la desigualdad. Aproximación a las consecuencias de los recortes en servicios sociales a partir de la experiencia en Navarra. *Rev. Española Del Terc. Sect.* **2015**, *31*, 65–88. Available online: https://academica-e.unavarra.es/bitstream/handle/2454/33264/101_Mart%C3%ADnez_ExclusionSocial.pdf?sequence=1&isAllowed=y (accessed on 14 June 2021).
49. Martínez Virto, L.; Pérez Eransus, B. El modelo de atención primaria de servicios sociales a debate: Dilemas y reflexiones profesionales a partir del caso de Navarra. *Cuad. Trab. Soc.* **2018**, *31*, 333–343. [CrossRef]
50. Anaut-Bravo, S. (Ed.) *El Sistema de Servicios Sociales en España*; Aranzadi-Thomson Reuters: Cizur Menor, Spain, 2019.
51. Anaut-Bravo, S.; Carbonero-Muñoz, D. El rumbo de las políticas sociales de atención sociosanitaria de las situaciones de dependencia. In *Las Políticas Sociales que Vendrán*; Zambrano, C., Ed.; Aranzadi-Thomson Reuters: Cizur Menor, Spain, 2021.
52. Pérez Eransus, B.; Martínez Virto, L. (Eds.) *Políticas de Inclusión en España: Viejos Debates, Nuevos Derechos: Un Estudio de los Modelos de Inclusión en Andalucía, Castilla y León, La Rioja, Navarra y Murcia*; CIS: Madrid, Spain, 2020.
53. García, I.; Caballeira, J.D. Receta electrónica: Diferencias entre comunidades autónomas que afectan al acceso a los tratamientos y a la calidad de la atención farmacéutica. *An. Del Sist. Sanit. Navar.* **2020**, *43*, 297–306. [CrossRef]

54. Álvarez, A.; Aranda, A.M.; García, G.; Ramírez, J.M.; Revilla, A.; Velasco, L.; Fuentes, M. Índice DEC 2020, 2021. Available online: https://directoressociales.com/project/indice-dec-2020 (accessed on 1 September 2021).

55. Porcino, A.J.; Verhoef, M.J. The use of mixed methods for therapeutic massage research. *Int. J. Ther. Massage Bodyw.* **2010**, *3*, 15–25. [CrossRef]

56. Sánchez, M.C. La dicotomía cualitativo-cuantitativo: Posibilidades de integración y diseños mixtos. *Campo Abierto* **2015**, *1*, 11–30. Available online: https://www.researchgate.net/publication/317156352 (accessed on 1 June 2021).

57. World Health Organization—WHO. *Global Action Plan on the Public Health Response to Dementia 2017–2025*; WHO: Geneva, Switzerland, 2017. Available online: https://www.alzint.org/what-we-do/partnerships/world-health-organization/who-global-plan-on-dementia/ (accessed on 6 September 2021).

58. Derogatis, L.R. The psychosocial adjustment to illness scale (PAIS). *J. Psychosom. Res.* **1986**, *30*, 77–91. [CrossRef]

59. Bullinger, M.; Alonso, J.; Apolone, G.; Leplège, A.; Sullivan, M.; Wood-Dauphinee, S.; The IQOLA Project Group. Translating health status questionnaires and evaluating their quality: The IQOLA Project approach. *J. Clin. Epidemiol.* **1998**, *51*, 913–923. [CrossRef]

60. Portillo, M.C.; Senosiain, J.M.; Arantzamendi, M.; Zaragoza, A.; Navarta, M.V.; Díaz de Cerio, S.; Riverol, M.; Martínez, E.; Luquin, M.R.; Ursúa, M.E.; et al. Proyecto ReNACE. Convivencia de pacientes y familiares con la enfermedad de Parkinson: Resultados preliminares de la Fase, I. *Revista Científica Sociedad Española de Enfermería y Neurología* **2012**, *35*, 32–39. Available online: https://www.sedene.com/wp-content/uploads/2012/08/Revista-36.pdf (accessed on 14 June 2021).

61. Derogatis, L.R.; Derogatis, M.A. *PAIS & PAIS-SR: Administration, Scoring & Procedures Manual-II Clinical Psychometric Research*. 1990. Available online: https://eprovide.mapi-trust.org/instruments/psychosocial-adjustment-to-illness-scale (accessed on 14 June 2020).

62. García-Bellido, R.; González, J.; Jornet, J.M. *SPSS: Análisis de Fiabilidad Alfa de Cronbach*; Universität de Valencia: Valencia, Spain, 2010. Available online: https://www.uv.es/innomide/spss/SPSS/SPSS_0801B.pdf (accessed on 8 June 2020).

63. Lê, S.; Josse, J.; Husson, F. FactoMineR: An R Package for Multivariate Analysis. *J. Stat. Softw.* **2008**, *25*, 1–18. [CrossRef]

64. Peduzzi, P.; Concato, J.; Feinstein, A.R.; Holford, T.R. Importance of events per independent variable in proportional hazards regression analysis II. Accuracy and precision of regression estimates. *Accuracy and precision of regression estimates. J. Clin. Epidemiol.* **1995**, *48*, 1503–1510. [CrossRef]

65. Miles, M.; Huberman, A.M.; Saldaña, J. *Qualitative Data Analysis: A Methods Sourcebook*; SAGE: Phoenix, AZ, USA, 2014.

66. Garzón, M.; Pascual, Y. Relación entre síntomas psicológicos-conductuales de pacientes con enfermedad de Alzheimer y sobrecarga percibida por sus cuidadores. *Rev. Cuba. Enfermería* **2018**, *34*, 2. Available online: http://revenfermeria.sld.cu/index.php/enf/article/view/1584/346 (accessed on 1 September 2021).

67. Tartaglini, M.F.; Ofman, S.D.; Stefani, D. El sentimiento de sobrecarga y afrontamiento en cuidadores familiares principales de pacientes con demencia. *Rev. Argent. Clínica Psicológica* **2010**, *19*, 221–226. Available online: https://www.redalyc.org/pdf/2819/281921798003.pdf (accessed on 14 June 2021).

68. Brodaty, H.; Donkin, M. Family caregivers of people with dementia. *Dialogues Clin. Neurosci.* **2009**, *11*, 217–228. [CrossRef]

69. Raggi, A.; Tasca, D.; Panerai, S.; Neri, W.; Ferri, R. The burden of distress and related coping processes in family caregivers of patients with Alzheimer's disease living in the community. *J. Neurol. Sci.* **2015**, *358*, 77–81. [CrossRef]

70. Kim, S.K.; Park, M.; Lee, Y.; Choi, S.H.; Moon, S.Y.; Seo, S.W.; Park, K.W.; Ku, B.D.; Han, H.J.; Park, K.H.; et al. Influence of personality on depression, burden, and health-related quality of life in family cargivers of persons with dementia. *Int. Psychogeriatr.* **2017**, *29*, 227–237. [CrossRef]

71. Martínez-Santos, A.E.; Facal, D.; Vicho de la Fuente, N.; Vilanova-Trillo, L.; Gandoy-Crego, M.; Rodríguez-González, R. Gender impact of caring on the health of caregivers of persons with dementia. *Patient Educ. Couns.* **2021**, *104*, 2165–2169. [CrossRef]

72. Del Río-Lozano, M.; García-Calvente, M.M.; Marcos-Marcos, J.; Entrena-Durán, F.; Maroto-Navarro, G. Gender identity in informal care: Impact on health in spanish caregivers. *Qual. Health Res.* **2013**, *23*, 1506–1520. [CrossRef]

73. García, F.E.; Manquián, E.; Rivasa, G. Bienestar psicológico, estrategias de afrontamiento y apoyo social en cuidadores informales. *Psicoperspectivas* **2016**, *15*, 101–111. [CrossRef]

74. Antelo, P.; Espinosa, P. La influencia del apoyo social en cuidadores de personas con deterioro cognitivo o demencia. *Rev. De Estud. E Investig. En Psicol. Y Educ.* **2017**, *14*, 14–21. [CrossRef]

75. Rubio, M.; Márquez, F.; Campos, S.; Alcayaga, S. Adaptando mi vida: Vivencias de cuidadores familiares de personas con enfermedad de Alzheimer. *Gerokomos* **2018**, *29*, 54–58. Available online: https://scielo.isciii.es/pdf/geroko/v29n2/1134-928X-geroko-29-02-00054.pdf (accessed on 1 October 2021).

76. Camarero-Rioja, L.; Cruz, F.; Oliva, J. Rural sustainability, inter-generational support and mobility. *Eur. Urban Reg. Stud.* **2016**, *23*, 734–749. [CrossRef]

77. Losada, A.; Márquez, M.; Vara-García, C.; Gallego, L.; Romero, R.; Olazarán, J. Impacto psicológico de las demencias en las familias: Propuesta de un modelo integrador. *Rev. Clínica Contemp.* **2017**, *8*, 1–27. [CrossRef]

78. Comas D'Argemir, D.; Soronellas-Masdeu, M. Men as carers in long-term caring: Doing gender and doing kinship. *J. Fam. Isuues* **2018**, *40*, 315–339. [CrossRef]

79. Abellán, A.; Ayala, A.; Pérez, J.; Puyol, R.; Sundström, G. *Los Nuevos Cuidadores*; Observatorio La Caixa: Barcelona, Spain, 2018. Available online: https://observatoriosociallacaixa.org/es/-/los-nuevos-cuidadores (accessed on 5 October 2021).

80. Durán, M.A. Dependientes y cuidadores: El desafío de los próximos años. *Rev. Del Minist. Trab. Y Asun. Soc.* **2006**, *60*, 57–73. Available online: https://digital.csic.es/handle/10261/100683 (accessed on 10 October 2021).
81. Navarro, M.; Jiménez, L.; García, M.C.; De Perosanz, M.; Blanco, E. Los enfermos de Alzheimer y sus cuidadores: Intervenciones de enfermería. *Gerokomos* **2018**, *29*, 79–82. Available online: https://scielo.isciii.es/pdf/geroko/v29n2/1134-928X-geroko-29-02-00079.pdf (accessed on 28 December 2021).
82. Lorenzo, T.; Millán-Calenti, J.C.; Lorenzo-López, L.; Maseda, A. Caracterización de un colectivo de cuidadores informales de acuerdo a su percepción de la salud. *Aposta Rev. Cienc. Soc.* **2014**, *62*, 1–20. Available online: http://www.apostadigital.com/revistav3/hemeroteca/jcmillan1.pdf (accessed on 16 June 2021).
83. Agulló-Tomás, M.S.; Zorrilla-Muñoz, V. Technologies and Images of Older Women. In *Human Aspects of IT for the Aged Population. Technology and Society. HCII 2020. Copenhagen, Dermark. Proceedings, Part III Lecture Notes in Computer Science*; Gao, Q., Zhou, J., Eds.; LNCS 12209; Springer: Cham, Switzerland, 2020. [CrossRef]
84. Goerlich, F.J.; Reig, E.; Cantarino, I. Construcción de una tipología rural/urbana para los municipios españoles. *Investig. Reg. J. Reg. Res.* **2016**, *35*, 151–173. Available online: http://hdl.handle.net/10017/27144 (accessed on 20 October 2021).

 *land*

Article

Gender Differences in Environmental Correlates of Cycling Activity among Older Urban Adults

Lanjing Wang [1], Xiayidan Xiaohelaiti [1], Yi Zhang [1,*], Xiaofei Liu [2], Xumei Chen [2], Chaoyang Li [1], Tao Wang [1] and Jiani Wu [1]

[1] State Key Laboratory of Ocean Engineering, China Institute for Urban Governance, Shanghai Jiao Tong University, Shanghai 200240, China; lanjing.wang@sjtu.edu.cn (L.W.); xayda.x@sjtu.edu.cn (X.X.); cyljjf@sjtu.edu.cn (C.L.); wangtao127@sjtu.edu.cn (T.W.); JiaNiZi@sjtu.edu.cn (J.W.)
[2] Key Laboratory of Advanced Public Transportation Science, China Academy of Transportation Sciences, MOT, Beijing 100029, China; liuxf@motcats.ac.cn (X.L.); chenxm@motcats.ac.cn (X.C.)
* Correspondence: darrenzhy@sjtu.edu.cn

Abstract: Cycling is a form of active transport that can improve the level of health among the elderly population. However, little is known about the environmental correlates of bicycle use among older adults. This study investigated the relationship between the built and social environment and the gender differences in cycling frequency among older urban adults in China. The data were derived from a household travel survey in 2012 and covered thirty-three urban neighborhoods in Zhongshan. The results suggest that denser intersections are negatively related to cycling trips among both older men and women. Reverse associations for either gender, however, are observed between the average income in a neighborhood and cycling frequency. For older women, living far from a bus stop is positively correlated to an increase in daily cycling trips. For older men, social environment, including the proportions of employed or elderly people in a neighborhood, is significantly associated with cycling activity. The findings facilitate the understanding of the gender gap in cycling activity among older urban adults, and help towards designing effective planning strategies as health interventions.

Keywords: gender differences; built environment; social environment; cycling activity; older urban adults

Citation: Wang, L.; Xiaohelaiti, X.; Zhang, Y.; Liu, X.; Chen, X.; Li, C.; Wang, T.; Wu, J. Gender Differences in Environmental Correlates of Cycling Activity among Older Urban Adults. *Land* **2022**, *11*, 52. https://doi.org/10.3390/land11010052

Academic Editors: Vanessa Zorrilla-Muñoz, Eduardo Fernandez, Blanca Criado Quesada, Sonia De Lucas Santos, Jesus Cuadrado Rojo and Maria Silveria Agulló-Tomás

Received: 29 November 2021
Accepted: 29 December 2021
Published: 30 December 2021

Publisher's Note: MDPI stays neutral with regard to jurisdictional claims in published maps and institutional affiliations.

1. Introduction

The elderly population is the fastest-growing group worldwide [1]. The world is expected to experience a significant demographic shift over the next few decades. By 2050, the world's elderly population (65 years and older) is estimated to reach 1.5 billion, a drastic growth from 727 million in 2020 [2]. Between 2020 and 2050, the proportion of the global older adult population will nearly double from 9.3% to 16% [3]. By 2050, 80% of older adults will be living in low- and middle-income countries. This raises major challenges to ensure that both health and social systems are ready to face this demographic shift [3].

Research findings indicate that active transport provides significant health benefits to the elderly population [4]. As a crucial form of active transport, cycling helps to prevent chronic diseases and bone fractures in older adults [5]. A recent review suggests that the health benefits of cycling outweigh the risks [6]. Older adults who choose cycling over driving for short trips have the most statistically significant estimated gain in life, compared to other age subgroups [6]. Specifically, for older women, a small amount of daily bicycling reduces the risk of atrial fibrillation and is beneficial for muscle strength and functional abilities [7,8]. Therefore, cycling may be a viable option for older adults where a favorable policy exists and a bicycle-friendly social and built environment is created [9].

Considering the benefits, policies have been implemented to enhance bicycle use among the general population. For example, in the Netherlands and Denmark, the cycling transportation infrastructure is sufficient and the modal choice of cycling is high, indicating the potential of bicycle use among older individuals [10]. However, few policies targeting the elderly population can be found, possibly due to the risks (exposure to traffic and air pollution) and physical barriers preventing older adults from cycling. The environmental correlates of cycling activity have been revealed among the general population. Nevertheless, the potential of cycling among older adults has been neglected for research and practice [11]. The social and built environment may facilitate or restrict older adults' participation in cycling activity [1]. There was limited evidence revealing the factors significantly associated with cycling activity among older adults [12]. Furthermore, older women partake in less bicycle use than their male counterparts. However, few studies have investigated the gender differences in the effects of the built and social environment on bicycle use among older adults.

Which environmental features are associated with cycling activity among older urban adults? Are there any gender differences in the environment-cycling activity relationships? Which planning strategies offer the most promising effects for increasing cycling among older urban men and women? This study attempts to answer these questions by investigating the gender-specific environmental correlates of bicycle use among older urban adults. Using data from thirty-three urban neighborhoods in Zhongshan, China, we investigated the environmental correlates of older urban adults' daily cycling trips for transportation and recreation, controlling for attitudinal and sociodemographic attributes. The findings will facilitate planners' and policymakers' attempts to increase bicycle use among both older urban men and women by optimizing land use planning and improving environmental conditions.

This paper is organized as follows: Section 2 reviews relevant literature. Section 3 details the data and method. Section 4 presents findings from negative binomial regression. Section 5 includes discussion and policy implications, and Section 6 concludes with limitations and strengths.

2. Literature Review

The planning [13–15] and public health [16–18] fields have mutually contributed to "environment-active transport" studies, with a focus on older adults. However, most research emphasized walking, with very little on cycling [19,20]. Generally, previous studies categorized the environmental factors as two types: The built environment and the social environment [21].

The built environment is defined as "the human-made space in which people live, work, and recreate on a day-to-day basis" [22] (p. 24) and "encompasses places and spaces created or modified by people including buildings, parks, and transportation systems" [23] (p. 1446). Prior studies have examined the built environment factors relating to older adults' cycling in different contexts [24–27]. However, few studies have explored the associations among the aging population in developing countries. The built environment factors commonly utilized are defined as the "five Ds": density, design, diversity, distance to transit, and destination accessibility [28]. A denser population is related to a higher propensity to cycle among older adults in China [29]. However, the urban density shows negativity in explaining bicycle use among older adults in the Netherlands [30]. Well-designed cycling infrastructure has shown positive effects, as expected [25,31–34]. Mixed development is linked to an increase in cycling frequency among older adults [24,25,29,31,35]. Being adjacent to services and destinations is also attributed to a higher propensity to cycle [25,26,36,37]. Inadequate transit services are related to more cycling trips, as older adults may opt to cycle for medium-distance trips [25,35,38].

The social environment "includes the culture that the individual was educated or lives in, and the people and institutions with whom they interact" [39] (p. 465). Role models and neighborhood social cohesion were associated with walking duration, and peer support may facilitate physical activity among older adults [21,22]. The role models in previous studies included both the young and the old who favor active transport most. Another study in Zhongshan, China, found that when the proportion of the elderly population exceeds 15% in a neighborhood, bus use among older adults decreases [40]. In denser areas with a higher ratio of younger adults, young travelers may choose the bus as part of an active lifestyle. They will act as role models for older adults. Therefore, in areas with a higher ratio of older people, the effects of role models will decrease, leading to a decrease in the modal share of bus use. However, little is known about the social environment correlates of older adults' cycling activity. A negative association was observed between the proportion of the aged in a community and cycling among rural older adults in China [25]. Factors related to the social image of cycling also influenced the cycling among older adults [9].

Gender-specific differences between environmental factors and physical activity among older adults have been preliminarily examined. However, most studies focused on walking. Women were less active than men, and environmental factors were significantly related to women's physical activity and walking [41]. Self-efficacy, density, and design were related to older men's walking, and self-efficacy and destinations were related to older women [42].

Most "environment-active transport" research has been conducted in developed countries. The findings are sometimes not transferable to developing countries with ultra-high population densities, including China [18]. Recently, scholars have begun to explore how environmental variables are related to active transport and health promotion among different age groups [43]. Nonetheless, little research has been carried out to explore the gender-specific environment correlates of cycling frequency among older urban adults throughout the world. The cycling activity of older urban adults is essential for efficacious interventions on health promotion. This study addresses these gaps by specifically examining the built and social environment correlates of bicycle use among older urban men or women in Zhongshan, China. In this paper, "cycling" includes utilitarian and recreational trips. Older adults focus on the group aged above 59.

It is worth noting the dataset and modeling approach in the present study are the same as in our previous study [44]. However, the two studies are significantly different regarding research focus and contribution. Our previous study [44] focused on the general older urban population and employed only built environment variables. It is among the earliest attempts to explore the environmental correlates of cycling among older adults [44]. This time we concentrated on the gender differences in the influences of both built and social environment on cycling trips of older urban adults. The findings will help the design of gender-specific interventions to facilitate cycling among older urban men and women and contribute to the land use-travel literature.

3. Data and Methods

3.1. Study Area

In China, the elderly population is defined as adults aged over 59, according to the Law of the People's Republic of China on Protection of the Rights and Interests of the Elderly. By 2020, the elderly Chinese population was over 200 million, reaching nearly 15% of the general population. It is necessary to encourage active transport among older Chinese adults as a health intervention. In 2007, the "China Healthy Lifestyle for All" initiative launched a campaign of "Ten Thousand Steps a Day" [45], aiming to promote walking among the general population. Cycling was once a prevailing travel mode in China; however, the modal split of cycling has decreased due to rapid motorization [46]. The factors facilitating cycling among Chinese older adults have barely been studied, and few interventions have been initiated to revive bicycle use in China. We chose Zhongshan to disentangle the gender-specific cycling activity among older urban adults in China [47,48]. Zhongshan is a medium-sized city in Guangdong Province, China (Figure 1). In China's urban agglomerations, there are over two dozen cities with similar urban transport patterns to Zhongshan [49]. Therefore, the findings of the study may transfer to those cities.

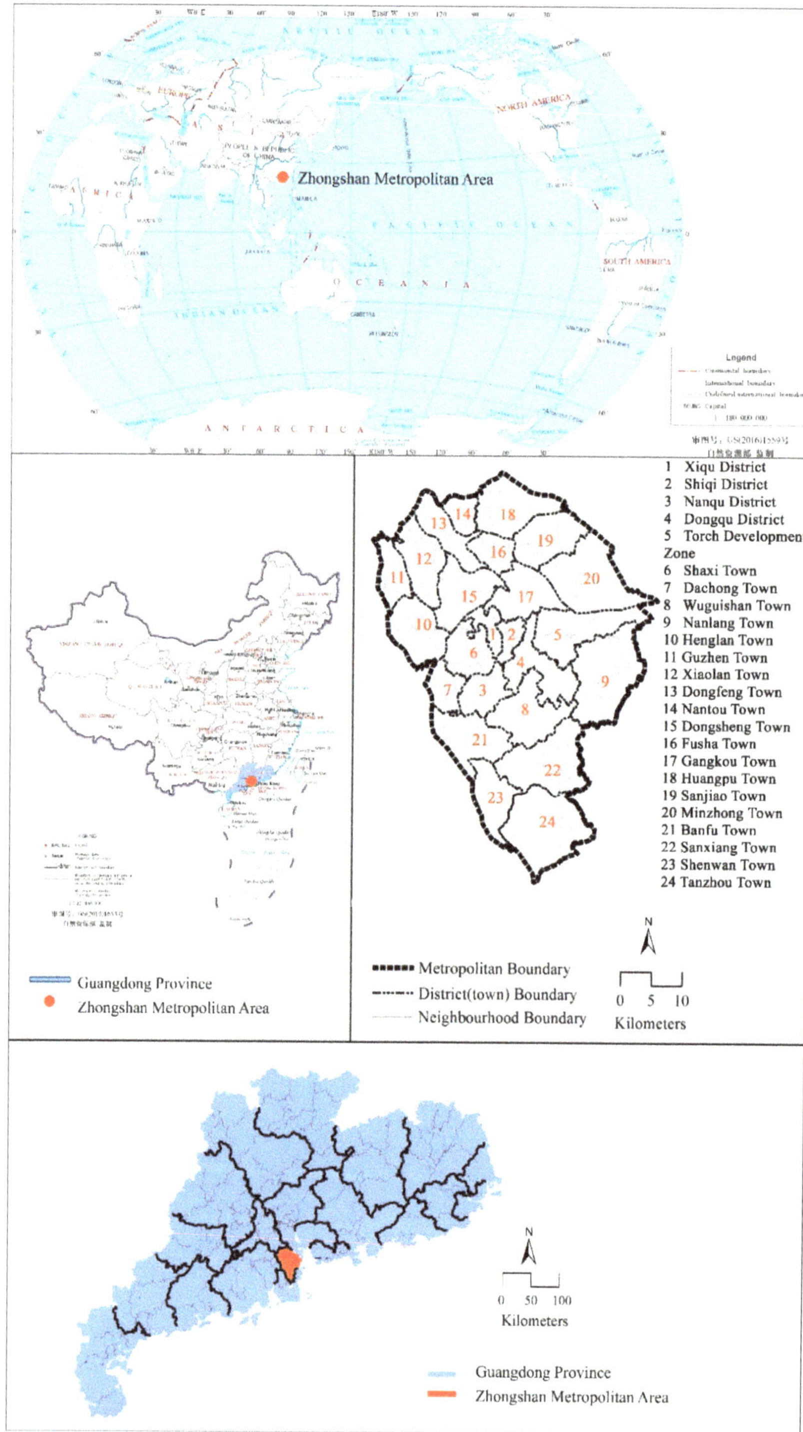

Figure 1. Study area.

3.2. Data Collection

The Zhongshan Household Travel Survey (ZHTS) in 2012 provided the bicycle use data [50]. The survey covered the entire Zhongshan Metropolitan Area. The survey included a self-reported one-day travel diary and the sociodemographics of the urban adults. The sampling method used was stratified random sampling. The sample size was 616 older urban women and 648 men from thirty-three neighborhoods. The sample rate was 2.0%.

The built and social environment data were provided by Zhongshan Municipal Bureau of Urban Planning in 2012 [47,48]. We imported the data into ArcGIS for analysis. The data included: (1) neighborhood boundaries and neighborhood-level socio-demographics (population, dwelling units, and employment); (2) five types of land uses (residential land, commercial land, industrial land, green space land, and other land); (3) road networks; and (4) bus stops.

3.3. Characterization of Social Environment and Built Environment Variables

The environmental variables were characterized based on neighborhood level [51]. Based on the administrative divisions of Zhongshan, thirty-three urban neighborhoods were selected for analysis. We characterized three neighborhood-level social environment variables, i.e., the proportions of the employed, the ratios of the aged, and the average household income.

We characterized five built environment variables according to the "five Ds" [52]. The dwelling unit density represents Density; the intersection density denotes Design; the distance between home and the closest bus stop defines Distance to Transit; the distance to the CBD stands for Destination Accessibility; and the land use mixture demonstrates Diversity [52]. The first four variables are self-explanatory. The land use mixture denotes the degree to which different land uses in a neighborhood are mixed. The land use mixture was calculated by the Entropy Index (*EI*) [53], wherein 0 denotes a single-use environment and 1 represents the equalization of different land uses in area coverage. *EI* is expressed by:

$$EI = \sum_{i=1}^{n} P_i \log(1/P_i) \tag{1}$$

where n = number of different land uses, $n \geq 1$; Pi = percentage of land use i's coverage in the entire land use coverage. In the Chinese standard, the officially recommended proportion of residential, industrial, commercial, green space, and other types of land use is around 2:2:1:1:1 [54], generating an *EI* of 0.67. This proportion is employed in land use planning practices of Chinese cities, including Zhongshan. Therefore, each of the original *EI* of a neighborhood in this study is transformed into a criterion that 0.67 is the standard 1, and all other indexes are ranged between 0 and 1 based on the standard 1 [47].

3.4. Modeling Approach

Older urban adults' cycling trips are non-negative count-dependent variables. Considering the statistical characteristics of the cycling trips (Appendix A), we first chose negative binomial regression. We then applied a Vuong model selection test, and the result preferred a standard one over a zero-inflated one (Figure 2). Finally, we utilized a negative binomial regression model to examine the gender-specific sociodemographics and the attitudinal and environmental correlates of bicycle use among older urban adults. The multicollinearity of all the independent variables were checked by the variance inflation factor (VIF). None of the VIFs were more significant than ten, suggesting a low degree of multicollinearity.

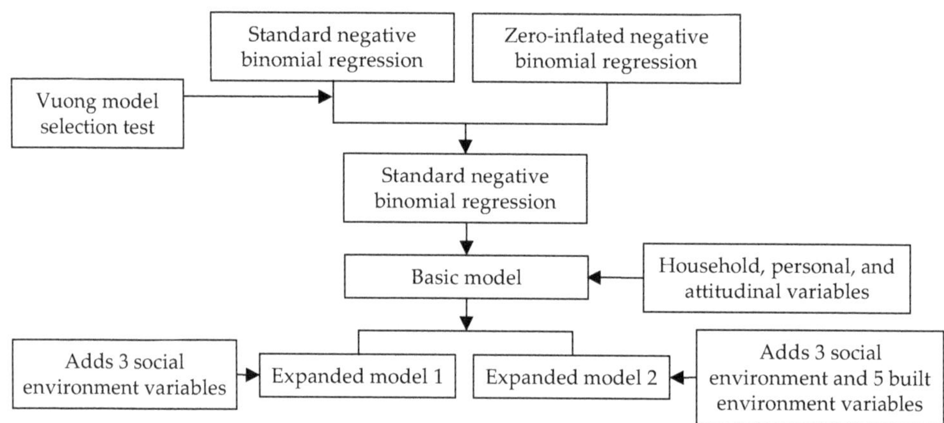

Figure 2. The modeling approach.

We built models separately for older men and women and used the same independent variable sets. We rewrote the negative binomial regression model with natural log transformation for calibrating the coefficient with Stata 12.0. The basic model was as follow:

$$N_{fr} = \beta_0 + \beta_1 \times \text{SIZE\_1} + \beta_2 \times \text{SIZE\_2} + \beta_3 \times \text{EMPLOYED} + \beta_4 \times \text{H\_HIGHINC} +$$
$$\beta_5 \times \text{H\_MEDINC} + \beta_6 \times \text{AGE} + \beta_7 \times \text{P\_HIGHINC} + \beta_8 \times \text{P\_MEDINC} + \quad (2)$$
$$\beta_9 \times \text{FAV\_BIKE} + \beta_{10} \times \text{FAV\_WALK} + \beta_{11} \times \text{FAV\_EBIKE} + \beta_{12} \times \text{FAV\_BUS}$$

where N_{fr} = frequency (times/day) of cycling of an older urban adult; SIZE_1 and SIZE_2 = dummies for the household size of one and two (with a household size of more than two as a reference category); EMPLOYED = number of family members employed; H_HIGHINC and H_MEDINC = dummies for high household monthly income over 6000 Chinese Yuan (Renminbi) (RMB, 6.4 Renminbi $\approx$ 1 US Dollar) and medium between 2500 and 6000 RMB (with a reference category of 0–2500 RMB); AGE = respondent's age in years; P_HIGHINC and P_MEDINC = dummies for high personal monthly income over 3000 RMB and medium between 1200 and 3000 RMB (with 0–1200 RMB as a reference category), FAV_BIKE, FAV_WALK, FAV_EBIKE, or FAV_BUS = whether the respondent's favorite travel mode is bicycle, walking, e-bike, or bus.

The regression proceeded in two expanded models based on the basic model (Figure 2). The expanded model 1 adds three neighborhood-level social environment attributes as independent variables, where N_EMPLOY and N_AGED demonstrate the proportions of the employed and the aged in a neighborhood, respectively; N_AVGINC represents the average monthly income of a neighborhood. The expanded Model 2 includes five built environment variables, in which DWELL_DEN, INTER_DEN, DIST_BUS, DIST_CBD, and LAND_MIX denote dwelling unit density, intersection density, the distance between home and the closest bus stop, Euclidean distance from the centroid of the neighborhood to the CBD, and land use mixture. Among the five attributes, DIST_CBD is a household-level attribute, and the other four are neighborhood-level.

4. Results

4.1. Descriptive Statistics

In Zhongshan, older urban men and women make 3.06 and 2.84 trips per day, respectively, among which 0.37 (men) and 0.18 (women) trips were cycling (Tables 1 and A1) [55]. One in five older urban adults lives alone, and 40% live with one partner. One-fifth live in high-income households. About 16% of males preferred cycling to other modes, whereas that figure was only 9% for females. The respondents' average age was 67. The social and built environment of both men and women was very similar due to the random sampling method.

Table 1. Description of variables.

Category	Variable	Description
Dependent Variable	**Frequency**	**Daily Cycling Trips, Count**
Household variables	SIZE_1	One member in a household, binary, 1 = yes
	SIZE_2	Two members in a household, binary, 1 = yes
	SIZE_2+	Three or more members in a household, binary, 1 = yes
	EMPLOYED	Number of household members employed, count
	H_HIGHINC	High household income (>6000 RMB/month), binary, 1 = yes
	H_MEDINC	Medium household income (2500–6000 RMB/month), binary, 1 = yes
	H_LOWINC	Low household income (<2500 RMB/year), binary, 1 = yes
Personal variables	AGE	Age in years, count
	P_HIGHINC	High personal income (>3000 RMB/month), binary, 1 = yes
	P_MEDINC	Medium personal income (1200–3000 RMB/month), binary, 1 = yes
	P_LOWINC	Low personal income (<1200 RMB/month), binary, 1 = yes
Attitudinal variables	FAV_BIKE	The respondent's favorite travel mode is bicycle, binary, 1 = yes
	FAV_WALK	The respondent's favorite travel mode is walking, binary, 1 = yes
	FAV_EBIKE	The respondent's favorite travel mode is e-bike, binary, 1 = yes
	FAV_BUS	The respondent's favorite travel mode is bus, binary, 1 = yes
	FAV_CAR	The respondent's favorite travel mode is car, binary, 1 = yes
Social environment variables	N_EMPLOY	Proportions of the employed in a neighborhood, continuous
	N_AGED	Proportions of the aged in a neighborhood, continuous
	N_AVGINC	Average monthly income of a neighborhood (in 100 RMB), continuous
Built environment variables	DWELL_DEN	Dwelling units' density, 1000 dwelling units/km^2, continuous
	INTER_DEN	Intersection density, five intersections/km^2, continuous
	DIST_BUS	Distance between home and the closest bus stop, km, continuous
	DIST_CBD	Euclidean distance from the neighborhood centroid to the CBD, in km, continuous
	LAND_MIX	Entropy Index of land use mixture, continuous

4.2. Analysis of Cycling Frequency among Older Urban Men and Women

Generally, the directions of the coefficients for sociodemographics and attitudes persisted across all models, and the coefficients indicated slight to moderate variation. The LR chi2 and Log-likelihood demonstrates the overall goodness of fit. The variations of pseudo-R2, LR chi2, and Log-likelihood in expanded models suggested that the environmental variables strengthened the explanatory power and predictability of the models (Table 2).

Two personal attributes (medium personal income and favoring bicycling over other modes), one social environment attribute (average neighborhood income), and one built environment (intersection density) attribute were significantly related to both males and females. The male respondents who were pro-bicycle would have 4.37 times more cycling trips than those who were not, while the number was even more prominent for female respondents, at 7.65 times. Regarding social environments, a 100 RMB increase in average neighborhood monthly income was related to 7.35% more cycling trips for older urban men, but 15.51% fewer for women. The cycling trips decreased by 7.39% (for men) or 18.88% (for women) when the intersection density increased by one unit, which is 5 intersections per km^2.

Table 2. Negative binomial regression analysis of cycling frequency among older urban men and women in Zhongshan.

Variable	Basic Model Coef.		Expanded Model 1 Coef.		Expanded Model 2 Coef.	
	Men	Women	Men	Women	Men	Women
Household socio-demographics (SIZE > 2 and H_LOWINC are reference categories)						
SIZE_1	−0.023	−1.273 **	−0.110	−1.050 ***	−0.041	−0.793 ***
SIZE_2	0.282	−1.853 *	0.191	−1.409 *	0.205	−1.463 *
EMPLOYED	−0.039	−1.195 *	−0.016	−1.136 *	−0.054	−1.258 *
H_HIGHINC	0.751 **	0.943	0.840 **	1.193	0.790 **	1.260
H_MEDINC	0.262	0.347	0.305	0.321	0.295	0.372
Personal socio-demographics (P_LOWINC is a reference category)						
AGE	−0.023 ***	−0.034	−0.024 ***	−0.044	−0.027 ***	−0.037
P_HIGHINC	−0.896 **	0.413	−1.110 **	0.696	−1.124 **	1.204
P_MEDINC	−0.835 *	−1.039 **	−0.992 *	−0.919 **	−0.991 *	−0.594 ***
Attitudes (FAV_CAR is a reference category)						
FAV_BIKE	1.681 *	2.157 *	1.739 *	2.030 *	1.817 *	2.103 *
FAV_WALK	−0.291	−0.315	−0.208	−0.362	−0.139	−0.251
FAV_EBIKE	−0.347	0.496	−0.272	0.419	−0.258	0.047
FAV_BUS	−2.403 *	−0.903	−2.281 *	−0.731	−2.192 *	−0.507
Social environment						
N_EMPLOY			0.308 *	−1.032	0.586 **	−0.700
N_AGED			−0.385 **	−1.756	−0.658 ***	−1.720
N_AVGINC			0.071 **	−0.168 *	0.087 *	−0.071 ***
Built environment						
DWELL_DEN					0.024	−0.005
INTER_DEN					−0.077 ***	−0.209 ***
DIST_BUS					−0.254	1.083 ***
DIST_CBD					−0.215 ***	0.017
LAND_MIX					−0.545	−1.483
Summary statistics						
_cons	−0.382	0.869	−5.264	5.690	−4.654	1.290
Number of obs.	648	616	648	616	648	616
LR chi2	283.63	128.07	296.46	138.89	302.97	148.38
Prob > chi2	0.0000	0.0000	0.0000	0.0000	0.0000	0.0000
Pseudo-R2	0.2306	0.2470	0.2411	0.2679	0.2464	0.2861
Log-likelihood	−473.0756	−195.2382	−466.6582	−189.8262	−463.4049	−185.0829

Note: *** represents significance at $p < 0.01$, ** represents significance at $p < 0.05$, and * represents significance at $p < 0.1$. Blank cells denote variables were not included in the model. Obs = observations; LR = likelihood ratio; chi2 = chi-square; prob = probability.

Seven attributes were significantly associated solely with older urban men's cycling activity, covering all four categories. At a household and personal level, household or personal income, age, and a positive attitude towards public transportation showed significance at 90% confidence. Being one year older was related to a 2.26% reduction in cycling trips. Male respondents with high or medium incomes had 59.17% or 56.62% fewer cycling trips, respectively, than those with low incomes. Older urban men who preferred the bus to other modes made 90.96% fewer cycling trips. Regarding social environment variables, male respondents living in a neighborhood with the highest proportion of employed or aged people made 11.86% more or 8.04% fewer cycling trips, respectively, compared to those residing in areas with the lowest proportions. The only built environment attribute significantly related to male respondents was the distance to the CBD. As the results showed, living one kilometer farther from the CBD was associated with 19.35% fewer cycling trips.

Four attributes were significantly associated solely with older urban women's cycling activity, among which three were related to household and one to built environment. Living alone or with a partner, female respondents made 72.01% or 84.32% fewer cycling trips, respectively, compared to those with more household members. Having one more family member employed was related to a 69.73% reduction in cycling trips. At the built environment level, older urban women make 19.53% more cycling trips if the closest bus stop is one hundred meters farther from home.

5. Discussion and Policy Implications

5.1. Discussion

One built environment variable, intersection density, is significantly related to bicycle use among both older urban men and women. In a neighborhood with denser intersections, both men and women make significantly fewer cycling trips. To be specific, the negative association of intersection density with female respondents' cycling trips was much stronger. The reasons for this finding are complex. On the one hand, this may be due to safety concerns, as a recent study suggested that cyclists are exposed to a higher risk of accidents when crossing intersections [56]. Another study in Italy, however, observed a decreased risk of pedestrian accidents at road intersections, as both pedestrians and drivers pay more attention when approaching the street nodes. [57] On another hand, denser intersections are positively related to a higher traffic density, represented by more vehicles on the roads [33,34]. The safety issues related to the conflicts of bicycles with other modes, especially vehicles, may increase the perceived barriers towards cycling activity among older adults [32]. The average monthly income of the neighborhood, as a social environment variable, shows reverse impacts on older urban men and women. Residing in a more affluent neighborhood, older urban men make more cycling trips, while older urban women make fewer. The underlying reasons for the results call for further study. As expected, favoring cycling over other modes is associated with an increase in bicycle use among both male and female respondents. The results indicate the potential of disseminating a healthy lifestyle regarding cycling among older urban adults.

The attributes related solely to male respondents include household and personal characteristics, the social environment, and the built environment. Being older, richer, or favoring the bus over cycling is linked to fewer cycle trips. Living in a more dynamic social environment with more employed people and fewer older people is related to an increase in cycling trips for older urban men. The results are probably connected to social norms concerning active transport [58]. Previous research has observed positive relationships between social promoters, such as having neighbors that bicycle and greater physical activity [33]. In this study, physically active older urban men may opt to reside in communities with younger or employed neighbors. Presumably, the younger or employed population tend to choose more active travel modes, including cycling. As role models, they may influence older adults to cycle as a part of an active lifestyle. However, the in-depth reasons require future study. The findings indicate that forming neighborhoods with relatively balanced age or employment structures may facilitate cycling among older urban men. The built environment variable regarding commercial accessibility demonstrated a significant association. Better accessibility to the CBD was related to more bicycle use. Presumably, older urban men will opt to cycle for short-to-medium commercial trips if they live adjacent to the CBD [59]. This finding is in line with existing literature that suggests that access to destinations appeared to be important for promoting cycling among older adults [24].

The attributes related solely to female respondents included household characteristics and the built environment. For example, household size is positively associated with older urban women's cycling activity. Living with two or more family members was linked to more cycling trips. The distance between home and the closest bus stop is the only significant built environment factor for older urban women. Specifically, living farther from a bus stop is correlated with an increase in cycling trips. Presumably, when the closest bus stop is beyond walking distance from home, older urban women may choose cycling to access the bus stop, or even shift from bus to bicycle for the whole trip [40].

5.2. Policy Implications

To encourage bicycle use among older urban men and women in China, planning and public health policies should consider the gender differences in the effects of environmental factors. For both men and women, a safe environment at intersections and a positive attitude towards cycling may be effective, albeit to varying degrees. Therefore, we recommend two interventions: (1) enhancing safety for older cyclists at intersections, and (2) disseminating an active lifestyle oriented towards cycling activity. Safety issues at intersections may be a significant reason for older urban adults to make fewer cycling trips. Possible interventions include the improvement of road intersection design, traffic signals, and traffic management measures. Regarding attitudes, we recommended various initiatives (health-focused campaigns [60], public lectures, specialized websites, etc.) that have shown success in the "Ten Thousand Steps a Day" program [61].

For older women, less dense bus stops are associated with more cycling trips, implying that cycling interventions targeting older urban women may incorporate transit. However, this invention may harness bus use among older urban women. Therefore, the threshold effects and effective range of bus stop accessibility for both bicycle and bus use require further non-linear analysis.

For older men, interventions should consider the age structure and employment structure of a neighborhood and CBD accessibility. It may be effective to form vibrant neighborhoods with well-balanced age and employment structures. Locating more commercial establishments and service destinations adjacent to neighborhoods might also increase bicycle use among older urban men.

6. Conclusions

This study contributes to the land use–travel literature by disentangling the gender-specific relationship between environmental attributes and cycling activity among older urban adults, with data from Zhongshan, China. First, the study characterized built environment and social environment variables, together with sociodemographics and attitudes. Second, the study employed negative binomial regression to investigate the gender-specific environmental correlates of daily cycling trips among older urban adults. The results indicate that intersection density and attitudes towards different travel modes are significantly correlated to bicycle use among both genders in Zhongshan. For older urban men, the proportions of the aged or employed in a neighborhood and commercial accessibility show significance. However, bicycle use among older urban women is more correlated to transit service accessibility.

The findings facilitate the health promotion interventions and urban planning approach to accommodate older urban men and women equally from the perspective of built and social environment and attitudes. For both older urban men and women, we suggest creating a safe environment for cyclists at intersections and disseminating an active lifestyle relating to bicycle use. For older women, we suggest further discovering the threshold effects and the effective range of bus stop accessibility in favor of both bicycle and bus use. For older men, we suggest maintaining a relatively balanced age and employment structure in neighborhoods and providing abundant commercial and service destinations adjacent to residences.

This study has some limitations regarding data and method. First, the analysis employed cross-sectional data. The causal effects of different variables on bicycle use will require longitudinal data. Second, the study used linear modeling. Recent studies have begun to examine the non-linear relationships and threshold effects of environmental variables. The study yields several strengths. First, the study addressed the gender gap in the land use–travel literature. The findings provide nuanced policy implications for healthy aging. Second, the results indicate gender-specific environmental correlates of bicycle use among older urban adults in China. The findings will facilitate comparative studies among different contexts.

Author Contributions: L.W. led the survey design and manuscript preparation. Y.Z. conceived the research. X.L., X.C., J.W. and X.X. participated in data collection. C.L. and T.W. contributed to modeling. All authors have read and agreed to the published version of the manuscript.

Funding: This research was funded by the National Social Science Foundation of China (Grant No. 18BSH143).

Institutional Review Board Statement: Ethical review and approval were not required for the study on human participants in accordance with the local legislation and institutional requirements.

Informed Consent Statement: The participants provided their written informed consent to participate in this study.

Data Availability Statement: The data employed in this study are not readily available because they belong to ongoing projects (Grant No. 20692109900 and Grant No. 21692106700 of Shanghai Science and Technology Program, and Grant No. 2020-APTS-04 of APTSLAB).

Conflicts of Interest: The authors declare no conflict of interest.

Appendix A

Table A1. Descriptive statistics for variables.

Variable	Description	Male (Sample Size = 648)				Female (Sample Size = 616)			
		Mean	S. D.	Min.	Max.	Mean	S. D.	Min.	Max.
	Dependent variable								
Frequency	Daily cycling trips, count	0.37	0.91	0	8	0.18	0.63	0	5
	Household variables								
SIZE_1	One member in a household, binary, 1 = yes	0.20	0.40	0	1	0.20	0.40	0	1
SIZE_2	Two members in a household, binary, 1 = yes	0.41	0.49	0	1	0.40	0.49	0	1
SIZE_2+	Three or more members in a household, binary, 1 = yes	0.39	0.49	0	1	0.40	0.49	0	1
EMPLOYED	Number of household members employed, count	0.70	0.96	0	5	0.71	0.93	0	4
H_HIGHINC	High household income (>6000 RMB/month), binary, 1 = yes	0.19	0.40	0	1	0.21	0.41	0	1
H_MEDINC	Medium household income (2500–6000 RMB/month), binary, 1 = yes	0.54	0.50	0	1	0.55	0.50	0	1
H_LOWINC	Low household income (<2500 RMB/year), binary, 1 = yes	0.26	0.44	0	1	0.24	0.43	0	1
	Personal variables								
AGE	Age in years, count	67.62	6.68	60	95	67.25	6.52	60	93
P_HIGHINC	High personal income (>3000 RMB/month), binary, 1 = yes	0.07	0.26	0	1	0.02	0.14	0	1
P_MEDINC	Medium personal income (1200–3000 RMB/month), binary, 1 = yes	0.74	0.44	0	1	0.76	0.43	0	1
P_LOWINC	Low personal income (<1200 RMB/month), binary, 1 = yes	0.19	0.40	0	1	0.22	0.42	0	1
	Attitudinal variables								
FAV_BIKE	The respondent's favorite travel mode is bicycle, binary, 1 = yes	0.16	0.37	0	1	0.09	0.28	0	1
FAV_WALK	The respondent's favorite travel mode is walking, binary, 1 = yes	0.27	0.45	0	1	0.39	0.49	0	1
FAV_EBIKE	The respondent's favorite travel mode is e-bike, binary, 1 = yes	0.07	0.25	0	1	0.04	0.19	0	1
FAV_BUS	The respondent's favorite travel mode is bus, binary, 1 = yes	0.21	0.41	0	1	0.27	0.44	0	1
FAV_CAR	The respondent's favorite travel mode is car, binary, 1 = yes	0.19	0.37	0	1	0.21	0.40	0	1
	Social environment variables								
N_EMPLOY	Proportions of the employed in a neighborhood, continuous	0.60	0.08	0.49	0.82	0.60	0.07	0.49	0.82
N_AGED	Proportions of the aged in a neighborhood, continuous	0.17	0.07	0.04	0.29	0.18	0.07	0.04	0.29
N_AVGINC	Average monthly income of a neighborhood (in 100 RMB), continuous	15.19	3.90	9.62	25.87	15.36	3.82	9.62	25.87
	Built environment variables								
DWELL_DEN	Dwelling units' density, 1000 dwelling units/km^2, continuous	9.78	5.86	1.11	29.07	10.10	5.72	1.11	29.07
INTER_DEN	Intersection density, five intersections/km^2, continuous	5.75	3.38	0.40	12.36	6.11	3.41	0.40	12.36
DIST_BUS	Distance between home and the closest bus stop, km, continuous	0.37	0.27	0.10	1.20	0.37	0.27	0.10	1.20
DIST_CBD	Euclidean distance from the neighborhood centroid to the CBD, in km, continuous	2.04	0.68	0.22	3.09	2.02	0.72	0.22	3.09
LAND_MIX	Entropy Index of land use mixture, continuous	0.68	0.18	0.33	1.00	0.67	0.19	0.33	1.00

Note: S. D. = Standard Deviation; Min. = minimum; Max. = maximum.

References

1. Wu, J.; Zhao, C.; Li, C.; Wang, T.; Wang, L.; Zhang, Y. Non-linear Relationships Between the Built Environment and Walking Frequency Among Older Adults in Zhongshan, China. *Front. Public Health* **2021**, *9*, 686144. [CrossRef] [PubMed]
2. World Health Organization. *World Report on Ageing and Health*; WHO: Geneva, Switzerland, 2015.
3. United Nations. *World Population Ageing 2020-Highlights*; Department of Economic and Social Affairs, United Nations: New York, NY, USA, 2020.
4. Nelson, M.E.; Rejeski, W.J.; Blair, S.N.; Duncan, P.W.; Judge, J.O.; King, A.C.; Macera, C.A.; Castaneda-Sceppa, C. Physical activity and public health in older adults: Recommendation from the American College of Sports Medicine and the American Heart Association. *Med. Sci. Sports Exerc.* **2007**, *39*, 1435. [CrossRef] [PubMed]
5. Keysor, J.J. Does late-life physical activity or exercise prevent or minimize disablement? A critical review of the scientific evidence. *Am. J. Prev. Med.* **2003**, *25*, 129–136. [CrossRef]
6. De Hartog, J.J.; Boogaard, H.; Nijland, H.; Hoek, G. Do the Health Benefits of Cycling Outweigh the Risks? *Environ. Health Perspect.* **2010**, *118*, 1109–1116. [CrossRef] [PubMed]
7. Drca, N.; Wolk, A.; Jensen-Urstad, M.; Larsson, S.C. Physical activity is associated with a reduced risk of atrial fibrillation in middle-aged and elderly women. *Heart* **2015**, *101*, 1627–1630. [CrossRef] [PubMed]
8. Macaluso, A.; Young, A.; Gibb, K.S.; Rowe, D.A.; De Vito, G. Cycling as a novel approach to resistance training increases muscle strength, power, and selected functional abilities in healthy older women. *J. Appl. Physiol.* **2003**, *95*, 2544–2553. [CrossRef]
9. Winters, M.; Sims-Gould, J.; Franke, T.; McKay, H. "I grew up on a bike": Cycling and older adults. *J. Transp. Health* **2015**, *2*, 58–67. [CrossRef]
10. Handy, S.; van Wee, B.; Kroesen, M. Promoting Cycling for Transport: Research Needs and Challenges. *Transp. Rev.* **2014**, *34*, 4–24. [CrossRef]
11. Metz, D. Transport policy for an ageing population. *Transp. Rev.* **2003**, *23*, 375–386. [CrossRef]
12. Fraser, S.D.S.; Lock, K. Cycling for transport and public health: A systematic review of the effect of the environment on cycling. *Eur. J. Public Health* **2011**, *21*, 738–743. [CrossRef]
13. Handy, S.L.; Boarnet, M.G.; Ewing, R.; Killingsworth, R.E. How the built environment affects physical activity: Views from urban planning. *Am. J. Prev. Med.* **2002**, *23*, 64–73. [CrossRef]
14. Saelens, B.E.; Sallis, J.F.; Black, J.B.; Chen, D. Neighborhood-based differences in physical activity: An environment scale evaluation. *Am. J. Public Health* **2003**, *93*, 1552–1558. [CrossRef]
15. Cao, X.Y.; Mokhtarian, P.L.; Handy, S.L. Neighborhood Design and the Accessibility of the Elderly: An Empirical Analysis in Northern California. *Int. J. Sustain. Transp.* **2010**, *4*, 347–371. [CrossRef]
16. Brownson, R.C.; Hoehner, C.M.; Day, K.; Forsyth, A.; Sallis, J.F. Measuring the Built Environment for Physical Activity State of the Science. *Am. J. Prev. Med.* **2009**, *36*, S99–S123. [CrossRef] [PubMed]
17. McCormack, G.R.; Shiell, A. In search of causality: A systematic review of the relationship between the built environment and physical activity among adults. *Int. J. Behav. Nutr. Phys. Act.* **2011**, *8*, 125. [CrossRef] [PubMed]
18. Van Cauwenberg, J.; De Bourdeaudhuij, I.; De Meester, F.; Van Dyck, D.; Salmon, J.; Clarys, P.; Deforche, B. Relationship between the physical environment and physical activity in older adults: A systematic review. *Health Place* **2011**, *17*, 458–469. [CrossRef]
19. Saelens, B.E.; Handy, S.L. Built environment correlates of walking: A review. *Med. Sci. Sports Exerc.* **2008**, *40*, S550. [CrossRef]
20. Willis, D.P.; Manaugh, K.; El-Geneidy, A. Cycling Under Influence: Summarizing the Influence of Perceptions, Attitudes, Habits, and Social Environments on Cycling for Transportation. *Int. J. Sustain. Transp.* **2015**, *9*, 565–579. [CrossRef]
21. Heinen, E.; van Wee, B.; Maat, K. Commuting by Bicycle: An Overview of the Literature. *Transp. Rev.* **2010**, *30*, 59–96. [CrossRef]
22. Roof, K.; Oleru, N. Public health: Seattle and King County's push for the built environment. *J. Environ. Health* **2008**, *71*, 24–27.
23. Srinivasan, S.; O'Fallon, L.R.; Dearry, A. Creating healthy communities, healthy homes, healthy people: Initiating a research agenda on the built environment and public health. *Am. J. Public Health* **2003**, *93*, 1446–1450. [CrossRef]
24. Van Cauwenberg, J.; Clarys, P.; De Bourdeaudhuij, I.; Van Holle, V.; Verte, D.; De Witte, N.; De Donder, L.; Buffel, T.; Dury, S.; Deforche, B. Physical environmental factors related to walking and cycling in older adults: The Belgian aging studies. *BMC Public Health* **2012**, *12*, 142. [CrossRef] [PubMed]
25. Zhang, Y.; Yang, X.G.; Li, Y.; Liu, Q.X.; Li, C.Y. Household, Personal and Environmental Correlates of Rural Elderly's Cycling Activity: Evidence from Zhongshan Metropolitan Area, China. *Sustainability* **2014**, *6*, 3599–3614. [CrossRef]
26. Verhoeven, H.; Simons, D.; Van Dyck, D.; Van Cauwenberg, J.; Clarys, P.; De Bourdeaudhuij, I.; de Geus, B.; Vandelanotte, C.; Deforche, B. Psychosocial and Environmental Correlates of Walking, Cycling, Public Transport and Passive Transport to Various Destinations in Flemish Older Adolescents. *PLoS ONE* **2016**, *11*, e0147128. [CrossRef]
27. Ji, Y.J.; Fan, Y.L.; Ermagun, A.; Cao, X.N.; Wang, W.; Das, K. Public bicycle as a feeder mode to rail transit in China: The role of gender, age, income, trip purpose, and bicycle theft experience. *Int. J. Sustain. Transp.* **2017**, *11*, 308–317. [CrossRef]
28. Ewing, R.; Cervero, R. Travel and the built environment: A meta-analysis. *J. Am. Plan. Assoc.* **2010**, *76*, 265–294. [CrossRef]
29. Li, Z.; Wang, W.; Yang, C.; Ding, H. Bicycle mode share in China: A city-level analysis of long term trends. *Transportation* **2017**, *44*, 773–788. [CrossRef]
30. Kemperman, A.; Timmermans, H. Influences of Built Environment on Walking and Cycling by Latent Segments of Aging Population. *Transp. Res. Rec.* **2009**, *2134*, 1–9. [CrossRef]

31. Leger, S.J.; Dean, J.L.; Edge, S.; Casello, J.M. "If I had a regular bicycle, I wouldn't be out riding anymore": Perspectives on the potential of e-bikes to support active living and independent mobility among older adults in Waterloo, Canada. *Transp. Res. Part A-Policy Pract.* **2019**, *123*, 240–254. [CrossRef]
32. Mertens, L.; Van Dyck, D.; Deforche, B.; De Bourdeaudhuij, I.; Brondeel, R.; Van Cauwenberg, J. Individual, social, and physical environmental factors related to changes in walking and cycling for transport among older adults: A longitudinal study. *Health Place* **2019**, *55*, 120–127. [CrossRef]
33. Grimes, A.; Chrisman, M.; Lightner, J. Barriers and Motivators of Bicycling by Gender Among Older Adult Bicyclists in the Midwest. *Health Educ. Behav.* **2020**, *47*, 67–77. [CrossRef]
34. Van Cauwenberg, J.; De Bourdeaudhuij, I.; Clarys, P.; De Geus, B.; Deforche, B. Older adults' environmental preferences for transportation cycling. *J. Transp. Health* **2019**, *13*, 185–199. [CrossRef]
35. Moudon, A.V.; Lee, C.; Cheadle, A.D.; Collier, C.W.; Johnson, D.; Schmid, T.L.; Weather, R.D. Cycling and the built environment, a US perspective. *Transp. Res. Part. D-Transport. Environ.* **2005**, *10*, 245–261. [CrossRef]
36. Siman-Tov, M.; Jaffe, D.H.; Peleg, K.; Israel Trauma, G. Bicycle injuries: A matter of mechanism and age. *Accid. Anal. Prev.* **2012**, *44*, 135–139. [CrossRef] [PubMed]
37. Prins, R.G.; Pierik, F.; Etman, A.; Sterkenburg, R.P.; Kamphuis, C.B.M.; van Lenthe, F.J. How many walking and cycling trips made by elderly are beyond commonly used buffer sizes: Results from a GPS study. *Health Place* **2014**, *27*, 127–133. [CrossRef]
38. Van Cauwenberg, J.; De Bourdeaudhuij, I.; Clarys, P.; de Geus, B.; Deforche, B. Environmental Preferences for Transportation Cycling Among Older Adults: An Experiment with Manipulated Photographs. *J. Transport. Health* **2018**, *9*, S4. [CrossRef]
39. Barnett, E.; Casper, M. A definition of "social environment". *Am. J. Public Health* **2001**, *91*, 465.
40. Wang, L.; Zhao, C.; Liu, X.; Chen, X.; Li, C.; Wang, T.; Wu, J.; Zhang, Y. Non-Linear Effects of the Built Environment and Social Environment on Bus Use among Older Adults in China: An Application of the XGBoost Model. *Int. J. Environ. Res. Public Health* **2021**, *18*, 9592. [CrossRef]
41. Lee, Y.-S. Gender Differences in Physical Activity and Walking Among Older Adults. *J. Women Aging* **2005**, *17*, 55–70. [CrossRef]
42. Gallagher, N.A.; Clarke, P.J.; Gretebeck, K.A. Gender Differences in Neighborhood Walking in Older Adults. *J. Aging Health* **2014**, *26*, 1280–1300. [CrossRef]
43. Zhang, Y.; Chen, L.; Zhu, W.; Liu, X.; Chen, D. Relationship between physical activity and environment in Shanghai, China: Analysis and evaluation in adults aged 45–80. *Med. Sport* **2011**, *64*, 269–284.
44. Zhang, Y.; Li, C.; Ding, C.; Zhao, C.; Huang, J. The Built Environment and the Frequency of Cycling Trips by Urban Elderly: Insights from Zhongshan, China. *J. Asian Archit. Build. Eng.* **2018**, *15*, 511–518. [CrossRef]
45. National Health and Family Planning Commission of China. *Promotion of Healthy Living Style in China*; National Health and Family Planning Commission of China (NHFPC): Beijing, China, 2007.
46. Zhang, Y.; He, Q.; Wu, W.; Li, C.Y. Public transport use among the urban and rural elderly in China: Effects of personal, attitudinal, household, social-environment and built-environment factors. *J. Transp. Land Use* **2018**, *11*, 701–719. [CrossRef]
47. Zhang, Y.; Li, Y.; Yang, X.; Liu, Q.; Li, C. Built Environment and Household Electric Bike Ownership. *Transp. Res. Rec. J. Transp. Res. Board* **2013**, *2387*, 102–111. [CrossRef]
48. Zhang, Y.; Li, Y.; Liu, Q.; Li, C. The Built Environment and Walking Activity of the Elderly: An Empirical Analysis in the Zhongshan Metropolitan Area, China. *Sustainability* **2014**, *6*, 1076–1092. [CrossRef]
49. National Bureau of Statistics of China. *China Statistical Yearbook 2012*; China Statistics Press: Beijing, China, 2012.
50. Zhongshan Municipal Bureau of Urban Planning. *Zhongshan Household Travel Survey*; Zhongshan Municipal Bureau of Urban Planning: Zhongshan, China, 2012.
51. Martinez, L.M.; Viegas, J.M.; Silva, E.A. A traffic analysis zone definition: A new methodology and algorithm. *Transportation* **2009**, *36*, 581–599. [CrossRef]
52. Ding, C.; Wang, D.; Liu, C.; Zhang, Y.; Yang, J. Exploring the influence of built environment on travel mode choice considering the mediating effects of car ownership and travel distance. *Transp. Res. Part A-Policy Pract.* **2017**, *100*, 65–80. [CrossRef]
53. Kockelman, K.M. Travel behavior as function of accessibility, land use mixing, and land use balance: Evidence from San Francisco Bay Area. *Transp. Res. Rec. J. Transp. Res. Board* **1997**, *1607*, 116–125. [CrossRef]
54. Ministry of Housing and Urban-Rural Development of the People's Republic of China. *Code for Urban Land Use Classes and Standards of Planning Construction Land (GB50137-2011)*; Ministry of Housing and Urban-Rural Development of the People's Republic of China: Beijing, China, 2011. (In Chinese)
55. Zhongshan Municipal Bureau of Urban Planning. *Zhongshan Transportation Development Planning*; Zhongshan Municipal Bureau of Urban Planning: Zhongshan, China, 2012.
56. Liu, P.; Marker, S. Evaluation of contributory factors' effects on bicycle-car crash risk at signalized intersections. *J. Transp. Saf. Secur.* **2020**, *12*, 82–93. [CrossRef]
57. Congiu, T.; Sotgiu, G.; Castiglia, P.; Azara, A.; Piana, A.; Saderi, L.; Dettori, M. Built Environment Features and Pedestrian Accidents: An Italian Retrospective Study. *Sustainability* **2019**, *11*, 1064. [CrossRef]
58. Ball, K.; Jeffery, R.W.; Abbott, G.; McNaughton, S.A.; Crawford, D. Is healthy behavior contagious: Associations of social norms with physical activity and healthy eating. *Int. J. Behav. Nutr. Phys. Act.* **2010**, *7*, 86. [CrossRef] [PubMed]

59. Wang, W.; Zhang, Y.; Zhao, C.; Liu, X.; Chen, X.; Li, C.; Wang, T.; Wu, J.; Wang, L. Nonlinear Associations of the Built Environment with Cycling Frequency among Older Adults in Zhongshan, China. *Int. J. Environ. Res. Public Health* **2021**, *18*, 10723. [CrossRef] [PubMed]
60. Gamble, T.; Walker, I.; Laketa, A. Bicycling campaigns promoting health versus campaigns promoting safety: A randomized controlled online study of 'dangerization'. *J. Transp. Health* **2015**, *2*, 369–378. [CrossRef]
61. Yang, Z. Behavior and health. *Chin. J. Behav. Med. Brain Sci.* **2009**, *18*, 61–63. [CrossRef]

land

Article

Short-Run Links in Ecological Footprint: A Dynamic Factor Analysis for the EU

María Jesús Delgado-Rodríguez [1,*], Sonia de Lucas-Santos [2] and Alfredo Cabezas-Ares [1]

[1] Department of Business Administration (ADO), Applied Economics II and Fundaments of Economic Analysis, Universidad Rey Juan Carlos de Madrid, Pº Artilleros s/n, 28670 Madrid, Spain; alfredo.cabezas@urjc.es

[2] Department of Applied Economics (Statistics), Ciudad Universitaria de Cantoblanco, Universidad Autónoma de Madrid, 28049 Madrid, Spain; sonia.delucas@uam.es

* Correspondence: mariajesus.delgado@urjc.es; Tel.: +34-91-488-7852

Abstract: The Ecological Footprint (EFP) is a useful indicator for assessing the progress of environmental performance and offers a solid basis for sustainability studies. In this paper, we contribute to the broadening of its possibilities of investigation by measuring the cross-country links in the EFP in global hectares per capita. The modeling framework is based on the dynamic factor analysis to estimate, in the parametric form, an index that provides information about the short-run dynamics of the EFP in the EU. Following this approach, we identify different patterns in the EFP behavior of the European countries during the period of 1962–2017. The results show stronger links across the EFP of the main European countries: France, Austria, Belgium, Germany, Denmark and the U.K. The proposed analysis gives a better understanding of the links behind environmental degradation in the EU and is applicable for the implementation and design of environmental policies.

Keywords: ecological footprint; dynamic factor model; cross-country links

Citation: Delgado-Rodríguez, M.J.; Lucas-Santos, S.d.; Cabezas-Ares, A. Short-Run Links in Ecological Footprint: A Dynamic Factor Analysis for the EU. *Land* **2021**, *10*, 1372. https://doi.org/10.3390/land10121372

Academic Editor: Benedetto Rugani

Received: 16 November 2021
Accepted: 9 December 2021
Published: 11 December 2021

Publisher's Note: MDPI stays neutral with regard to jurisdictional claims in published maps and institutional affiliations.

1. Introduction

As a world leader in taking action on climate change, the EU has constantly demanded environmental regulations to combat it. Currently, Europe faces unprecedented environmental, climate and sustainability challenges. These include biodiversity loss, climate change, resource use and pollution, which require strong commitments to respond to them. For this purpose, the European Commission published a proposal for an 8th Environment Action Programme in 2020 in order to ensure that the EU climate and environment laws are effectively implemented. It forms the EU's basis for achieving the United Nations´ 2030 Agenda and its Sustainable Development Goals. This proposal also supports the environment and climate action objectives of the European Green Deal.

The monitoring and assessment of these plans require the development of innovative and comprehensive indicators that are capable of providing information of interest for the implementation of policies related to global warming and climate change. Environmental variables are important as proxies to represent degradation and to track the efforts in the struggle against environmental threats. Traditionally, the variables of carbon emission and pollution type (degradation in land, forest, water, etc.) have been widely used in the literature to constitute policy rules. Recently, the ecological-footprint indicator has attracted attention in research and is highlighted as one of the most prominent environmental-variable and sustainability-evaluation tools. The EFP is an aggregate indicator that measures the environmental degradation caused by human activities (Ulucak and Lin, 2017 [1]). It helps individuals to understand the direct and indirect impact of their actions on the planet. According to the Global Footprint Network (2021) [2], on the demand side, the EFP represents the ecological assets that a population requires in order to produce all of the resources they consume, including livestock-based food, fishery products, forestry, and infrastructure. This variable is calculated in global hectares. The EFP is categorized by land type, and there are six types of ecological footprints: arable-land EFP, forest-land EFP, grazing-land EFP, fishing-land EFP, built-up-land EFP and carbon-uptaking-land EFP.

Much of the empirical analysis of EFPs has been dedicated to determining convergence in countries in order to provide administrators and societies with policy recommendations. It is expected that the countries that converge in environmental indicators will be able to implement their policies more effectively in the context of a common environmental framework. Therefore, the issue of whether or not countries converge has received great attention in the literature. The empirical findings of the EFP document the presence of a number of convergence clubs (Ulucak and Aperguis, 2018 [3]) and the existence of a convergence in income in groups of countries (Erdogan and Okumus, 2021 [4]). Thus, convergence is verified for countries with similar conditions, such as the growth process or the dependence on environmental resources. The existence of club convergence or groups of countries that converge suggests that environmental policies should consider the different convergence paths associated with them. Furthermore, these results demonstrate that certain countries have similar dynamics and conditions in regards to the drivers of environmental quality and environmental degradation.

Despite its undoubted interest, these works have focused only on the long-run side of the analysis, while the short-term analysis of the EFP has been ignored. The short-run approach can be of interest to supplementing and improving research into the efficiency of environmental policies. This makes it very interesting to delve into the short-term characteristics and dynamics of the EFP in the EU and the possible links between countries. In this paper, we estimate a Dynamic Factor Model in order to assess information about how the EFP co-moves across European countries. In our analysis, the fact that countries share the same short-run dynamics in the EFP indicates linkages across them that can be of interest to the design of the EU climate change policy.

This paper is motivated by these concerns and extends the research into the EFP by investigating the possibility of cross-country links of 21 European countries for the 1961–2017 period. We contribute to the existing literature on the EFP in two distinct ways. Firstly, we introduce the short-run perspective in the study of the EFP by estimating the short-run index of the EFP in the EU based on the dynamic factor model. Secondly, we assess the links across countries by measuring the degree to which European countries share the dynamic pattern of the EFP in Europe. In this paper, we offer an innovative empirical approach that may be a useful tool to evaluate the features of EFP performance across member states. The results could lead to environmental recommendations on which countries should make further adjustments in order to increase efforts to reduce EFP.

We arrange the remainder of this paper as follows: the next section presents the literature review on the recent papers on EFP analysis. The third section describes the data and the econometric strategy. The empirical results are reported in the fourth section. The final section presents the conclusion and the policy recommendations.

2. Literature

The most recent sustainability analyses have focused on the use of EFP indicators for the assessment of progress on environmental performance. The concept of the EFP is considered a more comprehensive indicator than other environmental measures (Al-Mulali et al., 2015, Solarin, 2019 [5,6]). Traditionally, papers on sustainability analysis have mostly used information about CO_2 emissions. However, CO_2 emissions only represent a fraction of the total fallout triggered by large-scale energy use (Al-Mulali et al., 2015, Solarin, 2019 [5,6]). Beyond CO_2 emissions, resource stocks including forestry soil, mining and oil stock also put nature under immense pressure. According to Stern (2014) [7], the developed countries have experienced a decline in many specific pollutants per unit of output due to technological innovations and progressively stringent environmental regulations. For this reason, the use of indicators that provide information on environmental degradation in terms of the EFP is increasingly in demand (Bello et al. 2018 [8]; Solarin and Al-Mulali 2018 [9]).

Some novel and recent studies focusing on the use of the EFP in environmental analyses are shown in Table 1, which briefly summarizes them. The relevant literature review demonstrates that the analysis of the EFP is a promising area of research. However, most of these studies have focused on convergence analysis. These papers have especially evaluated the long-term dynamics of per capita emissions and tried to determine whether the national data series show evidence of converging trends in the sense that nations that initially have lower levels of EFP per capita are encountering a higher growth in emissions and are therefore "catching up" with the nations that have a higher EFP per capita. Solarin (2019) [6] showed evidence of sigma convergence in the EFP for 27 OECD countries. Bilgili and Uluak (2018) [10] offered evidence in favor of environmental convergence by demonstrating the stochastic and deterministic convergences among the G20 countries. Yilanci and Pata (2020) [11] investigated the convergence process for the ASEAN-5 countries and provided strong support for the absolute convergence in the EFP per capita. Sarkodie (2021) [12] also confirmed long-run convergence for 242 countries using cross-country time series techniques.

Table 1. Recent empirical sustainability analysis based on environmental degradation variables.

Study	Variable	Methodology	Sample	Results
Solarin 2019 [6]	CO_2 emissions, Carbon Footprint pc and EFP pc	Stochastic convergence	27 OCDE Countries	Sigma convergence
Bilgili & Ulucak 2018 [10]	EFP	A bootstrap-based panel KPSS test	G-20 Countries	Stochastic and deterministic C.
Ulucak &Apergis 2018 [3]	EFP	Club-clustering approach	EU Countries	Convergence clubs
Solarin et al., 2019 [13]	EFP and its six components	Club-convergence approach	92 countries	Convergence clubs
Yilanci & Pata 2020 [11]	EFP	TAR panel unit root test	5 ASEAN countries	Absolute convergence
Ulucak et al., 2020 [14]	EFP and its sub-components	Log t regression	33 Sub-Saharan countries	Convergence clubs
Erdogan & Okumus 2020 [4]	EFP	Stochastic and club convergence approach	89 countries	Convergence clubs
Haider et al., 2021 [15]	Biomass material Footprint	Phillips–Sul approach	172 Countries	No convergence
Sarkodie 2021 [12]	EFP, Biocapacity, Carbon F., and Ecological Status	Cross-country time series techniques	245 Countries	Long-run convergence
Wu 2020 [20]	EFP	GWR and OLS models	Provinces of China	Main driving forces of EFP evolution
Wu & Liu 2020 [17]	EFP Intensity	Global Moran´s Index and LISA	Jiangsu's counties	Spatial distribution
Nathaniel et al., 2020 [19]	EFP	AMG estimation and panel co-integration	CIVETS countries	Relation economic variables
Guo et al., 2020 [21]	EFP and Ecological Capacity	Grey GM (1,1) prediction model	Quinghai Province (China)	Forecasting EFP
Caglar et al., 2021 [16]	EFP and its six components	SOR unit root test	France, Germany, Italy, Spain and UK	No convergence
Zambrano-Monserrate et al., 2020 [18]	EFP and Biocapacity	Dynamic spatial Durbin model	158 countries	Spatial effects

NOTE: Two-regime threshold Autoregressive (TAR); Convergence (C.); Ecological Footprint per capita (EFP pc); Geographically Weighted Regression model (GWR); Ordinary Least Square Regression (OLS); Association of Southeast Asian Nations (ASEAN-5); Local Spatial Association Index (LISA); Successive Overrelaxation Method (SOR); Grey Model First Order One Variable (Grey GM (1,1)): Augmented Mean Group (AMG); Colombia, Indonesia, Vietnam, Egypt, Turkey and South Africa (CIVETS).

Other papers have focused on the Convergence Club analysis. In this case, economies are grouped by common characteristics and each group has the same steady-state equilibrium, and each group reaches its own equilibrium. Solarin et al. (2019) [13] demonstrated 10 convergence clubs with regards to the EFP in their analysis of 92 countries. Ulucak and Apergis (2018) [3] employed the club-clustering-convergence method to examine the convergence in EU countries and document the presence of certain convergence clubs. while Erdogan and Okumus (2021) [4] provided a similar analysis for different income groups of countries using a panel-stationarity test and obtained several convergence clubs. Ulukak et al. (2020) [14] examined the convergence of 23 Sub-Saharan Africa countries using log t regression and found several clubs of convergence.

Contrary to these studies, the following research studies reached the opposite conclusion, namely that EFPs do not converge: Haider et al. (2012) [15] for 172 countries using the Phillips–Sul approach and Caglar et al. (2021) [16] for France, Germany, Italy, Spain and the U.K. employing the SOR unit root test.

There is a smaller number of papers that deal with other topics. These are the cases of Wu and Liu (2020) [17] and Zambrano-Monserrate et al. (2020) [18] that explored spatial distribution and the spatial effects of economic variables on the EFP. Other papers estimated relations of EFP with other economic variables (Nathaniel et al., 2020) [19] and offered evidence of the main driving forces of EFP evolution in China´s provinces (Wu, 2020) [20]. We can also find papers that offer forecasts (Guo et al., 2020) [21]. However, no studies were found that analyze the short-run dynamics of EFP and offer evidence of the cross-country links to this environmental variable.

3. Data and Methods

3.1. Data

Annual data on the EFP per capita (global hectares per capita) for the EU countries were obtained from the Global Footprint Network (http://data.footprintnetwork.org, accessed on 10 November 2021), spanning the period 1962–2017. The time period was consistent with data availability for Austria, Belgium, Sweden, France, Netherlands, Poland, Greece, Finland, Denmark, Italy, Ireland, Germany, United Kingdom, Luxembourg, Cyprus, Malta, Spain, Portugal, Hungary, Bulgaria and Romania. The annual series on EFP were log-transformed and differentiated ($\Delta = 1 - L$, being L the lag operator) in order to obtain the EFP short-run behavior of the European countries. The analysis focuses on the EU countries since the EU has adopted some of the highest environmental standards on a global basis, as well as common environmental policies.

3.2. Model

Our proposal to offer information about the dynamics of the EFP begins by applying dynamic factor analysis to estimate a short-run index of the EFP for the EU. The dynamic factor model is based on the assumption that a small number of unobserved latent factors, f_t, generate the observed time series through a stochastically perturbed linear structure. Formally, in the model it is assumed that the pattern of observed co-movements of a high-dimensional vector of time-series countries, $X_t = \Delta \ln EFP_{i.t}$, (where EFP is the growth rates of global hectares per capita by country) can be represented by a few unobserved, latent common dynamic factors. The latent factors follow the time series process and are commonly taken to be a vector autoregression (VAR). In equations, the dynamic factor model is

$$X_t = \Lambda f_t + e_t$$
$$f_t = \psi(L) f_{t-1} + \eta_t \tag{1}$$

where there are N countries, so X_t and e_t are $N \times 1$, there are m dynamic factors so f_t and η_t are $m \times 1$, $\Lambda = (\lambda_1, \lambda_2, \ldots, \lambda_m)$ is $N \times m$, L is the lag operator, and the lag polynomial matrix $\psi(L)$ is $m \times m$. The *i-th* λ_i are called factor loadings for the *i-th* countries, X_{it}. The idiosyncratic disturbances, $e_t = (e_{1,t}, e_{2,t}, \ldots, e_{N,t})'$, are the specific elements of each series contained in a vector; they are serially correlated and slightly cross-sectionally correlated with other variables in the model and are mutually uncorrelated at all leads and lags, that is, $Ee_{it}e_{js} = 0$ for all s if $i \neq s$. They are assumed to be uncorrelated with the factor innovations at all leads and lags, that is, $Ee_t\eta'_{t-k} = 0$ for all k. The *pth* order autoregressive polynomial, $\psi_i(L)$, is assumed to have stationary roots. As we did here, it is common to reduce the number of parameters by estimating the signal-to-noise ratios $q_{i,m} = \frac{\sigma^2_{\eta,i}}{\sigma^2_{e,i}}$ (see Harvey and Trimbur, 2008 [22], for its importance for spectral analysis).

Assuming that all the processes in (1) are stationary and not cointegrated, we used the GROCER's Econometric Toolbox written by Dubois and Michaux, 2019 [23], which provides the standard estimation method by maximizing the likelihood of the corresponding model and estimation accuracy via the Kalman filter, after a suitable reparameterization of the model in a state-space form. This method allowed us to explicitly model the dynamic of the factors $f_{i,t}$ that can take an autoregressive-moving-average ARMA(p,q) form as:

$$(1 + \phi_1 L + \ldots + \phi_p L^p)f_{i,t} = (1 + \theta_1 L + \ldots + \theta_q L^q)\eta_t \tag{2}$$

expressed in matrix form (1) and of the residuals $e_{i,t}$ than can take an autoregressive AR form as:

$$(1 + \delta_{1,1}L + \ldots + \delta_{1,l_i}L^i)e_{i,t} = \xi_{i,t} \tag{3}$$

where l_i is the order of the idiosyncratic AR governing $e_{i,t}$.

In our proposal, we first confirmed the existence of only one common factor, $\hat{f}_{1,t}$, employing the statistical criterion proposed by Bai and Ng, 2002 [24]. This single common factor represents the short-run index for the EFP of the EU. Furthermore, for an appropriate interpretation of results, we standardized the loading factors: $\lambda_i \frac{\sigma_{f_1}}{\sigma_{x_i}} = \frac{\text{cov}(x_i,f_1)}{\sigma^2_{f_1}} \frac{\sigma_f}{\sigma_{x_i}} = \frac{\text{cov}(x_i,f_1)}{\sigma_{x_i}\sigma_{f_1}} = \rho_i$, such that it provided an estimation of the correlation or linkages between the countries´ EFPs with the common factor. The values that were parametrically obtained, ρ_i, are the proportion of the total variation explained by the common factor, which offers a measure of the degree to which the country is co-moving following the short-run dynamic pattern of *EFP* in Europe (the short-run index of EFP) over the period.

To analyze the possible results obtained by ρ_i, we differentiated three types of results according to the percentage of variation explained by the short-run index of EFP:

- Group 1: $\rho_i \geq 0.5$—Denoting a strongly linked EFP. We interpret that this result is obtained by countries that share the short-run dynamics of EFP in Europe and we could expect them to exert influence on the neighboring countries.
- Group 2: $0 < \rho_i < 0.5$—Denoting emissions with weak links. In this case, countries are not so influenced by the short-run dynamics of the EFP common pattern.
- Group 3: $\rho_i = 0$—Denoting an independent EFP pattern. This type of result implies that these countries are not linked with the European pattern of EFP.

4. Results

The results of the estimation according to the dynamic factor model in (1) are shown in Figure 1 and Table A1 in the Appendix A. The AR idiosyncratic parameter and noise ratio confirm the suitability and dynamicity of the model. The significance of the loading factors indicates which countries are sharing a short-run dynamic behavior and which are not. Results confirm that all of the factor loadings are significant and statistically similar, with the exceptions of Luxembourg, the Netherlands, Portugal, Spain and Sweden. These countries were then excluded from the estimation of the model, which is why they do not appear in Figure 1. Luxembourg, the Netherlands, Portugal, Spain and Sweden follow independent short-run behavior in their EFP and thus, they are considered to be independent from the European EFP dynamic.

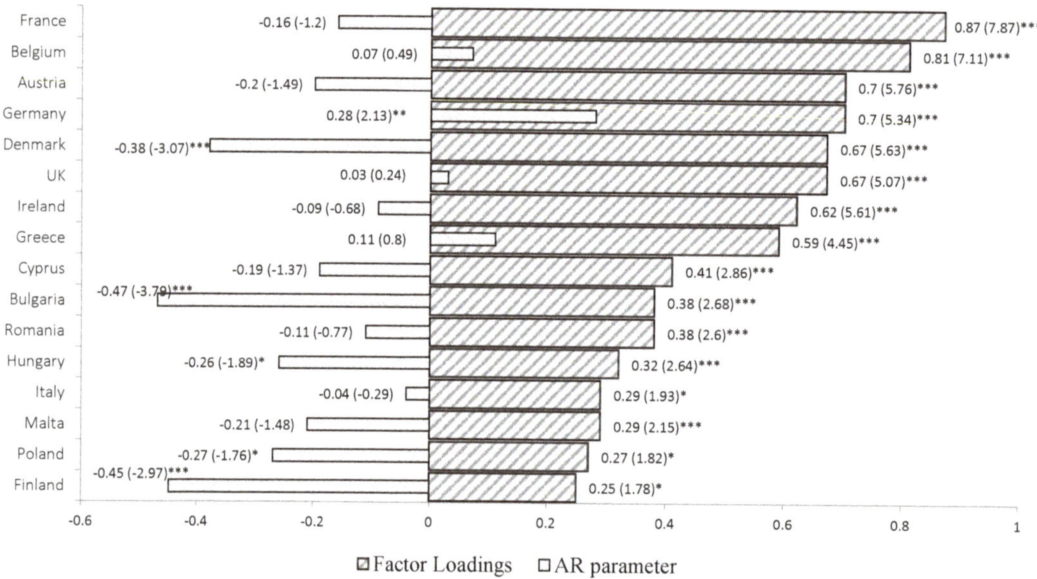

Figure 1. Countries ordered by factor loadings in model (1) and AR idiosyncratic parameters in (3). Period 1962–2017. Note: AR: Autoregressive. In () t-statistics at * 90%, ** 95% and *** 99% of significance. No significant countries in factor model: Luxembourg, the Netherlands, Portugal, Spain and Sweden.

Additionally, the proportion of total variation explained by the factor loading can be a measure of the degree of the cross-country links. Following the criteria established in the methodology section, we can identify countries that strongly share the common short-term dynamic pattern of the EU ($\rho_i \geq 0.5$). In Figure 1 the European countries are ordered by the value of their factor loadings from highest to lowest in order to facilitate analysis. We can observe that the main members of the EU, i.e., France, Germany, the UK, Austria, Belgium and Denmark, obtained values of $\rho_i \geq 0.5$. This group is joined by Ireland and Greece, which, although with lower values, also achieve the results of $\rho_i \geq 0.5$.

In our analysis, we also find the countries for which the emissions show weak linkages ($0 < \rho_i < 0.5$). This is the case of Cyprus, Bulgaria, Rumania, Hungary, Italy, Malta, Poland and Finland. The cases of Italy and Finland may be surprising, but their results may be related to the fact that they are closer to maintaining independent positions than to sharing the short-term behavior of the EU. For the rest of the countries, the enlargements to include the new member states in Central and Eastern Europe have given a greater dimension to the EU, but their progress towards a common EFP dynamic is in less-advanced stages. But the fact that they maintain significant correlations with the common factor is evidence of the positive progress of their environmental behavior.

Based on the previous results, we also provide a map in order to more intuitively follow the three groups of countries that were obtained (see Figure 2). The darker the shading of the map areas, the stronger the links across the EFP. It is straightforward to perceive that the core European countries are the ones that show more influence on the European EFP dynamic.

The information on the dynamics of the European EFP also allows for the analysis of their cyclical properties. Figure 3 shows the dynamic of the European EFP throughout the time period beginning in 1962 and ending in 2017. The characteristics of the changes in the short-run index of the EFP over the period show quite a symmetrical behavior, with the same average duration from peak to peak than from trough to trough. Although, we find that the averages of the duration (2.86 years) and amplitude (8.5 years) of the reduction in footprint are greater than the average duration (2.57 years) and average amplitude (6 years) of the expansions. Another result of interest is that the period of best EFP performance occurs during the period 1998–2008, during which period there was a decrease in environmental degradation.

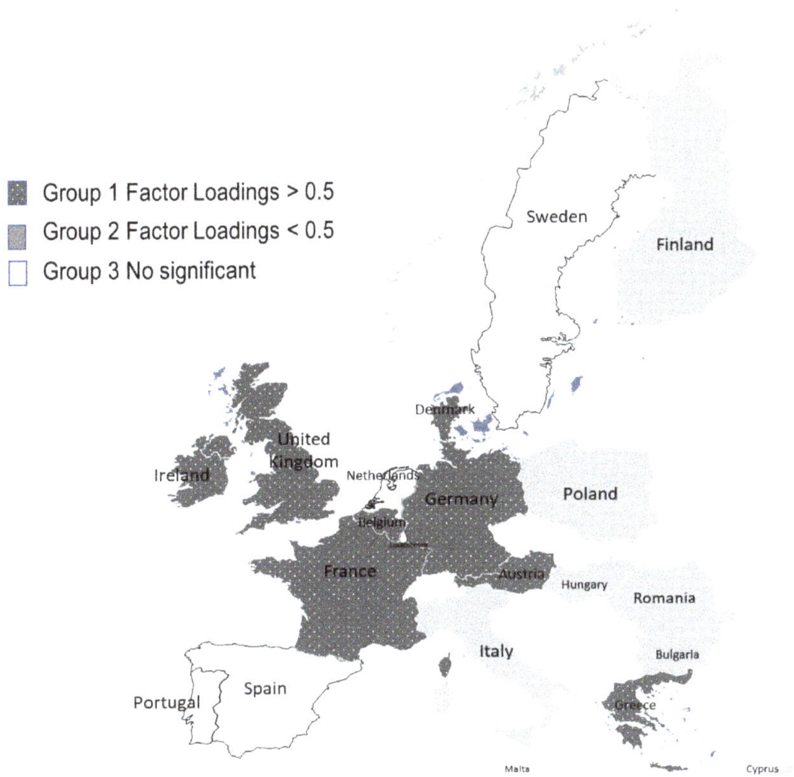

Figure 2. Spatial distribution of the European country´s correlation.

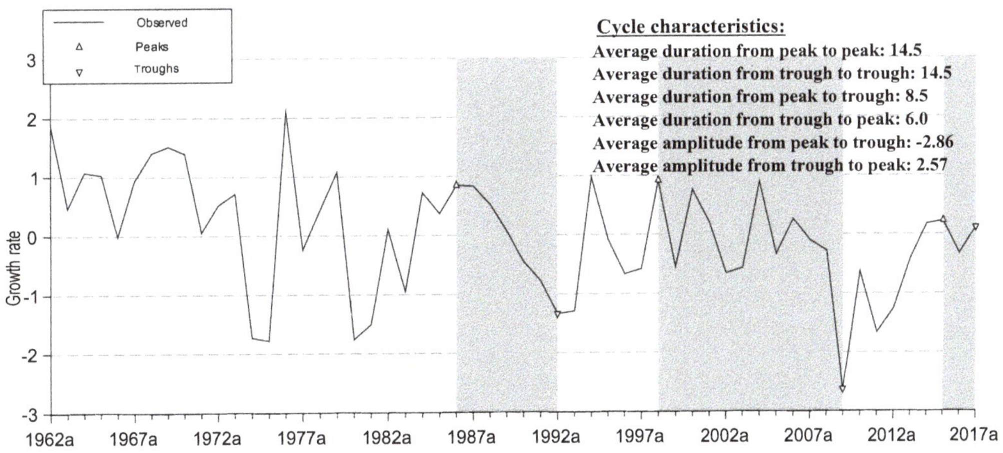

Figure 3. Dating the EFP Index and cycle characteristics. Note: Shaded areas correspond to the declining phases in the EU EFP. Source: own compilation using Harding and Pagan (2002) [25].

5. Robustness Checks

We performed a number of robustness checks in order to test the validity of our results. First, we confirmed the existence of only one common factor, $\hat{f}_{EFP,t}$, by employing the statistical criterion proposed by Bai and Ng (2002) [24]. The number of dynamic factors, p, according to these authors, is where $p \leq r$ is the number of static factors determined by Bai and Ng (2007) [26], and where $p = 1$ since $r = 1$ according to the following criteria:

$$IC_{p1}(q) = \log(\det(\Sigma)) + q\frac{(N+T)}{nT} + \log\left(\frac{nT}{N+T}\right)$$

$$IC_{p2}(q) = \log(\det(\Sigma)) + q\frac{(N+T)}{nT} + \log(\min(n,T)) \tag{4}$$

$$IC_{p3}(q) = \log(\det(\Sigma)) + q\frac{\log(\min(n,T))}{(\min(n,T))}$$

where $\Sigma = variance\ matrix\ of\ residual\ e_t$.

Next, we confirmed the robustness of our results by testing the stability of the estimated parameters in model (1) to check the existence of structural or temporal breaks that reflect changes in European behaviors that are related to the difficulties of maintain the EFP short-run behavior. Following Bueno et al. (2011) [27], if the date of a possible break is unknown, then a recursive testing procedure can be employed. The null hypothesis H_0 of the parameters' stability of each model is $H_0 : \rho_i(\tau) = 0$. Recursive estimations $\hat{\rho}_i(\tau)$ are obtained from:

$$x_{i,t} = \rho_i(\tau)\hat{f}_{EFP,t} + v_{i,t}(\tau) \tag{5}$$

and the F-type statistic (Wald-type statistic) to test for a break of an unknown date is

$$F_{SupWald,i} = max[F_i(\tau_0), F_i(\tau_0 + 1), \ldots, F_i(T - 1), F_i(T)] \tag{6}$$

where each of the $F(\tau)$ is defined as:

$$F(\tau) = \frac{1}{q}[R\rho_j(\tau) - \rho_j]'(R\hat{\Sigma}(\tau)R')^{-1}[R\rho_j(\tau) - \rho_j] \tag{7}$$

$\hat{\Sigma}(\tau)$ must be robust to heteroskedasticity and autocorrelation. For example, by employing the Newey–West (1987) [28] estimator.

The asymptotic distribution of the $F_{SupWald,i}$ statistic is not standard because the break date appears only under the alternative hypothesis. In this work, empirical critical values with no asymptotic sample sizes and autocorrelated errors were calculated by the Monte Carlo simulation according to our data characteristics, that is, by assuming autocorrelation disturbances in (1) and (5) and for the available sample sizes. For more details see Bueno et al. (2011) [27].

We show the results of the estimation of the breaks and the $F_{SupWald,i}$ (Wald type statistic) in Table 2 and Figure 4. Some temporal breaks were found in Romania and Hungary in 1989 which can be linked to the economic crisis in these countries.

Table 2. Estimation breaks of model (1) for EU countries, 1962–2017.

Countries	Trimming	90%	95%	99%	$F_{SupWald,i}$	Break Date	Breaks
France	20%	5.6	7.9	16.2	0.9	1972	
Belgium	20%	4.9	7	15.8	3.3	1985	
Austria	20%	5.2	7	12.5	1	1972	
Germany	20%	5.2	6.9	10.6	1.2	1974	
Denmark	20%	4.9	7.2	14	1.7	1992	
UK	30%	3.5	4.9	10.1	1.3	1989	
Ireland	30%	3.3	4.7	8.8	0.4	1978	
Greece	20%	5.1	7	12.8	1	1996	

Table 2. *Cont.*

Countries	Trimming	90%	95%	99%	$F_{SupWald,i}$	Break Date	Breaks
Cyprus	20%	5.7	7.7	13.5	1	1979	
Bulgaria	20%	4.9	6.7	15.7	2.1	1988	
Romania	20%	5.2	7.9	14.8	5.6	1989 *	Temporal break
Hungary	20%	5.1	7.3	15.9	8.8	1989 **	Temporal break
Italy	20%	5	7.3	15.2	0.6	1972	
Malta	20%	5.2	7.4	17.1	0.7	1978	
Poland	20%	5.1	6.9	10.6	0.8	1980	
Finland	20%	2.2	3.4	5.5	3.2	1980	

Note: significant $F_{SupWald,i}$ statistic * at 90% and ** at 95%.

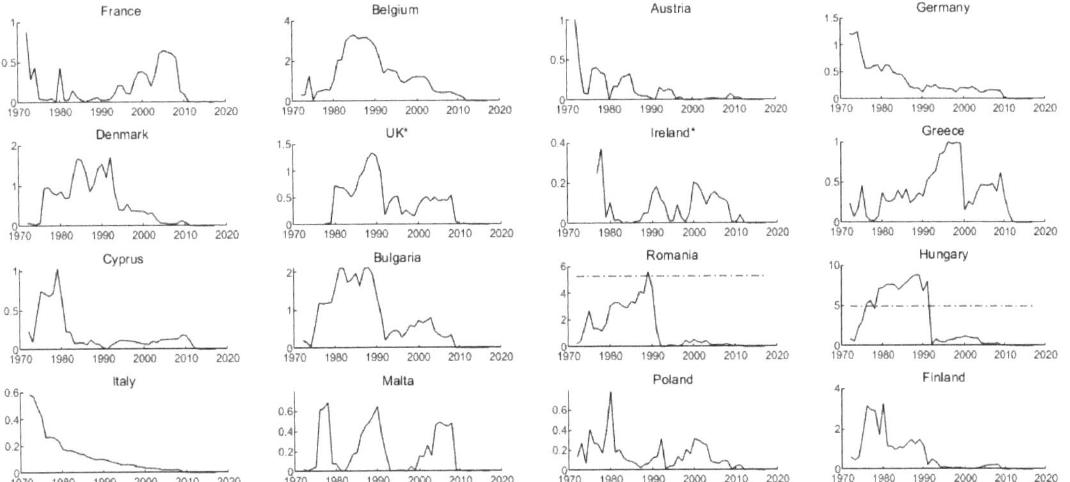

Figure 4. The asymptotic distribution of the $F_{SupWald,i}$ statistics of model (1) at 20% of trimming. Note: *y*-axis: value of $F_{SupWald,i}$; * at 30% of trimming. Source: Own compilation using Matlab.

6. Conclusions

The EFP is an appealing environmental factor that shows the human impact on the environment and is widely used for sustainability assessments. Most of the work on the EFP has been dedicated to determining the existence of convergence and has focused only on the long-run side of the analysis. In this paper, we broadened the possibilities of its investigation by measuring a short-run index of the EFP for the EU and testing the existence of cross-country links. To this end, we used a dynamic factor model which enabled estimation of the dynamics of the EFP in the EU and used the information about the proportion of variation explained by this index to characterize countries according to their cross-country links over the extended period of 1962–2017. So far, this type of analysis has not received enough attention in the literature.

The information on the dynamics of the EFP in Europe shows that the period of best EFP performance occurs over the period 1998–2008 during the crisis there, due to an increase in environmental degradation. Our results for the EU demonstrate the existence of three groups of countries according to the degree to which they share the EU short-run dynamics of EFP. The empirical findings reveal the existence of strong cross-country links across some of the main European countries: France, Austria, Belgium, Germany, Denmark and the U.K. These links allow us to extract information on the indirect effects that the dynamics of the EFP can have on their neighboring countries. We also find countries with weak links and countries with EFP-independent patterns. From the policy point of view, our findings recommend that policies related to the environment consider the distinctive short-run path of the EFP in Europe. The existence of different patterns of short-run EFP behavior shows that mitigation policies are required, and that the environmental degradation will not ease without exogenous intervention. For this, knowing the pattern of the EFP is crucial for evaluating the success of environmental actions.

This empirical study extends the literature on the EFP by serving as the first paper to investigate short-run characteristics of the EFP across European countries. Understanding short-run characteristics helps in assessing whether European sustainability policies can be effective and should therefore be reinforced. The methodology proposed in this article is a starting point for future work related to the understanding of the dynamics of the EFP. Its application will allow policies to be modulated for each country. Further research on how to contribute to the understanding of these cross-country links is needed in the environmental degradation literature.

Author Contributions: Conceptualization, M.J.D.-R.; Data curation, A.C.-A.; Formal analysis, S.d.L.-S.; Methodology, M.J.D.-R.; Resources, S.d.L.-S. and A.C.-A.; Writing—original draft, M.J.D.-R., M.J.D.-R. and A.C.-A. All the authors have equally contributed to the final version of this paper. All authors have read and agreed to the published version of the manuscript.

Funding: This research received no external funding.

Conflicts of Interest: The authors declare no conflict of interest.

Appendix A

Table A1. Estimation breaks of model (1) for EU countries, 1962–2017.

Countries	Factor Loadings	AR Idiosyncratic Parameters	Residual Variance
France	0.87 (7.87) ***	−0.45 (−2.97) ***	0.26 (3.77) ***
Belgium	0.81 (7.11) ***	−0.27 (−1.76) **	0.34 (4.35)
Austria	0.7 (5.76) ***	−0.21 (−1.48)	0.51 (4.82)
Germany	0.7 (5.34) ***	−0.04 (−0.29)	0.57 (4.86)
Denmark	0.67 (5.63) ***	−0.26 (−1.89) *	0.51 (4.87)
UK	0.67 (5.07) ***	−0.11 (−0.77)	0.62 (4.94)
Ireland	0.62 (5.61) ***	−0.47 (−3.79) ***	0.51 (4.91)
Greece	0.59 (4.45) ***	−0.19 (−1.37)	0.7 (5.06)
Cyprus	0.41 (2.86) ***	0.11 (0.8)	0.83 (5.18)
Bulgaria	0.38 (2.68) ***	−0.09 (−0.68)	0.87 (5.21)
Romania	0.38 (2.6) ***	0.03 (0.24)	0.87 (5.21)
Hungary	0.32 (2.64) ***	−0.38 (−3.07) ***	0.77 (5.23)
Italy	0.29 (1.93) *	0.28 (2.13) **	0.91 (5.23)
Malta	0.29 (2.15) **	−0.2 (−1.49)	0.89 (5.25)
Poland	0.27 (1.82) *	0.07 (0.49)	0.93 (5.25)
Finland	0.25 (1.78) *	−0.16 (−1.2)	0.92 (5.26)

Note: AR: Autoregressive; In () t-statistics at * 90%, ** 95% and *** 99% of significance. Note: Luxembourg, Netherlands, Portugal, Spain and Sweden are not significance.

Notes

1. Dynamic panel Model offers an alternative measure for cross-country links to the obtained through conventional input-output models or other types of analysis. In this case, the econometric model employed measure parametrically an indicator that captures the dynamics of EFP from the growth rates of global hectares per capita by country.
2. Croatia, Estonia, Latvia, Lithuania, Slovakia, Slovenia and Czech Republic are omitted from our analysis due to the lack of availability data for the same sample period.
3. This methodology is frequently used for business cycle estimations but the interest in the analysis of the cycle has led to the use of its tools in environmental studies. These are the case of McKitrick and Wood, 2013[29], Doda, 2014[30], Delgado et al., 2018[31], De Lucas et al, 2021[32] and Cabezas et al., 2020[33]. These pa-pers demonstrate the interest and suitability of the Dynamic Factor Model to understand the short-run behavior of environmental variables.

References

1. Recep, U.; Lin, D. Persistence of Policy Shocks to Ecological Footprint of the USA. *Ecol. Indic.* **2017**, *80*, 337–343. [CrossRef]
2. Global Footprint Network. Ecological Footprint. 2021. Available online: http://data.footprintnetwork.org/ (accessed on 10 November 2021).
3. Ulucak, R.; Apergis, N. Does convergence really matter for the environment? An application based on club convergence and on the ecological footprint concept for the EU countries. *Environ. Sci. Policy* **2018**, *80*, 21–27. [CrossRef]
4. Erdogan, S.; Okumus, I. Stochastic and club convergence of ecological footprint: An empirical analysis for different income group of countries. *Ecol. Indic.* **2020**, *121*, 107–123. [CrossRef]
5. Al-Mulali, U.; Weng-Wai, C.; Sheau-Ting, L.; Mohammed, A.H. Investigating the environmental Kuznets curve (EKC) hypothesis by utilizing the ecological footprint as an indicator of environmental degradation. *Ecol. Indic.* **2015**, *48*, 315–323. [CrossRef]
6. Solarin, S.A. Convergence in CO_2 emissions, carbon footprint and ecological footprint: Evidence from OECD countries. *Environ. Sci. Pollut. Res.* **2019**, *26*, 6167–6181. [CrossRef] [PubMed]
7. Stern, D.I.; The Environmental Kuznets Curve: A Primer. CCEP Working Paper 1404. 2014. Available online: https://ccep.crawford.anu.edu.au/ (accessed on 1 October 2021).
8. Bello, M.O.; Solarin, S.A.; Yen, Y.Y. The impact of electricity consumption on CO_2 emission, carbon footprint, water footprint and ecological footprint: The role of hydropower in an emerging economy. *J. Environ. Manag.* **2018**, *219*, 218–230. [CrossRef]
9. Solarin, S.A.; Al-Mulali, U. Influence of foreign direct investment on indicators of environmental degradation. *Environ. Sci. Pollut. Res.* **2018**, *25*, 24845–24859. [CrossRef]
10. Bilgili, F.; Ulucak, R. Is there deterministic, stochastic, and/or club convergence in ecological footprint indicator among G20 countries? *Environ. Sci. Pollut. Res.* **2018**, *25*, 35404–35419. [CrossRef]
11. Yilanci, V.; Pata, U.K. Convergence of per capita ecological footprint among the ASEAN-5 countries: Evidence from a non-linear panel unit root test. *Ecol. Indic.* **2020**, *113*, 106178. [CrossRef]
12. Sarkodie, S.A. Environmental performance, biocapacity, carbon & ecological footprint of nations: Drivers, trends and mitigation options. *Sci. Total. Environ.* **2020**, *751*, 141912. [CrossRef] [PubMed]
13. Solarin, S.A.; Tiwari, A.K.; Bello, M.O. A multi-country convergence analysis of ecological footprint and its components. *Sustain. Cities Soc.* **2019**, *46*, 101422. [CrossRef]
14. Ulucak, R.; Kassouri, Y.; Ilkay, S.; Altıntaş, H.; Garang, A.P.M. Does convergence contribute to reshaping sustainable development policies? Insights from Sub-Saharan Africa. *Ecol. Indic.* **2020**, *112*, 106140. [CrossRef]
15. Haider, S.; Akram, V.; Ali, J. Does biomass material footprint converge? Evidence from club convergence analysis. *Environ. Sci. Pollut. Res.* **2021**, *28*, 27362–27375. [CrossRef] [PubMed]
16. Caglar, A.E.; Balsalobre-Lorente, D.; Akin, C.S. Analysing the ecological footprint in EU-5 countries under a scenario of carbon neutrality: Evidence from newly developed sharp and smooth structural breaks in unit root testing. *J. Environ. Manag.* **2021**, *295*, 113155. [CrossRef] [PubMed]
17. Wu, D.; Liu, J. Spatial and Temporal Evaluation of Ecological Footprint Intensity of Jiangsu Province at the County-Level Scale. *Int. J. Environ. Res. Public Health* **2020**, *17*, 7833. [CrossRef]
18. Zambrano-Monserrate, M.A.; Ruano, M.A.; Ormeño-Candelario, V.; Sanchez-Loor, D. Global ecological footprint and spatial dependence between countries. *J. Environ. Manag.* **2020**, *272*, 111069. [CrossRef] [PubMed]
19. Nathaniel, S.; Nwodo, O.; Sharma, G.; Shah, M. Renewable energy, urbanization, and ecological footprint linkage in CIVETS. *Environ. Sci. Pollut. Res.* **2020**, *27*, 19616–19629. [CrossRef] [PubMed]
20. Wu, D. Spatially and temporally varying relationships between ecological footprint and influencing factors in China's provinces Using Geographically Weighted Regression (GWR). *J. Clean. Prod.* **2020**, *261*, 121089. [CrossRef]
21. Guo, J.; Ren, J.; Huang, X.; He, G.; Shi, Y.; Zhou, H. The Dynamic Evolution of the Ecological Footprint and Ecological Capacity of Qinghai Province. *Sustainability* **2020**, *12*, 3065. [CrossRef]
22. Harvey, A.; Trimbur, T. Trend Estimation and the Hodrick-Prescott Filter. *J. Jpn. Stat. Soc.* **2008**, *38*, 41–49. [CrossRef]
23. Dubois, É.; Michaux, E. Grocer 1.72: An econometric toolbox for Scilab 2019. 2019. Available online: http://dubois.ensae.net/grocer.html (accessed on 12 October 2021).
24. Bai, J.; Ng, S. Determining the Number of Factors in Approximate Factor Models. *Econometrica* **2002**, *70*, 191–221. [CrossRef]
25. Harding, D.; Pagan, A. Dissecting the cycle: A methodological investigation. *J. Monetary Econ.* **2002**, *49*, 365–381. [CrossRef]
26. Bai, J.; Ng, S. Determining the Number of Primitive Shocks in Factor Models. *J. Bus. Econ. Stat.* **2007**, *25*, 52–60. [CrossRef]
27. Bueno, J.L.C.; Santos, S.D.L.; Rodríguez, M.J.D.; Ayuso, I. Testing for structural breaks in factor loadings: An application to international business cycle. *Econ. Model.* **2011**, *28*, 259–263. [CrossRef]
28. Newey, W.K.; West, K.D. A Simple, Positive Semi-Definite, Heteroskedasticity and Autocorrelation Consistent Covariance Matrix. *Econometrica* **1987**, *55*, 703. [CrossRef]
29. McKitrick, R.; Wood, J. Co-fluctuation patterns of per capita carbon dioxide emissions: The role of energy markets. *Energy Econ.* **2013**, *39*, 1–12. [CrossRef]
30. Doda, B. Evidence on business cycles and emissions. *J. Macroecon.* **2014**, *40*, 214–227. [CrossRef]
31. Delgado-Rodríguez, M.J.; Ares, A.C.; Santos, S.D.L. Cyclical fluctuation patterns and decoupling: Towards common EU-28 environmental performance. *J. Clean. Prod.* **2018**, *175*, 696–706. [CrossRef]

32. De Lucas-Santos, S.; Delgado-Rodríguez, M.J.; Cabezas-Ares, A. Cyclical convergence in per capita carbon dioxide emission in US states: A dynamic unobserved component approach. *Energy* **2020**, *217*, 119349. [CrossRef]
33. Cabezas-Ares, A.; Delgado-Rodríguez, M.; Lucas-Santos, S. The Dynamics of Cyclical Convergence and Decoupling in the Environmental Performance of Spanish Regions. *Sustainability* **2020**, *12*, 8569. [CrossRef]

 *land*

Article

The Effect of Ageing, Gender and Environmental Problems in Subjective Well-Being

Manuela Ortega-Gil [1,*], Antonio Mata García [1] and Chaima ElHichou-Ahmed [2]

[1] Department of General Economy, University of Cadiz, 11202 Algeciras, Spain; antonio.mata@uca.es
[2] Facultad de Ciencias Económicas y Empresariales, Campus de Algeciras, University of Cadiz, 11202 Algeciras, Spain; chaima.elhichou@alum.uca.es
* Correspondence: manuela.ortega@uca.es

Abstract: This paper studies the relationship of factors such as ageing, gender and environmental problems included in the quality of life (QoL) with the subjective well-being represented by the life satisfaction (LS) indicator of the citizens of 33 European countries. To do so, it uses the LS of a country's citizens as the dependent variable; ageing, gender and environmental variables as independent variables; and other factors included in the QoL indicators and macroeconomic factors as control variables. Analysis uses data from the World Values Survey (WVS) and the European Values Study (EVS) for LS, from Eurostat for QoL indicators and World Bank for macroeconomic indicators. The values of LS have been treated based on the individual data from WVS and EVS in percentages according to their levels by country, and we present four robust models (two logit model and two OLS model). The results show that arrears of people aged 65 and over are a relation of positive significative in models with low levels of LS. The opposite is true for the income of people aged 65 and over. On the other hand, pollution, grime or other environmental problems and inequality show an inverse relationship with life satisfaction in models with high levels of satisfaction and a positive relationship in models with low levels of satisfaction. Nonetheless, the study has also shown contradictions in the gender gap and poverty indicators that should be studied further.

Keywords: ageing; gender; environment; satisfaction life; quality of life indicator

Citation: Ortega-Gil, M.; Mata García, A.; ElHichou-Ahmed, C. The Effect of Ageing, Gender and Environmental Problems in Subjective Well-Being. *Land* **2021**, *10*, 1314. https://doi.org/10.3390/land10121314

Academic Editors: Vanessa Zorrilla-Muñoz, Eduardo Fernandez, Blanca Criado Quesada, Sonia De Lucas Santos, Jesus Cuadrado Rojo and Maria Silveria Agulló-Tomás

Received: 14 October 2021
Accepted: 26 November 2021
Published: 28 November 2021

Publisher's Note: MDPI stays neutral with regard to jurisdictional claims in published maps and institutional affiliations.

1. Introduction

Every day, it is more usual to find citizens' well-being among the main objectives of the governments of developed countries and many international organizations. According to some authors, there is a belief that objective well-being (such as wealth) should be part of the overall indicator of well-being, and that GDP by itself is not adequate to describe the well-being of a country. In contrast, wealth on its own is not sufficient to describe a person's happiness [1]. The individual well-being of the citizens of a country "is highly subjective and person-specific, and thus policies should focus on making well-being possible by providing the freedoms and capabilities that allow each person to achieve what will contribute to his or her own well-being" [2] (p. 65). In that sense, the same thing happens with life satisfaction and happiness. The individual subjectivity of these variables makes it necessary to use polls for this study [3]. Among the most important surveys that include data on life satisfaction and happiness are the World Values Survey (WVS) and the European Values Study (EVS).

Many countries and researchers use psychological well-being, health, education, time use, cultural diversity and resilience, good governance, community vitality, ecological diversity and resilience, and living standards as measures of happiness or LS [4–6]. Many of the dimensions coincide with those considered by Eurostat and will be discussed in this study. Among them, there are four variables (income, wealth, civic commitment and governance) that report the highest quality of life indicators and the highest degree of happiness, and others (homicide rates, housing without basic services, poor working conditions and pollution) that present an inverse relationship with well-being [7].

Unlike life satisfaction and happiness, quality of life presents objective situations and subjective perceptions that combine to determine an individual's well-being [8]. Currently, there is an accepted consensus on the importance of both dimensions for the study of well-being. The QoL indicators are included such as objective income, and others, such as LS, that are entirely subjective. Among these measures of quality of life are the ones published by the OECD Better Life Index, which includes a total of 11 dimensions that are materialized in 24 indicators including five subjective indicators, one of which is LS and that is presented by the European Statistical System with nine dimensions: material living conditions, productive or other main activity, health, education, leisure and social interactions, economic security and physical safety, governance and basic rights, natural and living environment and overall experience of life. In this work, we will focus on the effect of some indicators that make up the European Statistical System's QoL on satisfaction with life from the World Values Survey (WVS) and the European Values Study (EVS). This work focuses especially on those indicators related to ageing, gender employment gap and the environment, and based on the studies analyzed on subjective well-being, other quality of life and macroeconomic indicators have been used as control variables. The LS indicator has been selected to reflect citizens' well-being since it is more stable over time to measure well-being [9]. In Pavot and Diener study, it showed a degree of the temporal stability of 0.54 for four years.

There are differences in studies between income and well-being. Higher incomes than others report higher levels of happiness [10], although dissatisfaction with income occurs when what you want exceeds what you get and when people feel they deserve more than they earn [11]. Other studies show that people with higher income tend to be happier. Still, as income increases, happiness does not increase in the same way [4] as higher-income aspirations reduce life satisfaction [12].

On the other hand, among the European Statistical System's QoL factors, there are also labour variables such as employment, unemployment, quality of work, inactivity and gender inequality. Unemployment and employment affect not only people's economic situation but also their social environment [13]. Employment causes an increase in subjective well-being. Just as it provides for the development of a person and acquiring new skills and knowledge, it makes people feel less monotonous and routine [14]. Temporary employees have generally lower job satisfaction, although it depends on the type of contract and job security [15]. Job loss and the resulting period of unemployment can lead to a high degree of stress, causing impacts on people's identities, depression due to the economic situation and the lack of satisfaction [16,17].

Unemployment affects people's social lives; their consumption patterns are altered, they show symptoms of isolation, loss of self-esteem, feelings of hopelessness that can lead to the deterioration of their physical well-being producing a prolonged duration of this state and an increase of negative emotions, which affect their mental health. Another relevant aspect is that the longer the duration and repetition of unemployment, the greater the dissatisfaction. Women, during the reaction process, once they were unemployed, presented fewer negative reactions than men. Similarly, repeated unemployment did not mean that responses were better [18]. It also affects the well-being of those close to you. The existence of high unemployment makes being employed difficult. Therefore, a key factor for well-being is the security of the labour market. Job instability affects men more than women [19].

It is difficult to find indicators that measure the effects of the gender gap on women's well-being [20]. Housework is mainly carried out by women, which prevents women's full incorporation into the labour market and affects women's well-being [21,22]. Therefore, the gender gap negatively affects the well-being of women [23]. More women than men are in the inactive population, although most of the existing studies on the inactive population focus on retirees.

The effect of retirement is positive with life satisfaction, but not to the same degree whether you are employed or unemployed at the time of retirement. The level of satisfaction with life when retiring is higher in employed people than in an unemployed person. Although, the well-being of the involuntarily unemployed increases considerably upon retirement [24,25]. Moreover, going from a work situation to being permanently ill implies lower levels of psychological well-being [26]. The psychosocial quality of labour and mental health is related to people's subjective well-being [27].

Physical and psychological well-being are also essential factors for well-being. The existence of a direct relationship between physical activity and well-being and inverse with sedentary behaviour has been proven [28–30]. Although [31] show that health does not directly affect the satisfaction of the quality of life in general but instead has an indirect effect on health satisfaction, [32] clarifies and indicates that the relative importance of satisfaction with one's health for QoL depends on the QoL measures used. For this author, people may perceive health as an essential factor about other vital areas of life. Still, satisfaction with their health may not necessarily be the most important determining factor in satisfaction with other critical life areas. In addition, a negative relationship has been confirmed between living in a deficient dwelling and health status [33].

Educational level also influences well-being; "Those individuals with a higher level of education are characterised by lower levels of emotional and physical distress than those with a lower level of education; those with higher levels of depression and discomfort among others; and who have lower levels of pleasure, hope, happiness, energy and physical fitness than more educated individuals. The latter is not satisfied with their work" [14] (p. 275). There is a positive relationship of educational level with QoL, associating secondary and tertiary education with a better quality of life [34,35].

Ageing, economic insecurity, payment arrears and physical safety also affects well-being. The first signs of ageing usually appear in people over 50 years of age and can negatively impact the individual's functioning in society and interpersonal relationships [30,36,37]. It has also been found that the life satisfaction of elderly people living in social housing is very low and that moving to better housing increases life satisfaction [38]. The economic insecurity and payment arrears contribute to reducing the well-being and health of households. Mortgage arrears are an additional burden on households facing financial difficulties, making it a social problem. They can cause health problems and homelessness, with indirect effects on state finances [39]. The frequency of arrears is relevant, as the occasional incidence had a significantly lower impact [40].

The relationship between well-being and physical safety has also been analysed. Life satisfaction is lower for victims and people who lived in areas with higher crime rates. Property crime (theft) affects all household members equally. The real nature of a murder is related to the quality of life perceived by the respondent [41]. The negative effects of both property crimes and homicides may be similar among all existing household members, making the perceived quality of life a valid measure of crime victimisation's well-being impact. Overall, all crimes and high crime rates reduce life satisfaction, and assaults quadruple the negative relationship with QoL, compared to robberies [42,43]. Personal well-being is more affected by fear of crime than by being a victim [44].

Environmental variables such as pollution, noise, and other environmental problems are also among the European Statistical System's QoL factors. Air pollution is a global health and environmental problem, especially in major developing countries [6,45]. There is consensus on a negative relationship between air pollution and satisfaction with life. There is also a binding relationship between well-being, income and air pollution concerning air quality [46] and negative effects on the cardiovascular and respiratory systems [47]. The exposure to PM10 may not have a direct health effect. Deficient air pollution levels can reduce subjective well-being and manifest itself through irritation, discomfort or high physical stress [48]. Besides, it directly influences people's lives through their visual perceptions, reducing their subjective well-being [6]. Climate change will have effects on physical and mental health and well-being. It is expected that extreme storms and droughts will occur, and this will result in human beings experiencing both short and long term decreases in quality of life, presenting: anxiety, depression, post-traumatic stress, substance abuse and even suicides [49].

These studies open a new path for us on the need to elaborate policies that pursue growth and development and consider the sustainability and well-being of citizens. For this reason, this work focuses on how the indicators of ageing, gender and environment included in the set of QoL indicators according to Eurostat affect the LS of citizens. This paper's main objective is to increase our understanding of the impact on life satisfaction of the environmental and social situations that affect the citizens of the 33 countries analysed, and the combined effects of these variables. To achieve this, we have used information from individual WVS and EVS surveys covering 33 countries from 1999 to 2018 and calculated the percentage of citizens per country in the ten levels of satisfaction with life that these surveys include. We tested four relationships by constructing the LS dependent variable differently for each, and two models were built.

The contribution of this work is to study the effect of QoL factors such as ageing, gender and environment issues of the European Statistical System on life satisfaction of citizens using several specifications based on the LS index as a dependent variable and using other QoL and macroeconomic indicators as control variables in order to visualize the results with different treatments of the life satisfaction variable and to deepen the study of these variables. We have found it appropriate to focus our research on these four different approaches to one variable—satisfaction with life (LS), which contains ten levels, and explore it using different models. We used these models to test the following hypotheses:

- H1: Countries with a higher percentage of the population aged 65 and over who have arrears (mortgage or rent, utility bills or hire purchase) present a higher percentage of citizens with low levels of LS (1–5).
- H2: As income increases for people aged 65 and over, the percentage of people with life satisfaction levels of 8 or above increases.
- H3: Increasing the percentage of the population aged 65 and over who perceive their health as "very good" raises the probability that a country will have more than 35% of its population at satisfaction levels 9 and 10.
- H4: Countries with a higher percentage of gender employment gap are more likely to have more than 25% of their population with LS levels from 1 to 5 than those with less gender employment gap.
- H5: Air pollution, noise from neighbours and other environmental problems have an inverse relationship with the percentage of citizens in countries with an LS level equal to or greater than 8.

2. Materials and Methods

Like many other studies on happiness and satisfaction with life, this work has used the World Values Survey (WVS) databases, European Values Study (EVS), Eurostat and the World Bank. The WVS and EVS databases provided data on the dependent variable, satisfaction with life. Both sources could be used as the WVS website indicates that interested users in European countries study EVS WVS's features [50–53]. The data obtained comes from citizens of 33 countries at the individual level. Eurostat provided the QoL indicator data and the World Bank the macroeconomic data on the countries under study. All previous data sets were entered into Stata statistical software (version 16) for analysis.

Previously, we have established the reasons for selecting life satisfaction as a variable to study subjective well-being and different indicators of quality of life and macroeconomics as independent and control variables. Some studies focused on these relationships, but this study goes further because it shows other models depending on the LS indicator's treatment. For this, the dependent variable (LS) has been constructed in four different ways using data from hundreds of surveys carried out between 1999 and 2018 by the World Values Survey and European Values Study groups in 33 countries to calculate the percentages by levels.

(1)   Satisf9_10_35. A dichotomous variable that takes a value of 0 when the condition is not met that more than 35% of its citizens have a satisfaction level equal to or more than 9, and the value of 1 when the condition is met.

(2)   Satist1_5_25. A dichotomous variable with value 0 when the condition is not met that in a country more than 25% of its citizens have a satisfaction level equal to or less than 5, and the value of 1 when the condition is met.

(3)   Satisf8_10. It is a variable that shows the percentage by country of citizens at the level equal to or more than 8.

(4)   Satisf1_5. It is a variable that shows the percentage by country of citizens at the level equal to or less than 5.

The independent and control variables were selected based on a preliminary analysis of all QoL indicators using Eurostat data and macroeconomic indicators such as unemployment (in percentage), GDP per capita, GDP growth (in percentage), life expectancy (years) and the Human Development Index (HDI, in percentage), the Gini index (Gini, in percentage) and the poverty rate at \$5.50 dollars per day (Pov55, in percentage). Finally, after reviewing the existing literature and analyzing all these variables and their correlations, those that were highly correlated were eliminated or replaced by others with similar characteristics. In addition, robust models have been built to control the observed heterogeneity. Therefore, the variables selected have been:

A.   Ageing QoL indicators

- q_mlc_mh65. Total population living in a dwelling with a leaking roof, damp walls, floors or foundation, or rot. (65 years or over). Percentage.
- q_safe_da65. Arrears (mortgage or rent, utility bills or hire purchase) from 2003 onwards—EU-SILC survey [ilc_mdes05]. One adult 65 years or over. Percentage.
- q_hlt_spvg65. Self-perceived health as very good (65 years or over). Percentage.
- logq_mlc_i65. Logarithm of median equivalised net income (65 years or over). Euro.

B.   Gender QoL indicator

- q_gov_dgg. Gender employment gap.

C.   Environment QoL indicator

- q_env_pol2.5. Exposure to air pollution by particulate matter. Particulates < 2.5 μm.
- q_env_polpa. Pollution, grime or other environmental problems. Above 60% of median equivalised income (% Total).
- q_env_polna. Noise from neighbours or from the street. Above 60% of median equivalised income.

D.   Other QoL indicators

- q_act_qlt. Temporary employees. From 15 to 64 years (percentage of employees).
- q_act_qtl. Long-term unemployment. From 15 to 74 years (percentage of active population).
- logq_mlc_c. Logarithm of main GDP aggregates per capita. Actual individual consumption. Euro.

E.   Macroeconomic indicators

- Gini. Gini Index.
- Pov55. Poverty headcount ratio at \$5.50 a day (2011 PPP) (% of the population).
- UnEmpl. Unemployment (as a percentage).

The model may be specified as

$$
\begin{aligned}
LS_{it} = \beta_0 &+ \sum_{l=1,\dots,e} \beta_l Ageing_{it} + \sum_{l=e+1,f} \beta_l Gender_{it} + \sum_{l=f+1,\dots,j} \beta_l Env_{it} \\
&+ \sum_{l=j+1\dots,i} \beta_l QoL_{it} + \sum_{l=i+1,\dots,m} \beta_l MAC_{it} + I(WAVE_t)\beta_{m+2} + \varepsilon_{it}
\end{aligned} \tag{1}
$$

Our dependent variable LS is "satisfaction with the life", measured per country "i" how satisfied their citizens are with their lives at WVS wave "t". We include five vectors whose variables have been mentioned previously. Ageing, gender and env are independent variables, QoL and macroeconomic are control variables. We also wave "t" fixed effects I (WAVE$_t$) and cluster standard errors at the country level.

Results were achieved using different models as the dependent variable was processed in four ways: two dichotomous variables, one shows the percentage by country of citizens at the level equal to or more than eight and other shows the percentage by country of citizens at the level equal to or less than 5. For dichotomous dependent variables Satisf9_10_35 (models 1) and Satisf1_5_25 (models 2), we used logit models [54–56]. For model 3 we used Satisf8_10 as the percentage of citizens whose satisfaction with life was equal to or greater than 8, and in model 4 Satisf1_5 as the percentages of citizens whose satisfaction with life was equal to or less than 5. These two models were performed by linear regression. The same control variables were used to build the models; those that best adapted for models 1 and 3 were selected while the same procedure was applied for models 2 and 4. All models used the same fixed effects as described above.

3. Results

We built two robust models with dummy variables to control wave t fixed effects in vectors and I (WAVEt). Results for models 1 and 2 (logit) and 3 and 4 (OLS) are listed in Table 1, and models 1 and 2 marginal effects are listed in Table 2.

Table 1. Logit and Linear Regression Model.

Satisfaction Life	Modelo (1) Logit	Modelo (2) Logit	Modelo (3) Linear Regression	Modelo (4) Linear Regression
Variables	**Satisf9_10_35**	**Satist1_5_25**	**Satisf8_10**	**Satisf1_5**
q_mlc_mh65		0.0747 (0.0641)		0.100 * (0.0519)
q_safe_da65		0.279 *** (0.0598)		0.571 *** (0.126)
q_hlt_spvg65	−0.163 ** (0.0641)		0.0263 (0.131)	
logq_mlc_i65	3.889 *** (0.890)		10.64 *** (1.328)	
q_gov_dgg	−0.227 *** (0.0692)	−0.141 *** (0.0441)	−0.123 (0.135)	−0.266 *** (0.0592)
q_env_pol2.5	0.166 *** (0.0491)		0.115 (0.124)	
q_env_polpa	−0.619 *** (0.235)	0.310 *** (0.0712)	−0.561 *** (0.175)	0.214 ** (0.104)
q_env_polna	−0.0191 (0.0845)	−0.358 *** (0.0868)	0.0689 (0.145)	0.0588 (0.114)
q_act_qlt	−0.0719 (0.0482)	−0.0913 * (0.0471)	0.272 *** (0.0866)	−0.361 *** (0.0588)
q_act_qtl	−0.437 ** (0.217)		−1.826 *** (0.287)	
logq_mlc_c		−5.405 *** (1.479)		−16.48 *** (2.011)
Gini	−0.191 * (0.107)	0.519 *** (0.112)	−0.924 *** (0.163)	0.421 *** (0.122)
Pov55	0.603 ***	−0.0666	0.951 ***	−0.588 ***

Table 1. *Cont.*

Satisfaction Life	Modelo (1) Logit	Modelo (2) Logit	Modelo (3) Linear Regression	Modelo (4) Linear Regression
Variables	Satisf9_10_35	Satist1_5_25	Satisf8_10	Satisf1_5
	(0.109)	(0.101)	(0.143)	(0.193)
UnEmpl		0.0764		0.535 ***
		(0.0908)		(0.203)
3.wave	−3.430 ***	−1.688 **	−2.823 *	1.211
	(0.934)	(0.809)	(1.532)	(0.963)
4.wave	−0.711	−5.276 ***	0.624	0.459
	(0.826)	(1.126)	(1.421)	(0.941)
Constant	−21.36 **	35.58 ***	−9.217	162.5 ***
	(8.407)	(12.11)	(14.19)	(19.73)
Observations	210	320	211	317
R-squared			0.779	0.734
Pseudo R^2	0.6034	0.7567		

Robust standard errors in parentheses,*** $p < 0.01$, ** $p < 0.05$, * $p < 0.1$.

Table 2. Margins y Odds Ratios.

Variables	dx/dy Satisf9_10_35	dx/dy Satist1_5_25	Odds Ratio Satisf9_10_35	Odds Ratio Satist1_5_25
q_mlc_mh65		0.0028191		1.077507
q_safe_da65		0.0105275 ***		1.321502 ***
q_hlt_spvg65	−0.0115144 **		0.8494863 **	
logq_mlc_i65	0.2745012 ***		48.85412 ***	
q_gov_dgg	−0.015995 ***	−0.0053183 ***	0.7972397 ***	0.8686387 ***
q_env_pol2.5	0.0117133 ***		1.180504 ***	
q_env_polpa	−0.0437032 ***	0.011691 ***	0.5384075 ***	1.362849 ***
q_env_polna	−0.0013499	−0.013538 ***	0.9810579	0.6987339 ***
q_act_qlt	−0.0050754	−0.0034484 *	0.9306211	0.9127323 *
q_act_qtl	−0.0308778 **		0.6456845 **	
logq_mlc_c		−0.2040985 ***		0.0044962 ***
Gini	−0.0135146 *	0.0196163 ***	0.8257518 *	1.681083 ***
Pov55	0.0425479 ***	−0.0025136	1.827178 ***	0.9356063
UnEmpl		0.0028862		1.079423

*** $p < 0.01$, ** $p < 0.05$, * $p < 0.1$.

Model 1 has treated the dependent variable as a dichotomous variable with the value of 0 when is not met that more than 35% of its citizens have a satisfaction level equal to or more than nine and the value 1 of when the condition is met. A robust logit model has been constructed with the aforementioned independent variables and has a correct classification of 91.43% (78.57% sensitivity and 96.10% specificity) (Figure 1).

Results show a positive significance ($p < 0.01$) and consequently an increase in the probability of that in a country more than 35% of its citizens have a satisfaction level equal to or more than 9 with the increase of net income equivalent at 65 years or over (logq_mlc_i65); exposure to air pollution by particulate < 2.5 µm (q_env_pol2.5); poverty index (Pov55).

Results further show an inverse significance at $p < 0.01$ with variables for gender employment gap (q_gov_dgg), pollution, grime or other environmental problems (q_env_polpa) and time period covered by variables 3wave and with lower significance and effect $p < 0.05$ with variables for self-perceived health as "very good" at 65 years or over (q_hlt_spvg65) and long-term unemployment (q_act_qtl).

```
Logistic model for Satisf9_10_35              Logistic model for Satist1_5_25

                    ─── True ───                                ─── True ───
Classified │      D          ~D       Total   Classified │      D          ~D       Total

     +     │     44           6         50         +      │     63           9         72
     -     │     12         148        160         -      │      8         240        248

   Total   │     56         154        210       Total    │     71         249        320

Classified + if predicted Pr(D) >= .5         Classified + if predicted Pr(D) >= .5
True D defined as Satisf9_10_35 != 0          True D defined as Satist1_5_25 != 0

Sensitivity                  Pr( +| D)  78.57% Sensitivity                  Pr( +| D)  88.73%
Specificity                  Pr( -|~D)  96.10% Specificity                  Pr( -|~D)  96.39%
Positive predictive value    Pr( D| +)  88.00% Positive predictive value    Pr( D| +)  87.50%
Negative predictive value    Pr(~D| -)  92.50% Negative predictive value    Pr(~D| -)  96.77%

False + rate for true ~D     Pr( +|~D)   3.90% False + rate for true ~D     Pr( +|~D)   3.61%
False - rate for true D      Pr( -| D)  21.43% False - rate for true D      Pr( -| D)  11.27%
False + rate for classified +  Pr(~D| +) 12.00% False + rate for classified +  Pr(~D| +) 12.50%
False - rate for classified -  Pr( D| -)  7.50% False - rate for classified -  Pr( D| -)  3.23%

Correctly classified                    91.43% Correctly classified                    94.69%
```

Figure 1. Classification statistics and table logit models (Models 1 y 2).

The study of the marginal effect on model 1 shows how the probability that more than 35% of its citizens have a satisfaction level equal to or more than 9 changes if the independent variables change in one unit. This positive effect is more powerful for net income equivalent at 65 years or over (0.2745 ***) and inverse or negative for grime or other environmental problems (−0.0437 ***). Overall, the marginal effect in model 1 is significant, positive and shows a result of 0.27 points; consequently, the probability that in a country more than 35% of its citizens have a satisfaction level equal to, or more than 9 when all variables are average is 26.66%, with a confidence of 95% that the probability is between 23.16% and 30.17% (Figure 2).

```
Predictive margins                         Number of obs      =        210
Model VCE     : Robust

Expression    : Pr(Satisf9_10_35), predict()

                         Delta-method
              Margin     Std. Err.      z      P>|z|     [95% Conf. Interval]

      _cons   .2666667   .0178787    14.92    0.000    .2316251    .3017082
```

Figure 2. Predictive Margins Model 1.

Model 2 has treated the dependent variable as a dichotomous variable with value 0 when the condition is not met that in a country more than 25% of its citizens have a satisfaction level equal to or less than 5, and the value of 1 when the condition is met. A robust logit model has been constructed with the aforementioned independent variables and has a correct classification of 94.69% (88.73% sensitivity and 96.39% specificity) (Figure 1).

Results show a positive significance ($p < 0.01$) and, consequently, an increase in the probability that in a country more than 25% of its citizens have a satisfaction level equal to or less than 5 with the increase of arrears (on mortgage or rent, utility bills or hire purchase) at 65 years or over (q_safe_da65); pollution, grime or other environmental problems (q_env_polpa); Gini index (Gini).

Results further show an inverse significance at $p < 0.01$ with variables for gender employment gap (q_gov_dgg), noise from neighbours or from the street (q_env_polna), of actual individual consumption (logq_mlc_c) and time period covered by variables 4wave and with lower significance and effect $p < 0.05$ with time period covered by variables 3wave.

The study of the marginal effect on model 2 shows how the probability that more than 25% of its citizens have a satisfaction level equal to or less than 5 changes if the independent variables change in one unit. This effect is stronger in positive Gini index (0.0196 ***) and inverse or negative on actual individual consumption (−0.2041 ***) Overall, the marginal effect in model 2 is significant, positive and shows a result of 0.22 points; subsequently, the probability that in a country more than 25% of its citizens have a satisfaction level equal to or less than 5 when all variables are average is 22.18%, with a confidence of 95% that the probability is between 20.04% and 24.34% (Figure 3).

```
Predictive margins                        Number of obs     =       320
Model VCE     : Robust

Expression    : Pr(Satist1_5_25), predict()


                          Delta-method
                Margin    Std. Err.      z     P>|z|     [95% Conf. Interval]

      _cons    .221875   .0109741     20.22   0.000     .2003662    .2433838
```

Figure 3. Predictive Margins Model 2.

Model 3 was performed as an ordinary least squares (OSL) regression with the dependent variable that shows the percentage by country of citizens equal to or more than 8 of LS. The results of this model showed a positive significance at $p < 0.01$ and consequently an increase in the percentage by country of citizens at the level equal to or more than 8 of LS with an increase of net income equivalent at 65 years or over (logq_mlc_i65); temporary employees (q_act_qlt) and poverty index (Pov55). Results further show an inverse significance at $p < 0.01$ with pollution, grime or other environmental problems (q_env_polpa); long-term unemployment (q_act_qtl) and Gini index (Gini). The biggest effects on the linear model of life satisfaction with one-unit variations in independent variables are found as positive for net income equivalent at 65 years or over (10.64 ***) and negative for long-term unemployment (−1.826 ***).

Model 4 was performed as an ordinary least squares (OSL) regression with the dependent variable that shows the percentage by country of citizens equal to or less than 5 of LS. The results of this model showed a positive significance at $p < 0.01$ and therefore, an increase in the percentage by country of citizens at the level equal to or less than 5 of LS with an increase of arrears (on mortgage or rent, utility bills or hire purchase) at 65 years or over (q_safe_da65); pollution, grime or other environmental problems (q_env_polpa); Gini index (Gini) and unemployment (UnEmpl).

Results further show an inverse significance at $p < 0.01$ with gender employment gap (q_gov_dgg), temporary employees (q_act_qlt); actual individual consumption (logq_mlc_c) and poverty index (Pov55). The biggest effects on the linear model of life satisfaction with one-unit variations in independent variables are found as positive for increase of arrears (on mortgage or rent, utility bills or hire purchase) at 65 years or over (0.571 ***) and negative for actual individual consumption (−16.48 ***).

4. Discussion

In relation to ageing, the income and arrears (mortgage or rent, utility bills or hire purchase) variables stand out. The study shows a positive significance ($p < 0.01$) of net income equivalent to 65 years and over (in logarithmic) and life satisfaction in models

1 and 3 (analyse high levels of life satisfaction). Thus, an increase of net income equivalent at 65 years or over (in logarithmic), leads to a rise in the probability that 35% of its citizens have a satisfaction level equal to or more than 9 with a probability of 27.45% (model 1), and also increases the percentage per country of citizens with a level of satisfaction level equal to or higher than 8 with β = 10.64 (model 3). These results are in line with studies by [7,10]. In the reverse direction, the individual consumption (the individual consumption variable has been used as income variables to better fit the model) shows significant inverse relationships ($p < 0.01$)in models 2 and 4 (analyse low levels of life satisfaction). This indicates that as consumption decreases, the probability increases in that more than 25% of a country's citizens have satisfaction levels equal to or below 5 (model 2), and that the percentage of people in a country with satisfaction levels between 0 and 5 increases (model 4).

Moreover, the results show a relation positive significative ($p < 0.01$) between arrears (mortgage or rent, utility bills or hire purchase) to adults 65 years or over and low levels of life satisfaction (models 2 and 4). Therefore, a rise in arrears increases the probability that more than 25% of a country's citizens have satisfaction levels equal to or below 5 (model 2) and increases the percentage of people in a country with satisfaction levels between 0 and 5 (model 4). These results are in line with [39,40] studies. In addition, in the results of model 1, self-perceived health as "very good" to adults 65 years or over variable, appear with a relation negative significative $p < 0.05$. This indicates that as this variable decreases the probability increases that more than 35% of a country's citizens have a level equal to or more than 9. These results are in line with the study of [33].

The analysis of the gender employment gap and well-being represented in our work with the variable satisfaction with life confirms that the gender employment gap negatively affects women's well-being [23]. Model 1 shows an inverse relationship between this variable and LS. Therefore, an increase of the gender employment gap leads to increase in the probability that 35% of the citizens in a country have a satisfaction level equal to or more than 9 with a probability of 1.6% (model 1). Although, the results of models 2 and 4 differ from the above because they present a relation negative significative ($p < 0.01$) since they are models with low levels of LS. This indicates that as the gender employment gap decreases, the probability increases that more than 25% of a country's citizens have satisfaction levels equal to or below 5 (model 2), and points out that the percentage of people in a country with satisfaction levels between 0 and 5 increases (model 4). The gender gap results of models 2 and 4 are not in accordance with other studies [23]. It would be interesting to delve into the analysis of gender variables.

Regarding the environmental variables analysed, this study shows a relation negative significative effect $p < 0.01$ of pollution, grime or other environmental problems with life satisfaction in models with high satisfaction levels (models 1 and 3), and the inverse sense in models with low satisfaction (models 2 and 4). This is in line with the study of [46,57]. However, the exposure to air pollution by particulate < 2.5 µm presents a relation positive significative in model 1. Concerning the relationship between the noise from neighbours or from the street variable and LS, the model 2 results show a relation inverse significative $p < 0.01$. Therefore, its decrease increases the probability that more than 25% of a country's citizens have satisfaction levels equal to or below 5. In line with the study conducted by [58], and in the opposite direction to the results of this work, are the studies by [59–61].

Concerning the variables of employment, this study shows a positive significative effect $p < 0.01$ of the percentage of the temporary employees (15 to 64 years) with LS in model 3, and inverse in model 4. Those results are in the same line as the studies by [15]. However, unemployment indicators show different behaviours depending on the duration of this situation. Our research shows how unemployment, in general, maintains a positive relationship in model 4 (model with low satisfaction 1–5). Its increase raises the percentage of people in a country with satisfaction levels between 0 and 5 (model 4). With respect to long-term unemployment, the results show a negative significative $p < 0.05$ in model 1 and $p < 0.01$ in model 3. Accordingly, a decrease of long-term unemployment leads to an increase in the probability that 35% of its citizens have a satisfaction level equal to or more than 9 (model 1), and also increases the percentage per country of citizens with a level of satisfaction equal to or higher than 8 (model 3). The results are similar to the study carried out by [16–18].

Relating the inequality (Gini variable), the results show a relation positive significative $p < 0.01$ in models 2 and 4, and negative in model 3. Therefore, the increase in inequality increases the probability that more than 25% of a country's citizens have satisfaction levels equal to or below 5 (model 2) and that the percentage of people in a country with satisfaction levels between 0 and 5 increases (model 4). In addition, its decrease increases the percentage per country of citizens with a satisfaction level equal to or higher than 8 (model 3). These results are in line with the study by [62].

Furthermore, the analysis shows that the poverty indicator in a country (Pov55) shows a significant positive relationship $p < 0.01$ in models 1 and 3, with a marginal effect in model 1 (4.25%) and in model 3 (0.951). It also shows an inverse relationship with the percentage per country of citizens at or below level 5 (model 4), but it is not significant in model 2. These results differ from the study conducted by [62,63] and open the debate about poverty and its effects on citizens in a developed country.

5. Conclusions

To conclude, it is possible to affirm from these results that countries with a higher percentage of the population aged 65 and over who have arrears (mortgage or rent, utility bills or hire purchase) present a higher percentage of citizens with low levels of LS (1–5). Therefore, hypothesis 1 is accepted.

We can also point out that as the income of people aged 65 and over increases, the percentage of people with life satisfaction levels equal to or higher than 8 increases (hypothesis 2 is accepted), and raises the probability that a country will have more than 35% of its population at satisfaction levels 9 and 10. In addition, as the percentage of the population aged 65 and over who perceive their health as "very good" increases, the probability of a country having more than 35% of its population in satisfaction levels 9 and 10 increases too, but with a lower significance $p < 0.05$. Consequently, hypothesis 3 is accepted considering its lower significance.

This paper also shows a significant inverse relationship $p < 0.01$ between the gender labour gap and LS indicators in three models (models 1, 2 and 4). Model 1 is in line with previous studies, but models 2 and 3 should be studied in more depth and detail as they show that as the gender employment gap decreases, this increases the probability that more than 25% of a country's citizens have satisfaction levels equal to or below 5 (model 2) (Hypothesis 4 is rejected), and indicates that the percentage of people in a country with satisfaction levels between 0 and 5 increases (model 4).

With regard to the analysis of the environmental variables, specifically the variable pollution, grime, or other environmental problems, the study reveals an inverse relationship between this indicator and life satisfaction in the models with high levels of satisfaction (model 1 and 3) and a positive relationship in models 2 and 4. These results are in line with previous studies and lead to the acceptance of hypothesis 5.

Finally, it is important to note the significant positive relationship of inequality (Gini index) with the models that represent low levels of satisfaction with life. These results go in the same direction as other studies analysed. Unlike this, the poverty indicator shows a relationship contrary to the expected one that should be studied further. What is more, we believe that it would be interesting to study these variables, including developing and underdeveloped countries in the study.

Author Contributions: Conceptualization, M.O.-G.; A.M.G. and C.E.-A.; methodology, M.O.-G.; software, M.O.-G.; validation, M.O.-G.; A.M.G. and C.E.-A.; formal analysis, M.O.-G.; investigation, M.O.-G.; A.M.G. and C.E.-A. resources, M.O.-G.; A.M.G. and C.E.-A.; data curation, M.O.-G. and C.E.-A.; writing—original draft preparation, M.O.-G.; A.M.G. and C.E.-A.; writing—review and editing, M.O.-G.; visualization, M.O.-G.; A.M.G. and C.E.-A.; supervision M.O.-G.; A.M.G. and C.E.-A. All authors have read and agreed to the published version of the manuscript.

Funding: This publication and research have been (partially) granted by Department of General Economy and Globalization and Territorial Dynamics Group of Universidad de Cádiz, Spain.

Institutional Review Board Statement: Not applicable.

Informed Consent Statement: Not applicable.

Data Availability Statement: As indicated, data used the World Values Survey (WVS), the European Values Study (EVS), the World Bank and Eurostat databases: https://www.worldvaluessurvey.org/wvs.jsp (accessed on 15 December 2020); https://europeanvaluesstudy.eu/methodology-datadocumentation/ (accessed on 14 May 2021); https://databank.worldbank.org/home.aspx; https://ec.europa.eu/eurostat/data/database (accessed on 26 May 2021).

Conflicts of Interest: The authors declare no conflict of interest.

References

1. D'Acci, L. Measuring Well-Being and Progress. *Soc. Indic. Res.* **2010**, *104*, 47–65. [CrossRef]
2. Rogers, D.S.; Duraiappah, A.K.; Antons, D.C.; Munoz, P.; Bai, X.; Fragkias, M.; Gutscher, H. A vision for human well-being: Transition to social sustainability. *Curr. Opin. Environ. Sustain.* **2012**, *4*, 61–73. [CrossRef]
3. Benjamin, D.J.; Heffetz, O.; Kimball, M.S.; Szembrot, N. Beyond Happiness and Satisfaction: Toward Well-Being Indices Based on Stated Preference. *Am. Econ. Rev.* **2014**, *104*, 2698–2735. [CrossRef]
4. Diener, E.; Lucas, R.; Schimmack, U.; Helliwell, J. *Well-Being for Public Policy*; Oxford University Press: Oxford, UK, 2009.
5. Brooks, J.S. Avoiding the Limits to Growth: Gross National Happiness in Bhutan as a Model for Sustainable Development. *Sustainability* **2013**, *5*, 3640–3664. [CrossRef]
6. Li, Y.; Guan, D.; Tao, S.; Wang, X.; He, K. A review of air pollution impact on subjective well-being: Survey versus visual psychophysics. *J. Clean. Prod.* **2018**, *184*, 959–968. [CrossRef]
7. Cortés, S.G.; Delgado, R.M.J.; Ortega, G.M. La medición de la felicidad y la construcción de indicadores de bienestar subjetivo. Aplicación del modelo de Rasch. In *Claves Para un Desarrollo Sostenible la Creatividad y Hapiness Ma-Nagement Como Portafolio de la Innovación Tecnológica, Empresarial y Marketing Social*; Rafael Ravina Ripoll, Luis Bayardo Tobar Pesántez, Araceli Galiano Coronil Coords: Granada, Spain, 2018; pp. 81–98. ISBN 978-84-9045-679-8.
8. Stiglitz, J.; Sen, A.; Fitoussi, J.P. *The Measurement of Economic Performance and Social Progress Revisited. Reflections and Overview*; Commission on the Measurement of Economic Performance and Social Progress: Paris, France, 2009.
9. Pavot, W.; Diener, E. Review of the Satisfaction With Life Scale. In *Assessing Well-Being*; Diener, E., Ed.; Springer: Dordrecht, The Netherlands, 2009; Volume 39, pp. 101–117. [CrossRef]
10. Stutzer, A. The role of income aspirations in individual happiness. *J. Econ. Behav. Organ.* **2004**, *54*, 89–109. [CrossRef]
11. Sweeney, P.D.; McFarlin, D.B.; Inderrieden, E.J. Research notes. Using relative deprivation theory to explain satisfaction with income and pay level: A multistudy examination. *Acad. Manag. J.* **1990**, *33*, 423–436. [CrossRef]
12. Layard, R. Happiness: Has Social Science a Clue? London School of Economics: London, UK, 2003; Volume 24.
13. Jahoda, M. *Empleo y Desempleo: Un Análisis Socio-Psicológico*; Ediciones Morata: Madrid, Spain, 1987; pp. 18–19.
14. Ross, C.E.; Van Willigen, M. Education and the Subjective Quality of Life. *J. Health Soc. Behav.* **1997**, *38*, 275–297. [CrossRef]
15. Dawson, C.; Veliziotis, M.; Hopkins, B. Temporary employment, job satisfaction and subjective well-being. *Econ. Ind. Democr.* **2014**, *38*, 69–98. [CrossRef]
16. Vansteenkiste, M.; Lens, W.; De Witte, S.; De Witte, H.; Deci, E.L. The'why' and'why not' of job search behaviour: Their relation to searching, unemployment experience, and well-being. *Eur. J. Soc. Psychol.* **2004**, *34*, 345–363. [CrossRef]
17. Binder, M.; Coad, A. Heterogeneity in the Relationship Between Unemployment and Subjective Wellbeing: A Quantile Approach. *Economica* **2015**, *82*, 865–891. [CrossRef]
18. Lucas, R.E.; Clark, A.E.; Georgellis, Y.; Diener, E. Unemployment Alters the Set Point for Life Satisfaction. *Psychol. Sci.* **2004**, *15*, 8–13. [CrossRef] [PubMed]
19. Clark, A.; Knabe, A.; Rätzel, S. Boon or bane? Others' unemployment, well-being and job insecurity. *Labour Econ.* **2010**, *17*, 52–61. [CrossRef]
20. Klasen, S. Gender-related Indicators of Well-being. In *Human Well-Being*; Palgrave Macmillan: London, UK, 2007; pp. 167–192.
21. Perrons, D. Final Comparative Report: Flexible Working and the Reconciliation of Work and Family Life-or a New form of Precariousness. In *Equality between Men and Women. European Commission*; Employment and Social Affairs: Brussels, Belgium, 1998.
22. Figart, D.M.; Mutari, E. Horarios largos y cortos: Qué puede aprender norteamérica de europa? *Investig. Econ.* **2000**, *60*, 53–71.

23. Offer, S.; Schneider, B. Revisiting the gender gap in time-use patterns: Multitasking and well-being among mothers and fathers in dual-earner families. *Am. Sociol. Rev.* **2011**, *76*, 809–833. [CrossRef]

24. Hetschko, C.; Knabe, A.; Schöb, R. Changing Identity: Retiring from Unemployment. *Econ. J.* **2014**, *124*, 149–166. [CrossRef]

25. Ponomarenko, V.; Leist, A.K.; Chauvel, L. Increases in well-being in the transition to retirement for the unemployed: Catching up with formerly employed persons. *Ageing Soc.* **2019**, *39*, 254–276. [CrossRef]

26. Flint, E.; Bartley, M.; Shelton, N.; Sacker, A. Do labour market status transitions predict changes in psychological well-being? *J. Epidemiol. Community Health* **2013**, *67*, 796–802. [CrossRef] [PubMed]

27. LaMontagne, A.D.; Milner, A.; Krnjacki, L.; Schlichthorst, M.; Kavanagh, A.; Page, K.; Pirkis, J. Psychosocial job quality, mental health, and subjective well-being: A cross-sectional analysis of the baseline wave of the Australian longitudinal study on male health. *BMC Public Health* **2016**, *16*, 33–41. [CrossRef]

28. Marans, R.W. Understanding environmental quality through quality of life studies: The 2001 DAS and its use of subjective and objective indicators. *Landsc. Urban Plan.* **2003**, *65*, 73–83. [CrossRef]

29. Steptoe, A.; Deaton, A. A Stone, A. Subjective wellbeing, health, and ageing. *Lancet* **2015**, *385*, 640–648. [CrossRef]

30. Rétsági, E.; Prémusz, V.; Makai, A.; Melczer, C.; Betlehem, J.; Lampek, K.; Ács, P.; Hock, M. Association with subjective measured physical activity (GPAQ) and quality of life (WHOQoL-BREF) of ageing adults in Hungary, a cross-sectional study. *BMC Public Health* **2020**, *20*, 1–11. [CrossRef] [PubMed]

31. Michalos, A.C.; Zumbo, B.D. Healthy Days, Health Satisfaction and Satisfaction with the Overall Quality of Life. *Soc. Indic. Res.* **2002**, *59*, 321–338. [CrossRef]

32. Hsieh, C.-M. The Relative Importance of Health. *Soc. Indic. Res.* **2008**, *87*, 127–137. [CrossRef]

33. Ayala, L.; José, M. Labeaga & Carolina Navarro, "undated". "Housing deprivation and health status: Evidence from Spain?". In *Working Papers*; Fedea: Madrid, Spain, 2005; p. 02.

34. Bowling, A.; Windsor, J. Towards the Good Life: A Population Survey of Dimensions of Quality of Life. *J. Happiness Stud.* **2001**, *2*, 55–82. [CrossRef]

35. Skevington, S.M. Qualities of life, educational level and human development: An international investigation of health. *Soc. Psychiatry Psychiatr. Epidemiol.* **2009**, *45*, 999–1009. [CrossRef] [PubMed]

36. Chen, C. Aging and Life Satisfaction. *Soc. Indic. Res.* **2001**, *54*, 57–79. [CrossRef]

37. Abbasimoghadam, M.A.; Dabiran, S.; Safdari, R.; Djafarian, K. Quality of life and its relation to sociodemographic factors among elderly people living in Tehran. *Geriatr. Gerontol. Int.* **2009**, *9*, 270–275. [CrossRef]

38. Addae-Dapaah, K.; Juan, Q.S. Life Satisfaction among Elderly Households in Public Rental Housing in Singapore. *Health.* **2014**, *6*, 1057–1076. [CrossRef]

39. Gerlach-Kristen, P.; Lyons, S. Determinants of mortgage arrears in Europe: Evidence from household microdata. *Int. J. Hous. Policy* **2017**, *18*, 545–567. [CrossRef]

40. Białowolski, P. Hard Times! How do Households Cope with Financial Difficulties? Evidence from the Swiss Household Panel. *Soc. Indic. Res.* **2017**, *139*, 147–161. [CrossRef]

41. Powdthavee, N. Unhappiness and Crime: Evidence from South Africa. *Economica* **2005**, *72*, 531–547. [CrossRef]

42. Staubli, S.; Killias, M.; Frey, B.S. Happiness and victimisation: An empirical study for Switzerland. *Eur. J. Criminol.* **2014**, *11*, 57–72. [CrossRef]

43. Spencer, N.; Liu, Z. Victimisation and life satisfaction: Evidence from a high crime country. *Soc. Indic. Res.* **2019**, *144*, 475–495. [CrossRef]

44. Møller, V. Resilient or Resigned? Criminal Victimisation and Quality of Life in South Africa. *Soc. Indic. Res.* **2005**, *72*, 263–317. [CrossRef]

45. Cooper, C.D.; Alley, F.C. *Air Pollution Control: A design Approach*; Waveland Press: Long Grove, IL, USA, 2010.

46. Welsch, H. Environment and happiness: Valuation of air pollution using life satisfaction data. *Ecol. Econ.* **2006**, *58*, 801–813. [CrossRef]

47. Giovanis, E. Worthy to lose some money for better air quality: Applications of Bayesian networks on the causal effect of income and air pollution on life satisfaction in Switzerland. *Empir. Econ.* **2018**, *57*, 1579–1611. [CrossRef]

48. Orru, K.; Orru, H.; Maasikmets, M.; Hendrikson, R.; Ainsaar, M. Well-being and environmental quality: Does pollution affect life satisfaction? *Qual. Life Res.* **2016**, *25*, 699–705. [CrossRef] [PubMed]

49. Manning, C.; Clayton, S. Threats to mental health and well-being associated with climate change. In *Psychology and Climate Change*; Academic Press: Cambridge, MA, USA, 2018; pp. 217–244.

50. Systems Institute; WVSA Secretariat. In *World Values Survey: All Rounds—Country-Pooled Datafile*; Inglehart, R.; Haerpfer, C.; Moreno, A.; Welzel, C.; Kizilova, K.; Diez-Medrano, J.; Lagos, M.; Norris, P.; Ponarin, E.; Puranen, B.; et al. (Eds.) Systems Institute & WVSA Secretariat: Madrid, Spain, 2020; Available online: http://www.worldvaluessurvey.org/WVSDocumentationWVL.jsp (accessed on 15 February 2021).

51. European Values Study. Integrated Dataset (EVS 2017). In *ZA7500 Data File Version 4.0*; GESIS Data Archive: Cologne, Germany, 2017. [CrossRef]

52. Blanchflower, D.G.; Oswald, A.J. International Happiness: A New View on the Measure of Performance. *Acad. Manag. Perspect.* **2011**, *25*, 6–22. [CrossRef]

53. Letelier-S, L.E.; Sáez-Lozano, J.L. Expenditure Decentralization: Does It Make Us Happier? An Empirical Analysis Using a Panel of Countries. *Sustainability* **2020**, *12*, 7236. [CrossRef]
54. Aldrich, J.; Nelson, F. *Quantitative Applications in the Social Sciences: Linear Probability, Logit, and Probit Models*; SAGE Publications, Inc.: Thousand Oaks, CA, USA, 1984.
55. Hosmer, D.W.; Lemeshow, S. *Applied Logistic Regression*; John Wiley & Sons: New York, NY, USA, 2000.
56. Cramer, J.S. *Logit Models from Economics and Other Fields*; Cambridge University Press: Cambridge, UK, 2003. [CrossRef]
57. Luechinger, S. Life satisfaction and transboundary air pollution. *Econ. Lett.* **2010**, *107*, 4–6. [CrossRef]
58. Ortega-Gil, M.; Cortés-Sierra, G.; ElHichou-Ahmed, C. The Effect of Environmental Degradation, Climate Change, and the European Green Deal Tools on Life Satisfaction. *Energies* **2021**, *14*, 5839. [CrossRef]
59. Clark, C.; Paunovic, K. WHO Environmental Noise Guidelines for the European Region: A Systematic Review on Environmental Noise and Quality of Life, Wellbeing and Mental Health. *Int. J. Environ. Res. Public Health* **2018**, *15*, 2400. [CrossRef] [PubMed]
60. JarosiŃska, D.; Héroux, M.-È.; Wilkhu, P.; Creswick, J.; Verbeek, J.; Wothge, J.; Paunović, E. Development of the WHO Environmental Noise Guidelines for the European Region: An Introduction. *Int. J. Environ. Res. Public Health* **2018**, *15*, 813. [CrossRef] [PubMed]
61. Shepherd, D.; Dirks, K.; Welch, D.; McBride, D.; Landon, J. The Covariance between Air Pollution Annoyance and Noise Annoyance, and Its Relationship with Health-Related Quality of Life. *Int. J. Environ. Res. Public Health* **2016**, *13*, 792. [CrossRef] [PubMed]
62. Cooper, D.; McCausland, W.D.; Theodossiou, I. Income Inequality and Wellbeing: The Plight of the Poor and the Curse of Permanent Inequality. *J. Econ. Issues* **2013**, *47*, 939–958. [CrossRef]
63. Lawless, N.M.; Lucas, R. Predictors of Regional Well-Being: A County Level Analysis. *Soc. Indic. Res.* **2010**, *101*, 341–357. [CrossRef]

9 783036 582023

MDPI

St. Alban-Anlage 66

4052 Basel

Switzerland

Tel. +41 61 683 77 34

Fax +41 61 302 89 18

www.mdpi.com

Land Editorial Office

E-mail: land@mdpi.com

www.mdpi.com/journal/land